AF552827

Biotechnology

BIOTECHNOLOGY

By

MANJU YADAV

Lecturer

Department of Zoology

M.M.H. College

Ghaziabad (U.P.)

(India)

DISCOVERY PUBLISHING HOUSE

NEW DELHI-110002

First Published – 2003

Reprinted – 2025

ISBN: 978-81-7141-712-4

Biotechnology

Published by:

DISCOVERY PUBLISHING HOUSE
4383/4B, Ansari Road, Darya Ganj
New Delhi-110 002 (India)
Phone: +91-11-23279245; 23253475; 43596065
Mobile: +91 9811179893 / +91 9871656464
E-mail: discoverybooksindia@gmail.com
orderdphbooks@gmail.com
namitwasan9@gmail.com
web: www.discoverypublishinggroup.com

Printed at:
Infinity Imaging Systems
Delhi (INDIA)

Preface

Biotechnology impinges on every one's lives. It is one of the major technologies of the twenty first century. In fact, it is an old field that has been rejuvenated in recent years because of the development of genetic engineering techniques. At present, biotechnology is an amazing growth phase whose end is nowhere in sight. Its, huge, wide ranging, multi-disciplinary activities include recombinant DNA technique, cloning and genetics, and the application of microbiology to the production of goods as prosaic as bread, beer, cheese and antibiotics. It continuous to revolutionize treatments of many diseases, and uses to provide clear technologies and to deal with environmental problems.

Biotechnology is one of the fast moving fields of science that has undergone a variable revolution over the last two decades, leading to the major advances in the understanding and manipulating gene at molecular level. The way that a biotechnologist thinks and works is influenced by his knowledge of events on two levels: the biotechnical and the transmission genetic level. It provides an overview of scientific perspectives. By intent, the chapters are not exhaustic reviews. The chapters are written with the expectation that the readers will have a general understanding of the cell biology and genetics, as well as a fun background in the major disciplines of modern biology.

It has been constant endeavour of the author to furnish maximum substance, keeping in view the limitations of the size of the book. Effects have been made of condense the matter as far as practicable. It is hoped that this book will not only meet the requirement of students but will also be useful as a guidelines to the teachers in their teaching.

There can be no claim of originality except in the manner of treatment and presentation and much of the information has been gathered from the books and scientific journals available in different libraries.

The author expresses his thanks to her friends and colleagues whose constant inspiration have initiated her in bringing out this book.

Special thanks are given to Mr. Wasan and staff of M/s Discovery Publishing House for their whole hearted co-operation in the publication of this book.

Author

CONTENTS

1

INTRODUCTION

In the second half of the twentieth century, biology has entered what some scientists have referred to as its 'Golden Age'. From the discovery of the structure of DNA in 1953 to the ability to write the genetic code for a human being by the end of the century, the relatively new science of molecular biology has combined with genetics to give us a new and powerful biotechnology. It will have applications in industry, medicine and agriculture and many other fields. Its importance is reflected in the fact that the United States now spends about half its academic research budget on the life sciences. Like all new knowledge, it can be used for the benefit of humankind, but also presents new dangers. Physicists faced the same problem in the first half of the twentieth century when knowledge of the structure of the atom led to our ability to use nuclear power, with its potential for peaceful or destructive use.

Francis Cricks and *Maurice Wilkins*, who with *James Watson* won the Nobel prize for their work in discovering the structure of DNA, had been physicists and had worked on weapons research during the Second World War before moving on to the science of life. Past experience has made scientists very aware of the social and ethical implications of their research and, as we shall see later, they even slowed down their work in the late 1970s while strict regulations and guidelines for genetic engineering were worked out. As a student of biology you will share in the responsibility for discussing the new issues that will certainly arise from our expanding knowledge of molecular genetics. The more informed our opinions are, and the more people are prepared to discuss the issue, the more likely we all are to benefit.

Molecular biologists also discovered unconventional processes for transferring genetic information through two different biological entities, *plasmids* (1965) and *viruses* (1960s and 1970s). Neither are considered to be true organisms in a biological sense because they cannot reproduce independently outside of the host cells they normally inhabit. Nevertheless, it become clear that plasmids and viruses were capable of transmitting genetic information from one species to another in way that were eventually harnessed to have an enormous impact on biology and society. The new methods and techniques developed in the 1970s provided deeper insights into the chemistry of DNA and genes.

A spectacular accomplishment occurred in 1973 when DNA from two different organisms was spliced together–the first successful experiment in which DNA from two different species was "recombined." It later became possible to clone eukaryotic genes, carried by plasmids, inside of bacterial (prokaryotic) cells (1974) and to determine the exact DNA nucleotide sequences of entire genes and longer segments of DNA (1977). Companies were founded in the 1980s to take advantage of the new biological technologies to produce valuable substances such as vaccines and hormones. Like several other major breakthroughs in biology, manipulating genes for human purposes has led to a number of issues and questions that are of concern to informed citizens. The new molecular tools gave rise to novel techniques referred to as *genetic engineering* and products known as *recombinant DNA*. For the first time, it became possible for molecular biologists to isolate genes from one organism, clone them in another, and have them expressed in bacteria, manufacturing the protein encoded by the gene. The pace of discovery in genetic engineering remains rapid, and useful products from this field continue to be developed.

Recombinant DNA Technology

DNA can be viewed as a master set of instructions that, once expressed, regulate the development, growth, functions, and ageing of organisms throughout their lives. The instructions of DNA are translated through a genetic code into proteins, the workhorses of biochemistry. The common feature of all life on Earth that makes genetic engineering possible is the genetic code; it is virtually the same for all organisms, from bacteria to humans.

The Early Development of Genetics

For the first 30 years of its life this new science grew at an astonishing rate. The idea that genes reside on *chromosomes* was proposed by *W. Sutton* in 1903, and received experimental backing from *T.H. Morgan* in 1910. Morgan and his colleagues then developed the techniques for *gene mapping*, and by 1922 had produced a comprehensive analysis of the relative positions of over 2000 genes on the four chromosomes of the fruit fly, *Drosophila melanogaster*. Despite the brilliance of these classical genetic studies, there was no real understanding of the molecular nature of the gene until the 1940s. Indeed, it was not until the experiments of Avery, MacLeod and McCarty in 1944, and of Hershey and Chase in 1952, that anyone believed DNA to be the genetic material; up to then it was widely thought that genes were made of protein.

Table 1.1. Terms used in the Field of Genetic Engineering.

Bacteria	: One-celled organisms; prokaryotes lacking a nucleus; capable of carrying on many chemical reactions (for example, protein synthesis) that are similar to those of higher organisms. *Strains* of bacteria are a group of organisms within a species that is characterized by some particular quality (for example, the rough and smooth strains of *Diplococcus pneumoniae*.
Bacteriophage	: Viruses that infect bacteria; commonly referred to as *phages*.
Biotechnology	: A broad field that employs an industrial technology based on the biological synthesis of important chemical compounds, especially proteins (for example, insulin), genetic engineering of plants and animals, and other, related technologies.
Clone	: A large number of *cells* or *molecules* identical to an ancestral cell or molecule. Also, a specific gene sequence, isolated and replicated.
Cloning vectors	: Small *plasmid*, *phage*, or animal *virus* DNA molecules used to transfer a DNA fragment from a test tube into a living cell. Cloning vectors are capable of multiplying inside of living cells.

Complementary DNA (*cDNA*) : DNA copied from a messenger RNA (mRNA) molecules using an enzyme called *reverse transcriptase.* The DNA sequence is thus *complementary* to that of the mRNA.

DNA ligase : An enzyme that connects two separate DNA molecules, end to end.

Escherichia coli (*E.coli*) : The bacterium most widely used in recombinant DNA research; commonly found in the digestive tracts of mammals.

Expression vector : A *plasmid* designed to permit expression (transcription) of a foreign gene inside a cell.

Genetic code : The sequence of mRNA base triplets that specify the exact series of amino acids for a protein.

Genetic engineering : The manipulation of genetic information (DNA or RNA) of an organism to alter the characteristics of that organism. The basic process consists of inserting foreign DNA carrying instructions for a valuable enzyme, hormone, or other protein into the DNA of some other organism so that the host organism makes the desired product at the same time as it makes its own proteins.

Genomic clone : A fragment of genomic DNA from an organism.

Genomic DNA : All DNAs equences of an organism.

Host cell : A cell (usually a bacterium) in which a *cloning vector* can be propagated.

λ (lambda) phage : A particular *bacteriophage* used extensively in gene cloning.

Library : A set of cloned fragments together reprsenting a part or all of the genome of a species.

Plasmids : Small, circular DNA molecules found inside bacterial cells. Plasmids repro-

		duce every time the bacterial cell reproduces.
Recombinant DNA (rDNA)	:	A DNA molecule containing two or more regions of DNA from two different genes or species (for example, a fragment of human DNA spliced into plasmid DNA).
Restriction enzymes	:	Enzymes that cut DNA molecules at specific nucleotide sequences.
Reverse transcriptase	:	An enzyme purified from retroviruses that makes DNA from RNA.
Transduction	:	The transfer of DNA (genes) from one bacterium to another by bacteriophages.
Transformation	:	The insertion of any foreign into any organism.

The discovery of the role of DNA a tremendous stimulus to genetic research, and many famous biologists contributed to the second great age of genetics. In the 14 years between 1952 and 1966 the structure of DNA was elucidated, the genetic code cracked, and the processes of transcription and translation described.

The Advent of Gene Cloning

These years of activity and discovery were followed by a lull, a period of anticlimax when it seemed to some molecular biologists (as the new generation of geneticists styled themselves) that there was little of fundamental importance that was not understood. In truth there was a frustration that the experimental techniques of the late 1960s were not sophisticated enough to allow the gene to be studied in any great detail. Then, in the years 1971-1973 genetic research was thrown back into gear by what at the time was described as a revolution in experimental biology.

A whole new methodology was developed, enabling previously impossible experiments to be planned and carried out, if not with ease, then at least with success. These methods, referred to as *recombinant DNA technology* or *genetic engineering*, and having at their core the process of gene cloning, sparked the third great age of genetics. Twenty-five years later we are still riding the rollercoaster set in motion by the gene cloning revolution, and there is no end to the excitement in sight.

What is Gene Cloning?

The basic step in a gene cloning experiment are as follows:

1. A fragment of DNA, containing the gene to be cloned, is inserted into a circular DNA molecule called a *vector*, to produce a *chimaera* or *recombinant DNA molecule*.
2. The vector acts as a *vehicle* that transports the gene into a host cell, which is usually a bacterium, although other types of living cell can be used.
3. Within the host cell the vector multiplies, producing numerous identical copies not only of itself but also of the gene that it carries.
4. When the host cell divides, copies of the recombinant DNA molecule are passed to the progeny and further vector replication takes place.
5. After a large number of cell divisions, a colony, or *clone*, of identical host cells is produced. Each cell in the clone contains one or more copies of the recombinant DNA molecule; the gene carried by the recombinant molecule is now said to be cloned.

Obtained DNA to be Cloned

Recall that genes are specific sequences of DNA that encode proteins. One way to obtain relevant genes is to isolate them from the *genome* of the organism; that is, to identify and remove one gene from all of the DNA present in the chromosomes of a single somatic cell. This is a daunting task when one considers that the genome of an average mammal consists of approximately 1 billion nucleotide base pairs and that a typical gene contains 5,000 base pairs. To reduce the scope of this "needle in a haystack" problem, two general approaches have been developed for isolating small segments of DNA.

Genomic DNA Isolation and Restriction Enzymes

DNA molecules can be removed from cells and cut into small segments that can be manipulated and analyzed. Two basic methods are used to remove genomic DNA from cells and divide it into small segments. One employs mechanical techniques such as exposing cells and DNA to high-frequency sound (sonication), causing the rupture and release of fragmented DNA. The other uses a two-step approach. First, enzymes are used to digest membranes, which liberates the DNA. Second, *restriction enzymes* obtained from bacteria are used to cut the DNA into small pieces. The normal function of *restriction enzymes* in bacteria is to destroy ("restrict") foreign DNA attempting to gain entry into the cell. Through this mechanism, bacteria are protected against the insertion of DNA by viruses that infect and usually kill them.

Restriction enzymes recognize specific, short nucleotide sequences and cleave this DNA sequence at a specific *restriction site.* For example, *BamHI,* a restriction enzyme from *Bacillus amyloliquefaciens,* specifically recognizes the DNA sequence GGATCC and cleaves it at a restriction site between the two guanines. Thus it recognizes the DNA sequence reading G¯GATCC and cuts between the two Gs, leaving G and GATCC. Approximately 200 restriction enzymes are now commercially available. By exposing a genome to suitable restriction enzymes, a set of fragments is obtained.

Cloning Vectors

Almost any genomic or cDNA fragment can be cloned after it is integrated into a cloning vector forming rDNA. The rDNA molecule is inserted into a bacterial cell, where it is *amplified.* Two submicroscopic agents normally associated with bacteria, *plasmids* and *phages,* are commonly used as vector systems for cloning DNA fragments.

Plasmids

Plasmids are small, circular, double-stranded DNA molecules that reside inside prokaryotic cells such as bacteria. Plasmids replicate independently of the bacterial chromosome, although they are dependent on the host cell's DNA-synthesizing machinery. Plasmid are retained throughout cell division, and they generally encode proteins that are of some benefit to the bacterial host cells. For example, many plasmids (so-called R factors) contain genes that confer drug resistance on their bacterial host cells. Several strains of bacteria are now resistant to specific antibiotics such as ampicillin and tetracycline because they harbor plasmids with antibiotic-resistant genes.

The fact that plasmids are able to confer drug resistance on a bacterial host has been exploited in identifying bacteria that contain such plasmids. The plasmid is opened through the action of a restriction enzyme, which cuts its DNA at a specific restriction site. The plasmid DNA and foreign DNA, which have been cleaved by the same restriction enzyme, are mixed together in the presence of the enzyme *DNA ligase,* which can bind them together to create an rDNA molecule. The manipulations described so far are carried out *in vitro* (in a test tube). For the rDNA molecules to be amplified, they are placed in the presence of a suitable bacterial strain. Only a very small percentage of the bacterial cells become *transformed,* meaning that they take up the plasmid containing the foreign DNA. The cells containing plasmid

DNA must then be distinguished from those that do not; this is done by growing the bacterial cells in a medium containing the antibiotic whose resistance is encoded by the plasmid. Only transformed bacterial cells containing the plasmids survive, untransformed bacterial cells perish.

Phages

Bacteriophages, commonly known as *phages*, are viruses that infect bacteria. They are more complicated than plasmids in that they contain genes that encode proteins used in their own structures and replication. Like plasmids, however, they depend on a bacterial host cell to carry out their DNA replication. Each phage type will replicate only in a specific bacterial species or strain. Using phages instead of plasmids as cloning vectors has three advantages: phages are easier to introduce into bacteria, larger foreign DNA fragments can be inserted into the phage DNA to create stable rDNA molecules, and phages have a greater efficiency for incorporating foreign DNA. The success for plasmids is usually 10 percent or less, whereas for phages it is greater than 50 percent.

Many phages have been used in gene-cloning operations, but one of the most frequently used has been the l (lambda) phage (bacteriophage l) that infects *E.coli.* The procedures for incorporating foreign DNA into the l phage genome follow the same lines as for plasmids. The viral DNA is opened using specific restriction enzymes; foreign DNA is then introduced, and the two DNAs are joined by DNA ligase to form an rDNA molecule. The l phage cloning vector then infects a bacterial cell where the rDNA may become integrated into the bacterial chromosome. In this phage, the rDNA will be replicated along with the bacterial DNA. Under other conditions, the l phage DNA remains separated from the host chromosome and replicates autonomously. When this happens, the phage DNA replicates at a much greater rate, which causes death of the bacterial host but also results in greater yields of the cloned gene.

Importance of Gene Cloning

Gene cloning is a relatively straightforward procedure. To understand exactly how this works, consider a gene cloning experiment drawn in a slightly different way. In this example the DNA fragment to be cloned is one member of a mixture of many different fragments, each carrying a different gene or part of a gene. This mixture could indeed be the entire genetic complement of an organism, a human for instance. All these fragments will become inserted into different vector

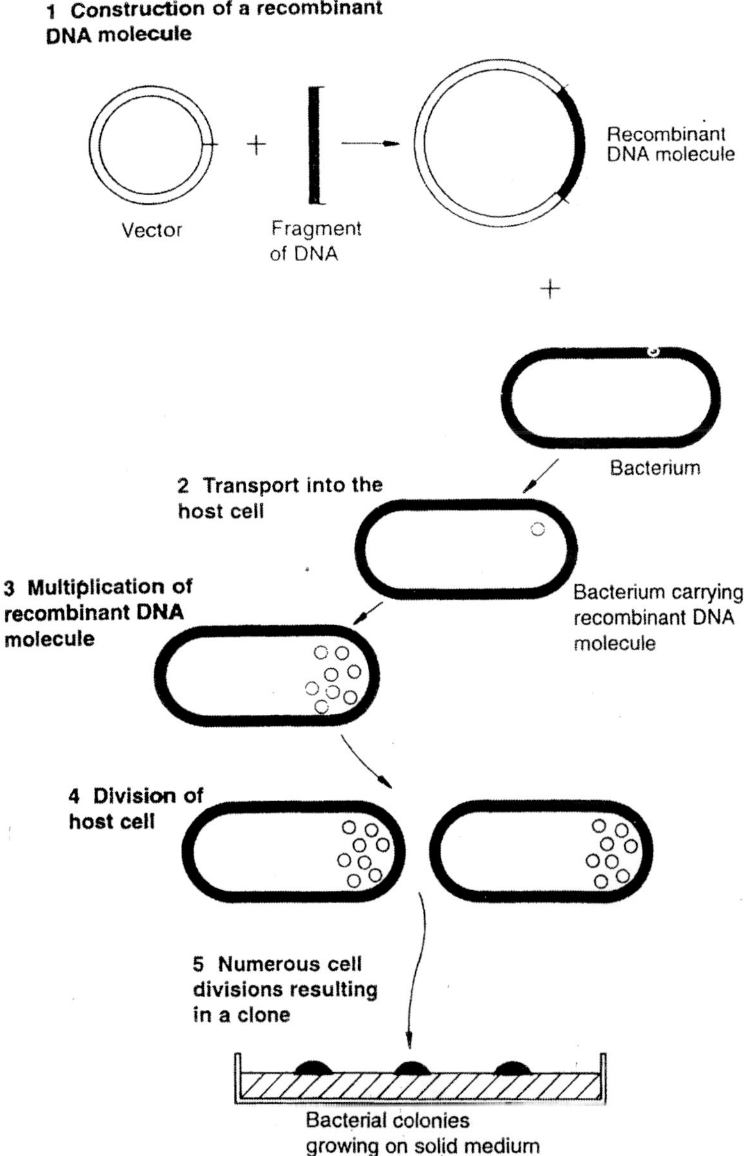

Fig. 1.1. The basic steps in gene cloning.

molecules to produce a family of recombinant DNA molecules, one of which carries the gene of interest. Usually only one recombinant

DNA molecule will be transported into any single host cell, so that although the final set of clones may contain many different recombinant DNA molecules, each individual clone contains multiple copies of just one molecule.

The gene is now separated away from all the other genes in the original mixture, and its specific features can be studied in detail. In practice, the key to the success or failure of a cloning experiment is

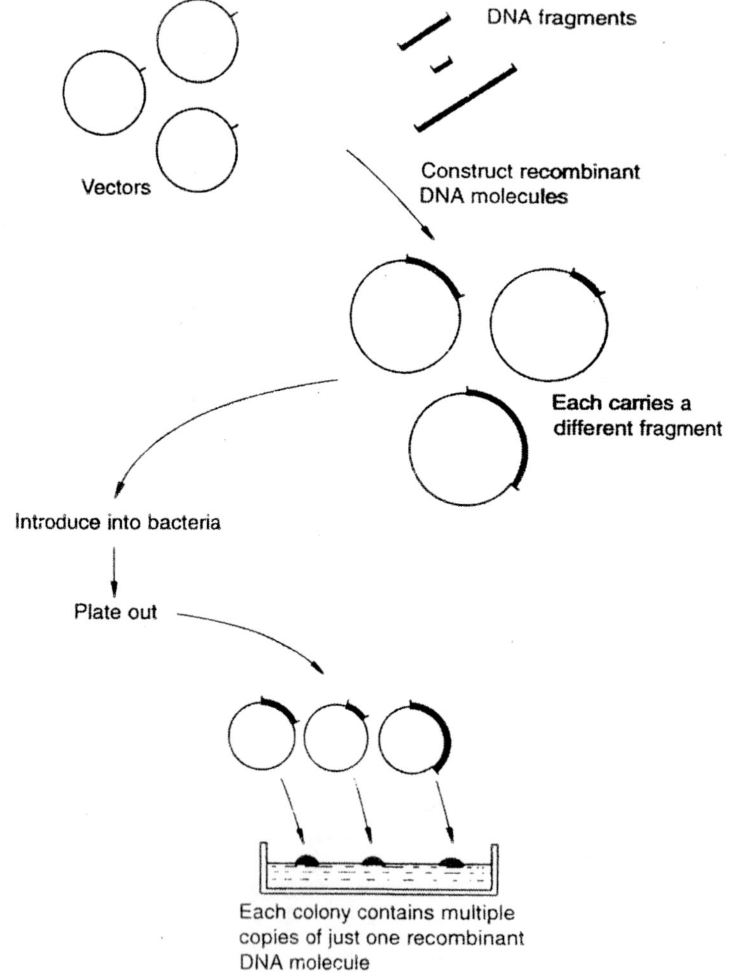

Fig. 1.2. Cloning allows individual fragments of DNA to be purified.

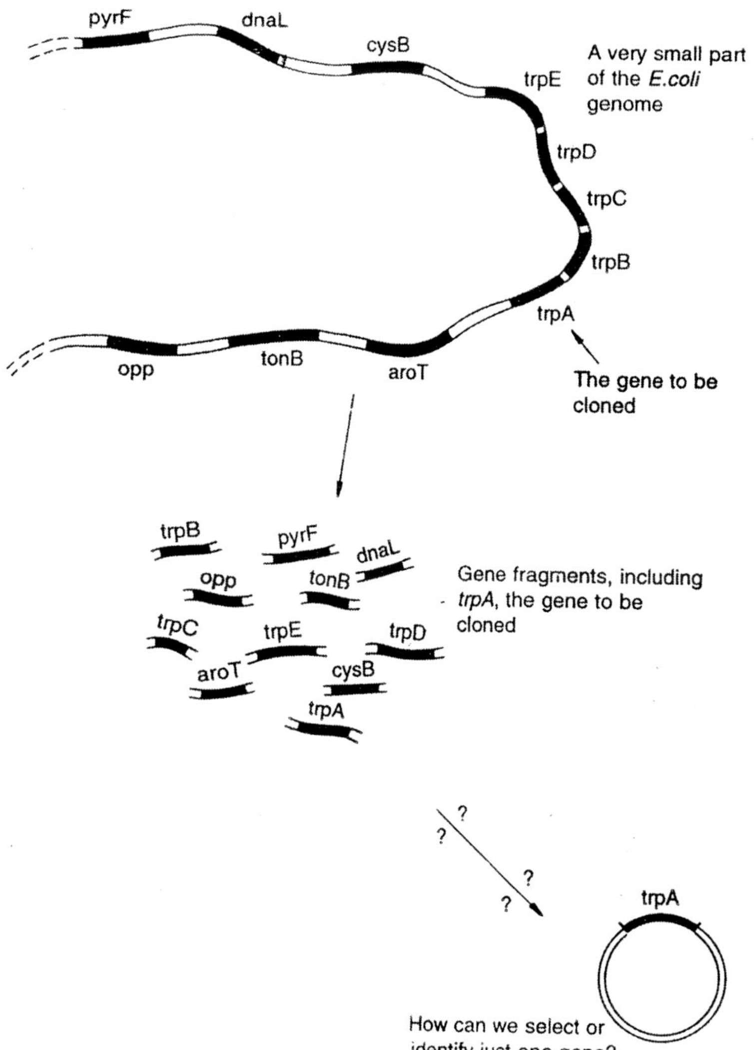

Fig. 1.3. The problem of selection.

the ability to identify the particular clone of interest from the many different ones that are obtained. If we consider the *genome* of the bacterium *Escherichia coli*, which contains something in the region of 4000 different genes, we might at first despair of begin able to find just one gene among all the possible clones. The problem becomes even more overwhelming when we remember that bacteria are relatively simple organisms and that the human genome contains about 20 times as many genes.

A variety of different strategies can be used to ensure that the correct gene can be obtained at the end of the cloning experiment. Some of these strategies involve modifications to the basic cloning procedure, so that only cells containing the desired recombinant DNA molecule can divide and the clone of interest is automatically *selected.* Other methods involve techniques that enable the desired clone to be identified from a mixture of lots of different clones. Once a gene has been cloned there is almost no limit to the information that can be obtained about the structure and expression of that gene. The availability of cloned material has stimulated the development of analytical methods for studying genes, with new techniques being introduced all the time.

Recombinant DNA and Biotechnology

There are very few areas of biological research that have not been touched by gene cloning, PCR, and the recombinant DNA techniques that these procedures have made possible. In industry, for example, the ability to clone genes has led to far-reaching advances in *biotechnology.* For many years microorganisms have been used as living factories for the production of useful compounds. Examples are provided by antibiotics, such as penicillin, which is synthesized by a fungus called *Penicillium*, and streptomycin, produced by the bacterium *Streptomyces griseus.*

Gene cloning has revolutionized biotechnology, most notably in providing a way in which mammalian proteins can be produced in bacterial cells. A remarkable property of a cloned gene is that it can often be made to function in an organism totally unrelated to that in which it is normally found. For example, an animal gene can be transferred by cloning into a bacterium and then induced by some careful modifications to carry on working as though nothing had happened. The implications are enormous. Genes controlling the synthesis of important pharmaceuticals, such as drugs and hormones, can be taken from the organism in which they occur naturally, but from which they may be costly and difficult to prepare, and placed in a bacterium or other type of organism, from which the product can be recovered conveniently and in large quantities.

A number of successes have been notched up by biotechnologists, with recombinant insulin being the most noteworthy achievement so far. Medical and agricultural research have also received important

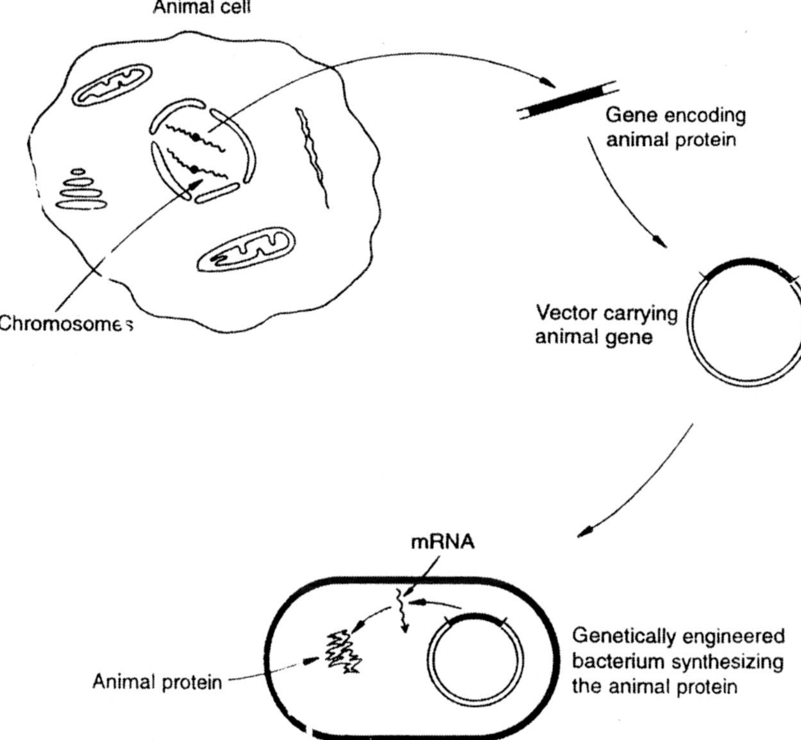

Fig. 1.4. A possible scheme for the production of an animal protein by a bacterium.

boosts from gene cloning. New types of vaccines, providing protection against diseases for which vaccination was previously impossible, have been developed thanks to the ability to clone genes. Many inherited diseases can now be diagnosed in an unborn child, and recent research has led to the hope that cystic fibrosis, breast cancer and other heart-reading diseases will soon be treatable. In agriculture, equally important problems are being addressed through the development of genetically engineered crops able to withstand the revages of insects.

Benefits and Hazards—The Ethical and Social Implications of Genetic Engineering

From the earliest days of genetic engineering scientists have been very aware of the need to consider the potential hazards and ethical issues associated with this new branch of biology. Original concerns in 1971 focused on plans to clone cancer genes from viruses into *E.coli.*

It was argued that if the genetically altered *E.coli* escaped from the laboratory, they might spread the gene into *E.coli* that live in the human gut by transferring the plasmids, for example by conjugation. It was also argued that human DNA, and the DNA of other mammals like mice, could contain cancer-causing genes (oncogenes) and that these might inadvertently get transmitted with neighbouring pieces of DNA being used for genetic engineering. In February 1975 a group of more than 100 internationally well-known molecular biologists met in California and decided that, until the risks could be more precisely estimated, certain restrictions should be placed on genetic engineering research.

This was a remarkable self-imposed brake on scientific progress, not imposed by the Government but by the scientists themselves. Work on cancer viruses was stopped. Non-scientists who had been part of the debate were invited to join the Advisory Committee that was set up by the American Government in 1976. Similar bodies were set up in Europe, including the Genetic Manipulation Advisory Group (GMAG) in the UK. After a two-year halt, during which safe procedures were established, research on cancer viruses and other work continued. Careful checks on the safety and implications of procedures remain to this day. Debate continues about how strict the rules and regulations should be. There has been more resistance to developments in Europe than in the USA. In the 1980s an explosion of activity and interest was unleashed. Manufacturing companies quickly began investing billions of dollars into not just genetic engineering but all the 'biotechnologies' which were emerging from molecular biology. *Biotechnology*, a new industry was born.

2

Nucleic Acids

The life in a cell is due to nucleic acid and the most interesting thing is that the nucleic acid itself is non-living. *Friedrich Miescher* in (1868) isolated a material from pus cells by digesting them for weeks with dilute HCl and called this material as *nuclein*. *Hoppe Segler* and his co-workers confirmed the work of *Miescher* and also proved the presence of nuclein in yeast and the erythrocytes of the birds and reptiles and various other tissues. The term *nucleic acid* was introduced by *Altmann*. *Albrecht Kossel*, *P.A. Levine* and *Walter Jones* described the chemical nature of nucleic acid. According to them the nucleic acid is composed of phosphoric acid, a sugar and nitrogenous bases. *Franklin W. Stahl* presented first evidence that nucleic acid forms the genetic material. *Chargaff* (1951) described the occurrence of nitrogenous bases in equal proportions. *Dotty* (1961) has emphasized much on the physical properties of DNA. In 1962 *Willkins*, *Klatson* and *Crick* proposed a model to explain the double helicular structure of DNA. They also shared the Nobel Prize for the same.

Biological Role

The functions of DNA may be summarized below:

1. It contains the genetic information that is transmitted from generation to generation, which is achieved by self-replication of DNA during cell growth and division so that two daughter double helical molecules of DNA are obtained, each identical to parent DNA.
2. It expresses its encoded genetic information for the synthesis of RNA and proteins for metabolic function and control of all cellular

activities. The genetic information is carried in the form of genes and expressed at appropriate times.

A gene is defined as the sequence of bases in DNA which specifies the complete amino acid sequence of a polypeptide chain or the base sequence of an RNA molecule (rRNA, tRNA). The sequence that specifies a polypeptide chain is commonly called a structural gene.

Most genes of eukaryotic organisms do not consist of a single continuous sequence which is transcribed into mRNA. Rather, they are interrupted by regions called introns which do not specify the protein product. The regions which are transcribed and specify the final protein product are called as *exons.*

Characteristics and Properties

1. *Genetic material.* DNA is the molecule of heredity and is responsible for the progeny to have the same characteristics as their parents.
2. *DNA content.* The DNA content of a cell is remarkably constant for each species (except in germ cells and when chromosomal variations occurs) and cannot be altered by environmental circumstances, with change in age or nutritional status.
3. *Base composition.* DNA isolated from different tissues of the same organisms has the same base composition. The base composition of DNA varies from one species to the other; while the DNA from closely related species has more or less similar base composition.

   In nearly all DNAs, the number of adenine residues is equal to the number of thymine residues i.e., A=T, and the number of guanine residues is equal to the number of cytosine residues, i.e., G=C or A+G = T+C or

$$\frac{A+G}{T+C} = 1$$

4. *Effect of pH.* DNA is a polybasic acid due to the presence of phosphate groups which are fully ionized at physiological pH. Because of negative charges present DNA binds strongly to histones and cations like Na^+ and Mg^{2+}. pH also affects the stability of the double helical structure of DNA. The hydrogen bonded base pairs are stable between pH 4.0 and pH 10.0. Outside these limits, their hydrogen bonds break and the complementary strands separate from each other, a process known as *denaturation.*
5. *Effect of temperature.* When highly polymerized double-stranded DNA is slowly heated, the double helix 'melts', as a result the

double-stranded structure is converted to a random coil over a range of a few degrees of temperature. This transition from a helix to a coil results in increase in absorbance. The midpoint temperature (T_m) is the melting temperature of the helix of a specific DNA polymer. The T_m's of different DNA's increase linearly as a function of the percentage of G-C base pairs.

6. *Absorbance.* The purine and pyramidine bases found in the DNA and also RNA, strongly absorb ultraviolet radiation of wavelength at 260 nm. This property is used to identify and estimate nucleic acids. The high molecular weight DNA typically has an optical density at 260 nm which is about 35-40 percent less than the optical density expected from adding up the individual absorbances of bases in the DNA. This phenomenon is called the hypochromic effect which is explained by the fact that in a helical structure, the bases are stacked one above the other. Interaction of π electrons between the bases then results in a decrease in absorbancy.
7. *Hydrolysis.* Gentle acid hydrolysis of DNA at pH 3.0 causes selective hydrolytic removal of all its purine bases without affecting the pyrimidine-deoxyribose bonds or the phosphodiester bonds of the backbone. The resulting DNA derivative devoid of the purine bases is called an apurine acid. Selective removal of the pyrimidine bases by hydrazine produces apyrimidinic acid. DNA is not hydrolyzed by dilute alkali unlike RNA because of no 2′-hydroxyl groups.

Enzymes that hydrolyze the phosphodiester bonds of nucleic acids are collectively known as nucleases. Nucleases can be classified by their point of attack upon the polynucleotide chain. Those that attack the polymer at either its 3′ or 5′ terminus and sequentially remove nucleotide residues one at a time or as small oligonucleotides are known as *exonucleases*; those that attack within the chain are called *endonucleases.*

Chemistry of Nucleic Acids

Having identified the genetic material as the nucleic acid DNA (or RNA), we need to examine the chemical structure of these molecules. Their structure will tell us a good deal about how they function.

Nucleic acids are made by joining *nucleotides* in a repetitive way into long, chainlike polymers. Nucleotides are made of three components: phosphate, sugar, and a nitrogenous base. When incorporated into a nucleic acid, a nucleotide contains one each of the

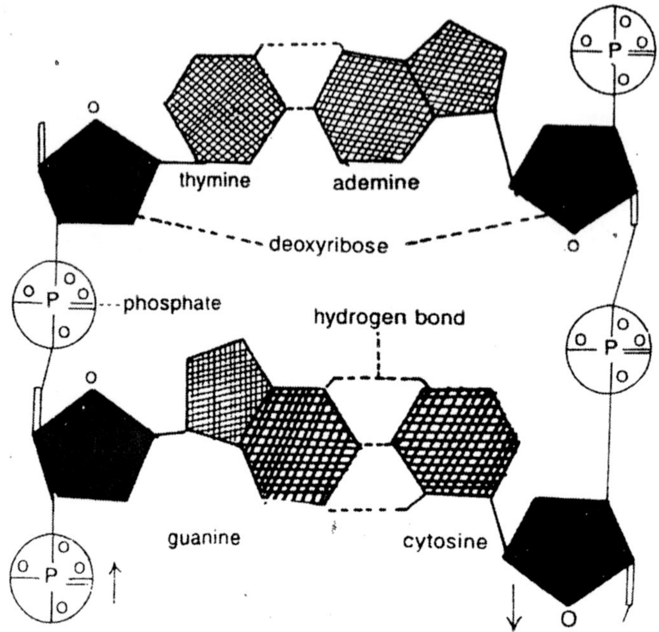

Fig. 2.1. Showing antiparallel nature of backbones of DNA.

three components. But, when free in the cell pool, nucleotides usually occur as triphosphates. The energy held in the extra phosphates is used, among other purposes, to synthesize the polymer. A *nucleoside* is a sugar-base compound. Nucleotides are therefore nucleoside phosphates. The sugars differ only in the presence (ribose in RNA) or absence (deoxyribose in DNA) of an oxygen in the 2′ position. The carbons of the sugars are numbered 1′ to 5′. The primes are used to avoid confusion with the numbering system of the bases. DNA and RNA both have four types of bases (two *purines* and *pyrimidines*) in their nucleotide chains. Both molecules have the purines *adenine* and *guanine* and the pyrimidine *cytosine*. DNA has the pyrimidine *thymine*, RNA has the pyrimidine *uracil*. Thus three of the nitrogenous bases are found in both DNA and RNA, whereas thymine is unique to DNA and uracil is unique to RNA.

A nucleotide is formed in the cell by attachment of a base to the 1′ carbon of the sugar and attachment of a phosphate to the 5′ carbon of the same sugar, the nucleotide takes its name from the base. Nucleotides are linked together (*polymerized*) by the formation of a bond between the phosphate of one nucleotide and the hydroxyl (OH) group at the 3′ carbon of an adjacent molecule. Very long strings of nucleotides can be polymerized by this ***phosphodiester bonding.***

Table 2.1. Components of Nucleic Acids

			Base	
	Phosphate	*Sugar*	*Purines*	*Pyrimidines*
DNA	Present	Deoxyribose	Guanine	Cytosine
			Adenine	Thymine
RNA	Present	Ribose	Guanine	Cytosine
			Adenine	Uracil

FUNCTIONAL STRUCTURE

Although the identity of the nucleotides that polymerized to form a strand of DNA or RNA was known, the actual structure of these nucleic acids when they function as the genetic material remained unknown until 1953. The general feeling was that the biologically active structure of DNA was more complex than a single string of nucleotides linked together by phosphodiester bonds and that several interacting strands were involved.

In 1953, Linus Pauling, a Nobel laureate who had discovered the a-helical structure of proteins, was investigating a three-stranded structure for the genetic material, whereas Watson and Crick decided that a two-stranded structure was more consistent with available evidence. Three lines of evidence directed Watson and Crick: the chemical nature of the components of DNA, X-ray crystallography, and Chargaff's ratios.

DNA X-Ray Crystallography

Maurice Wilkins, Rosalind Franklin, and their colleagues were using *X-ray crystallography* to analyze the structure of DNA. The molecules in a crystal are arranged in an orderly fashion, such that when a beam of X rays is passed through the crystal, the beam will be scattered. The pattern of the scatter can be recorded on photographic film. The nature of this pattern depends on the structure of the crystal. The cross in the center of the photograph indicates that the molecule is a helix; the dark areas at the top and bottom come from the bases, stacked perpendicularly to the main axis of the molecule.

Chargaff's Rule

Until Erwin Chargaff's work, scientists had laboured under the erroneous ***tetranucleotide hypothesis*** in which it was believed that

DNA was made up of equal quantities of the four bases; therefore, a subunit of this DNA consisted of one copy of each base. Chargaff carefully analyzed the base composition of DNA in various species. He found that although the relative amount of a given nucleotide differs among species, the amount of adenine equaled that of thymine and the amount of guanine equaled that of cytosine. That is, in the DNA of all the organisms studied, there is a 1:1 correspondence between the purine and pyrimidine bases. This is known as *Chargaff's rule.* Chargaff's observations disproved the tetranucleotide hypothesis; the four bases of DNA were not in a 1:1:1:1 ratio. His results were extremely important to Watson and Crick in the development of their model.

Table 2.2. Nucleotide Nomenclature

Base	Nucleotide (Nucleoside monophosphate)	*Abbreviation* Monophosphate *Ribose*	Monophosphate *Deoxyribose*	Diphosphate *Ribose*	Diphosphate *Deoxyribose*	Triphosphate *Ribose*	Triphosphate *Deoxyribose*
Guanine	Guanosine monophosphate	GMP		GDP		GTP	
	Deoxyguanosine monophosphate		dGMP		dGDP		dGTP
Adenine	Adenosine monophosphate	AMP		ADP		ATP	
	Deoxyadenosine monophosphate		dAMP		dADP		dATP
Cytosine	Cytidine monophosphate	CMP		CDP		CTP	
	Deoxycytidine monophosphate		dCMP		dCDP		dCTP
Thymine	Deoxythymidine monophosphate		dTMP		dTDP		dTTP
Uracil	Uridine monophosphate	UMP		UDP		UTP	

Table 2.3. Percentage Base Composition of Some DNAs

Species	*Adenine*	*Thymine*	*Guanine*	*Cytosine*
Human Being (Liver)	30.3	30.3	19.5	19.9
Mycobacterium tuberculosis	15.1	14.6	34.9	35.4
Sea Urchin	32.8	32.1	17.7	18.4

The Watson-Crick Model

With the information available, Watson and Crick began making molecular models. They found that a possible structure was one in which two helices coiled around one another (a *double helix*) with the sugar-phosphate back-bones on the outside and the bases on the inside. This structure would fit the dimension established for DNA by X-ray crystallography if the bases from the two strands were opposite each other and formed rungs in a helical ladder.

The diameter of the helix could only be kept constant (about 20Å–angstrom units) if there were one purine and one pyrimidine base per rung. Two purines per rung would be too big and two pyrimidines would be too small. After further experimentation with models of the bases, Watson and Crick found that the hydrogen bonding necessary to form the rungs of their helical ladder could occur readily between certain base pairs, the pairs that Chargaff found in equal frequencies. Thermodynamically stable hydrogen bonding occurs

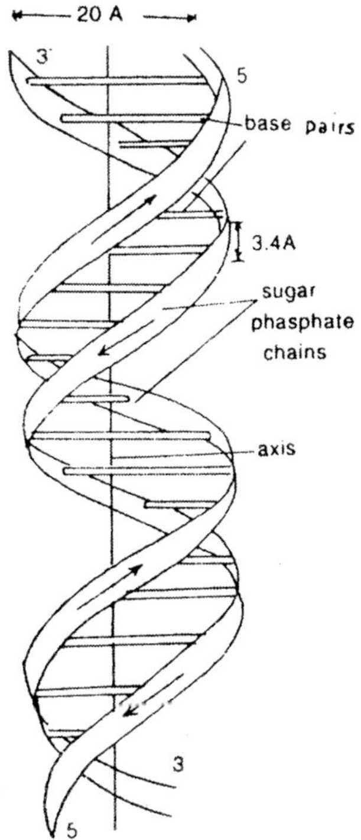

Fig. 2.2. Watson and Crick's model of DNA showing double helical structure.

between thymine and adenine and between cytosine and guanine. The relation is called *complementarity*. There are two hydrogen bonds between adenine and thymine and three between cytosine and guanine.

Another point about DNA structure relates to the fact that *polarity* exists in each strand. That is, one end of a DNA strand will have a 5´ phosphate and the other end will have a 3´ hydroxyl group. Watson and Crick found that hydrogen bonding could only occur if the polarity of the two strands ran in opposite directions; that is, the two strands were *antiparallel*.

Molecular Structure of DNA

The DNA molecule is a polymer consisting of several thousand pairs of nucleotide monomers. Each nucleotide consists of the pentose sugar–*deoxyribose*, a *phosphate* group, and a *nitrogenous base*, which may be either a *purine* or a *pyrimidine*.

Pentose Sugar

Levine (1909) identified ribose sugar as a main constituent of nucleic acid structure. *Mori* (1929) isolated sugar from the guanine nucleoside and showed that it is deoxyribose sugar. It is pentose type which in DNA molecule shows the absence of one oxygen molecule from carbon-2 position of ribose sugar. Both deoxyribose and ribose (pentose sugars of nucleic acids) have a pentagonal ring with five carbons, among which two (i.e., 3´ and 5´) are attached to phosphoric acid and three (1) to the base This sugar is called *deoxyribose* and it simply acts as a support column to which bases are attached.

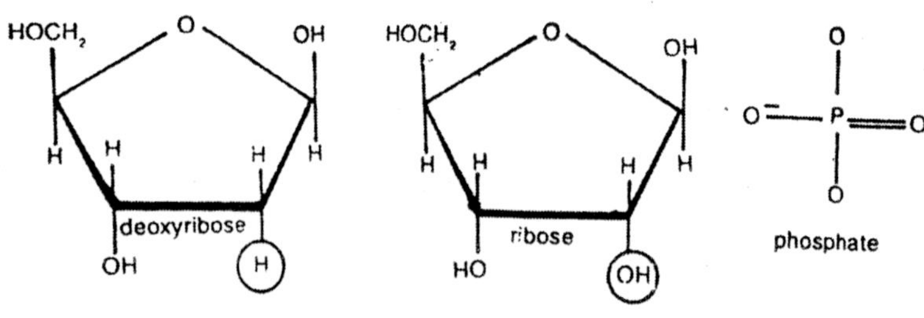

Fig. 2.3. Pentose sugars and phosphate.

Phosphate

In the DNA strand the phosphate groups alternate with deoxyribose. Each phosphate group is joined to carbon atom 3´ of one deoxyribose and to carbon atom 5´ of another. Thus each strnd has a 3´ end and a

5′ end. The two strands are oriented in opposite directions. The 3′ end of one strand corresponds to the 5′ end of the other. Consequently the oxygen atoms of deoxyribose point in opposite directions in the two strands. *Phosphodiester bonds* are formed between the sugars of two different nucleotides and a phosphate group.

Nitrogenous Bases

The nitrogenous bases of nucleic acid are of two types: *Purine* and *Pyrimidine*. Purine bases comprise mainly adenine (A) and guanine (G) which are common to both DNA and RNA while pyrimidine bases comprise cytosine (C) and thymine (T) in DNA and in the cases of RNA, thymine (T) is replaced uracil (U). These bases are arranged linearly as the back bones of chains which (back bones) are composed of alternating sugar and phosphate. The base have specific way of combining, if C is found in one chain G will be opposite to it and in other if A is on one T must be on the other. Or in other words adenine (A) is always paired with thymine (T) and guanine (G) with cytosine (C).

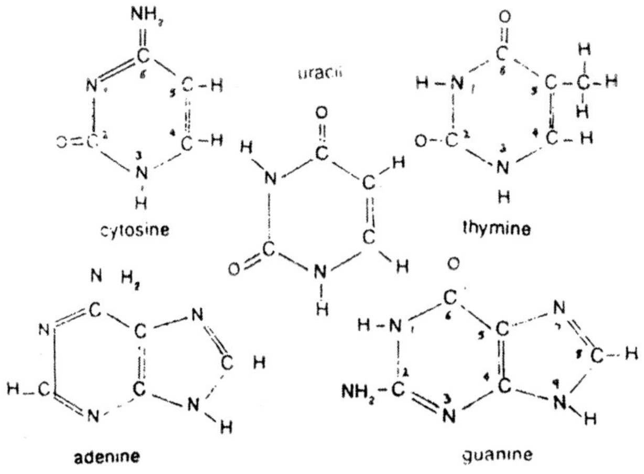

Fig. 2.4. Common nitrogenous bases of nucleic acids.

Among bases A-T and G-C, two hydrogen bonds are formed between A and three hydrogen bonds are formed between C and G. This hydrogen bond formation precludes A-C or G-T pairs. Only these arrangements A-T and G-C are possible, for two purines would occupy too much space to allow a regular helix and correspondingly, two pyrimidines would occupy too little. The stickness of these pairing rules result in a complementary relation between the sequences of bases on the two interwined chains. For example, if we have a sequence ATGTC on one chain, the opposite chain must have a sequence TACAG.

The hydrogen atom with its positive charge is shared between an oxygen atom and a nitrogen atom, both with slight negative charges. Although hydrogen bonds are weak, the fact that there are so many gives stability to the DNA molecule. The weak hydrogen bonding enables the two strands of the DNA to separate during replication.

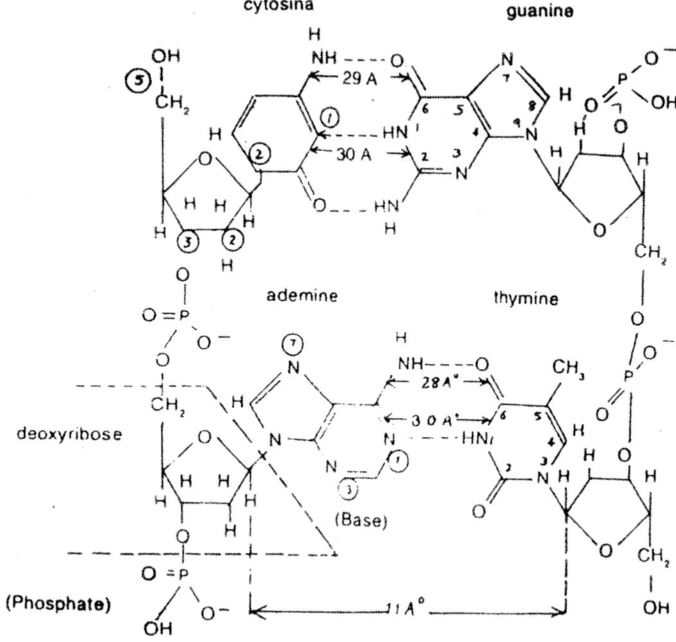

Fig. 2.5. Segment of a DNA molecule.

The bases are joined to the pentoses by N-C *glycosidic bonds.* For the purine, the glycosidic bond is between the C_1 position of the pentose sugar and the N_9 position of the bases. For the pyrimidines the linkage joins the C_1 and N_2 positions.

Nucleosides and Nucleotides

A sugar molecule and a nitrogenous base form a *nucleoside,* and a nucleoside plus a phosphate group form a *nucleotide.* In other words a nucleoside is a base-sugar combination and a nucleotide is a nucleoside phosphate. The nucleotides of RNA are called ribonucleotides, and those of DNA deoxyribonucleotides. Ribonucleotides contain the sugar ribose, and deoxyribonucleotides the sugar deoxyribose. The nucleosides are usually abbreviated as A, T, G or C, with *d*-as a prefix for deoxynucleosides (e.g., dG-is the abbreviation for deoxy guanosine).

Table 2.4. Four Nitrogen Bases, Nucleosides and Nucleotides of DNA Molecule.

Nitrogen base	Base deoxyribose =deoxyribo-nucleoside	Deoxyribonucleoside + Phosphoric acid = Deoxyribonucleotide	Abbreviation for nucleotide
1. Adenine (A)	Deoxyadenosine	Deoxyadenylic acid (Deoxyadenosine monophosphate)	3′-dAMP
2. Guanine (G)	Deoxyguanosine	Deoxyguanylic acid (Deoxyguanosine monophosphate)	3′-dGMP
3. Cytosine (C)	Deoxycytidine	Deoxycytidylic acid (Deoxycytidine monophosphate)	5′-dCMP
4. Thymine (T)	Deoxythymidine	Thymidylic acid (Deoxythymidine monophosphate)	5′-dTMP

Nucleosides

The nucleosides are compounds formed by linking purine and pyrimidine bases to either D-ribose or 2-deoxy-D-ribose in a N-β-glycosidic bond. The point of attachment of the base to the sugar is N-9 of the purines or N-1 of the pyrimidines to C-1 of D-ribose or 2-deoxy-D-ribose. The carbon atoms of the sugar are designated by prime numbers (i.e., C-1′, C-5′), while the atoms in the bases lack the prime sign.

Table. 2.5. Lists the trivial names of the purine and pyrimidine nucleosides which are related to the bases that occur in RNA and DNA.

Table 2.5. Names of nucleosides.

Base	*Ribonucleoside*	*Deoxyribonucleoside*
Adenine	Adenosine	2′-Deoxyadenosine
Guanine	Guanosine	2′-Deoxyguanosine
Uracil	Uridine	2′-Deoxyuridine
Cytosine	Cytidine	2′-Deoxycytidine
Thymine	Thymine ribonucleoside	2′-Deoxythymidine

Nucleotides

These are phosphoric acid esters of nucleosides in which phosphoric acid is esterified with one of the hydroxyl groups of D-ribose ($2'$, $3'$ and $5'$ hydroxyl group) or $2'$-deoxy-D-ribose ($3'$ and $5'$ hydroxyl group). Therefore, $2'$, $3'$ or $5'$ ribonucleoside monophosphate and $3'$ or $5'$ deoxyribonucleoside monophosphates can be formed. However, the $5'$ position is most commonly phosphorylated.

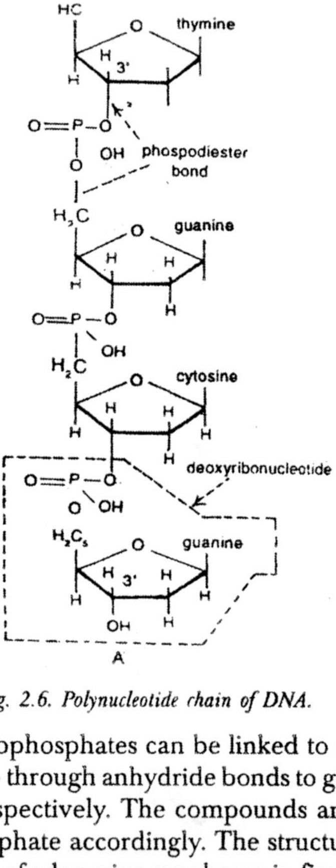

Fig. 2.6. Polynucleotide chain of DNA.

Nucleoside monophosphates can be linked to a phosphate or a pyrophosphate group through anhydride bonds to give nucleoside di- and triphosphate, respectively. The compounds are called acids or mono-, di- or triphosphate accordingly. The structure of the mono-, di- and triphosphates of adenosine are shown in figure as examples.

In the absence of oxygen atom at C-$2'$ of the above sugar ring, the structures are correspondingl known as deoxyadenosine-$5'$-monophosphate (dAMP), deoxyadenosine-$5'$-diphosphate (dADP) and deoxyadenosine-$5'$-triphosphate (dATP), respectively. A list of biologically important nucleotides is present in table.

Table 2.6. Biologicaly important nucleotides* and their nomenclature.

Ribonucleotides	*Deoxyribonucleotides*
Adenosine-5′-monophosphate (adenylic acid; AMP)	Deoxyadenosine-5′-monophosphate (deoxyadenylic acid; dAMP)
Guanosine-5′-monophosphate (guanylic acid; GMP)	Deoxyguanosine-5′-monophosphate (deoxyguanylic acid; dGMP)
Cytidine-5′-monophosphate (cytidylic acid; CMP)	Deoxycytidine-5′-monophosphate (deoxycytidylic acid; dCMP)
Uridine-5′-monophosphate (uridylic acid; UMP)	Deoxythymidine-5′-monophosphate (deoxythymidylic acid; dTMP)

* Each of the 5′-monophosphates exists as the 5′-diphosphate and 5′-triphosphate. Thus, as an example, there occurs GMP, GDP, GTP, dGMP, dGDP and dGTP.

An important discovery has been the identification of cyclic nucleotides. An important cyclic nucleotide is 3′5′-cyclic adenosine monophosphate (3′,5′-cyclic AMP or cAMP) which is called a second messenger plays a key role in the biochemical action of a number of hormones.

Metabolic Function of Nucleotides

All types of cells contain a wide variety of nucleotides and their derivatives. Nucleotides serve many functions some of which are as follows:

(i) Role in energy metabolism

ATP is the main form of chemical energy available to the cells. It is generated in cells by oxidative phosphorylation and substrate-level phosphorylation. ATP is utilized to drive metabolic reactions, as a phosphorylating agent, and is involved in such processes as muscle contraction, active transport and maintenance of cell membrane integrity. As a phosphorylating agent, ATP serves as the phosphate donor for the generation of the other nucleoside-5′-triphosphates (e.g. GTP, UTP, CTP).

(ii) Monomeric units of nucleic acids

The nucleic acids, DNA and RNA are composed of monomeric units of the nucleotides as building-blocks.

(iii) Physiological mediators

cAMP plays an important role as a 'second messenger' in epinephrine- and glucagon-mediated control of glycogenolysis and glycogenesis. cGMP acts as a mediator of cellular events. ADP is important for normal platelet aggregation and hence blood coagulation.

(iv) Components of coenzymes

Coenzymes such as NAD^+, FAD and coenzyme A are important metabolic constituents of cells and are involved in many metabolic pathways.

(v) Activated intermediates

The nucleotides also serve as carriers of activated intermediates required for a variety of reactions. UDP-glucose is a key intermediate in the synthesis of glycogen and glycoproteins. CTP is utilized to generate CDP-choline, CDP-ethanolamine which are involved in phospholipid metabolism.

(vi) Allosteric effectors

Many of the regulated steps of the metabolic pathways are controlled by the intracellular concentrations of nucleotides.

Table 2.7. Showing Amount of Various Bases in Different Tissues.

S.No.	*Source*	*Adenine*	*Guanine*	*Cytosine*	*Thymine*	$\frac{A+T}{G+C}$
1.	Human liver	30.3	19.5	19.9	30.3	1.53
2.	Human sperm	30.7	19.3	18.8	31.2	1.62
3.	Hen red cells	28.8	20.5	21.5	29.2	1.38
4.	Rat bone marrow	28.6	26.4	21.5	28.4	1.33
5.	Herring sperm	27.8	22.2	22.6	27.5	1.23
6.	*Paracentrotus lividus* (sea urchin) sperm	32.8	17.7	18.4	32.1	1.85
7.	Salmon	29.7	20.8	20.4	29.1	1.43
8.	Wheat germ	26.5	23.5	23.0	27.0	1.19
9.	Yeast	31.3	18.7	17.1	32.9	1.79
10.	*Diplococcus pneumoniae*	29.8	20.5	18.0	31.6	1.59
11.	K-12 *Escherchia coli*	26.0	24.9	25.2	23.9	1.00
12.	*Mycobacterium tuberculosis*	15.1	34.9	35.4	14.6	0.42
13.	Bacteriophage T_2	32.5	18.2	16.7	32.6	1.86

STRUCTURAL VARIATION IN DNA

Following three types of DNA have been recognized.

1. Double stranded DNA
2. Single Stranded DNA
3. Circular DNA

Double Stranded DNA

This is the common DNA present in all the living organisms. These contain two polynucleotide chain running antiparallel and twisted around each other in the form of regular double helix. The detailed structure of this type of DNA has already been explained.

Single Stranded DNA

At first it was thought that all DNA molecules are double stranded except during replication, when a small region around the replicating fork is temporarily in a non-hydrogen-bonded, single-stranded form. It therefore came as quite a surprise when experiments revealed that the DNA of several groups of small bacterial viruses exists as single-stranded molecules in which the amount of A is not equal to the amount of T and the amount of G does not equal the amount of C. Among these single-stranded phages are the spherically shaped ϕX174 and S13 viruses and the rod-shaped f1 and M13 viruses.

The chromosomes of the parvoviruses, minute viruses that infect many vertebrate organisms, are also single stranded DNA molecules. When these single-stranded viral chromosomes enter their host cells, they serve as templates for the formation of complementary strands. The resulting double helices in turn serve as templates for new single strands that then become incorporated into new virus particles. Thus, the fundamental mechanism for orderning nucleotides during the synthesis of single-stranded DNA is basically the same as that used for double-helical DNA. Nucleotide selection always occurs by attaction of the complementary base. Single-stranded DNA replication usually differs from double-helical replication in that it uses only one of the two complementary strans (the–strand) as a template for the progeny (+) strand.

Although single-stranded DNA can serve as the chromosome of a small virus, it would not be very effective in storing genetic information within the chromosomes of cells. In bacteria and in many eukaryotic cells (e.g., the haploid phase of yeast cells), there usually is only one copy of a given gene. If interphase chromosomes contained single-stranded DNA, with the double helix existing only briefly during mitosis, then the daughter cells would contain two completely different sets of genetic informations. This follows from the fact that the two complementary chains do not have identical base sequences and hence would code for entirely different amino acid sequences.

The double-strand ensures that each daughter cell will contain the same genetic information. The double-stranded form also permits effective DNA repair mechanisms to exist. For example, when one strand is damaged by exposure to X-rays, the remaining strand can provide the genetic information to rebuild the damaged section. It is probably no accident that only very small viral chromosomes are single-stranded. If they were to become, say, the size of T2, they would present too large a target for chain-damaging events.

Structure of Single-stranded DNA

Single-stranded DNA molecule have a strong tendency to fold back on themselves to form irregular double-helical hairpin loops whenever their sequences permit significant numbers of nucleotides to base-pair. Imperfect as these loops are, they nonetheless are energetically favoured under physiological salt conditions, since they allow much more effective stacking of the flat hydrophobic surfaces of the bases than is possible in any fully extended single-stranded structure.

Often, as many as half the bases of single-stranded DNA are so hydrogen-bonded, and their molecules are highly compacted. If, however, there are only minimal numbers of, neutralizing cations (e.g., less than or equal to 0.01 M Na^+) then the electrostatic repulsion between the partially unneutralized phosphates will keep the single-stranded chains highly extended, and there will be fewer hairpin loops. Single-stranded DNA is always denser than double-helical DNA. Much more complete hydrogen bonding to water is possible for the atoms of the highly regular double helix than for the atoms of the inherently irregularly shaped single-stranded DNA. In the absence of base-pairing hydrogen bonds, Van der Waal's interactions take over and bring the atoms of DNA closer together than would be the case if they formed regular hydrogen bonds with each other. An analogy can be made with behaviour of water, which becomes less dense as it forms the regular, hydrogen-bonded lattice of crystalline ice.

Rigorous Crystallography

X-ray diffraction patterns from parallel-oriented fibers of DNA provided the first clues that the polynucleotide chains of DNA have helical conformations. They also told us that DNA was a multichained molecule composed of two or three polynucleotide chains. However, the double-helical nature of DNA did not directly emerge from X-ray diffraction analysis alone. Equally important was the use of model building in which the question was asked. Given the chemical features of a single DNA chain, into what three-dimensional helical conformations could it fold? The simplest models were those in which the sugar-phosphate backbones were on the outside with the bases stacking in the center.

The double helix emerged when it was realised that thymine and guanine had keto, not enol, configurations and that by base-pairing adenine to thymine and guanine to cytosine, a stereochemically pleasing, regular helical molecule resulted. They were originally discovered through their antimicrobial and anticancer attributes. Each time these intercalating agents become inserted into a DNA molecule, they necessarily *increase* the spacing of successive base pairs along the helical axes to roughly 7Å, almost the distance between phosphate atoms in a fully extended polynucleotide chain. In this situation, very little rotation is possible around the helical axis, and so intercalating agents not only extend double helices, but extensively unwind them. For example, when a molecule of either ethidium bromide (an inhibitor of DNA synthesis) or actinomycin D (a powerful inhibitor of RNA synthesis) intercalates, the rotation angle between the two adjacent base pairs is reduced from 36° to 10°. The fact that intercalation occurs of readily indictes that it must be energetically favoured, with the van der Waals bonds holding the inserted molecules to the base pairs being stronger than those found between conventionally stacked base pairs. Intercalation is additional evidence for the *metastability* of the double-helical structure–its ability to temporarily assume many inherently unstable configurations that normally quickly revert back to the standard B conformation. One such metastable variant must be short stretches of extended chains with gaps separating adjacent base pairs, into which an intercalating agent may bind.

The Chromosomes of Viruses, E.coli, and Yeast are Single DNA Molecules

The first estimates of the average molecular weights of DNA centered at about a million, the size needed to encompass an average-size bacterial gene coding for some 300 to 400 amino acid. Therefore, it seemed natural to equate single DNA molecules with single genes. However, these early reports were inaccurate owing to DNA breakage during its isolation and study. Now it is clear that virtually all undegraded DNA molecules contain the information of at least several genes.

The most certain molecular weight values come from DNA-containing viruses. Regardless of whether the DNA content is relatively small or large, each virus particle contains a single DNA molecule. For example, all the DNA of the small monkey (simian) virus SV40 is present within a single molecule of molecular weight ~ 3×10^6 daltons (5×10^3 has pairs, or 5 kbp), while the DNA molecule of the large bacterial virus T2 has a molecular weight of 1.2×10^8 daltons (2×10^3 kbp). In these cases, the entire viral chromosome, rather than separate genes, corresponds to a single DNA molecules. Likewise, the chromosome of an *E.coli* cells is a single DNA molecule whose molecular weight is about 2.5×10^9 daltons (about 4×10^3 kbp) and whose extended length is, roughly 1 mm. DNA molecules of the yeast *Saccharomyces cerevistiae* range in size between the T2 and *E.coli* DNAs, with their average size being that expected if each of the yeast's 17 chromosomes contain one DNA molecule.

The centromeres of yeast chromosomes (if not those of all eukaryotic chromosomes) do not represent discontinuities between separate DNA molecules, but instead are specialized regions of DNA evolved to interact with the spindle bodies upon which chromosomal segreation occurs during cells division. As of yet, there is no direct proof that the very much larger chromosomes of higher plants and animals (often 50 times larger than the *E.coli* chromosomes) also contain only one DNA molecule. However, we now expect that the one chromosome–one DNA molecule rule will hold for the chromosomes of all organisms. In any case, some DNA molecules are larger than any other biological molecules by several powers of ten.

Circular Versus Linear DNA Molecules

Autoradiography and electron microscopy, initially suggested that all DNA molecules are linear and have two free ends. But when it became possible to take a better look at undergraded DNA, many DNA molecules were found to be circular. For instance, the small monkey DNA virus SV40 has a 5000-base pair circular double-helical chromosome; the similarly short chromosomes of single-stranded phages are likewise circular, as are almost all autonomously replicating plasmid DNAs.

Most, if not all, bacterial chromosomes are also circular We suspect that the DNA found in the rare circular chromosomes of higher cells are likewise circular molecules. Circular shapes were initially puzzling, since they seemed to present obstacles to the untwisting of double helices during DNA replication. Only when specific enzymes that first snip and then join DNA chains were discovered did the untwisting dilemma disappear. Equall important is the realization that linear DNA molecules do not have the uncomplicated structures first envisioned for them. As we see in the next chapter, replicating the ends of DNA molecules is not a straightforward process.

All linear DNA molecules have evolved special tricks to replicate their ends, as well as to prevent their free ends from either being nibbled back by cellular enzymes or being enzymatically ligated together to form even larger chromosomes. For example, the ends of the linear adenovirus chromosomes have specialized proteins covalently attached to the 5′ ends of strands; and poxvirus chromosomes have closed hairpin loops at their ends, with their basic structures being that of largely self-complementary circular single strands. Still other linear viral chromosomes (e.g., T2 and T7) have the same sequence at both ends, allowing the ends of the chromosome to recombine together during DNA replication to form very long, end-to-end aggregates (concatamers). Moreover, some DNA molecules that are linear when isolated from a virus particle (e.g., phage λ) are found as circles inside the host cell. This tells us that the linear and circular forms of such DNA molecules are interconvertible. Starting with a closed circle, a specific enzyme introduces two breaks, one in the + strand and one in the – strand. As these breaks are very close to each other, the intervening hydrogen bonds occasionally are broken by thermal agitation, and then the circle unfolds. Most importantly, the resulting linear form contains single-stranded ends that have complementary nucleotide sequences ("*stickly ends*").

As a later time the linear form can base-pair and resume a circular configuration. If the mission phosphodiester bonds are then reformed, the covalent circle is regenerated. There is solid evidence that the phage λ DNA interconverts between its circular and linear forms using precisely this mechanism. Later, we shall see that the duplication of such "sticky" DNAs involves circular replicative intermediates. Thus, the linear form is most likely an adaptation for injection of the viral chromosome through the narrow phase tail.

Supercoiling of Circular DNA

DNA is a very flexible structure, its exact molecular parameters are a function of both the surrounding ionic environment and the nature of the DNA-binding proteins with which it is complexed. Because their ends are free, linear DNA molecules can freely rotate to accommodate changes in the number of times the two chains of the double helix twist about each other. But once the two ends become covalently linked to form circular DNA molecules, the absolute number of times the chains twist about each other (the *linkage number*) cannot change. For the most part, changes in the average number of base pairs per turn of the double helix will necessarily be accommodated by the formation of the appropriate number of supercoils in the opposite direction.

Supercoiling refers to the further twisting of double-helical DNA molecules. Untwisting of the double helix usually leads to supercoiling in the negative (left-handed) direction while overtwisting leads to positive (right-handed) supercoiling. The supercoiled state is inherently less stable than uncoiled DNA, and the cutting, or *nicking*, of a single strand instantly converts a supercoiled molecule into its simple (*relaxed*) circular state. Because circular DNA molecules become increasingly compacted as they become more supercoiled, relaxed DNA is easily distinguished from supercoiled DNA by its slower sedimentation in a centrifugal field and by the longer time it takes to move through an agarose gel. At the present time gel electrophoresis affords the most direct way to measure supercoiling, since it can separate molecules that differ by only single turns.

Just because isolated DNA molecules would energetically prefer to be in the relaxed state does not mean that the "natural state" of DNA is relaxed. In fact, virtually all DNA within both prokaryotic and eukaryotic cells exists in the negative supercoiled state. As such it seldom, if ever, occurs as free DNA, but is complexed with specific DNA-binding proteins to form compacted molecules called *chromatin*. How this compaction occurs is best understood for eucaryotic chromatin, whose most prominent DNA-binding components are the histones. *Histones* are relatively small, positively charged, arginine- and lysine rich proteins that aggregate together to form discrete ellipsoid-shaped packets (histone cores) around which the DNA supercoils. The resulting *nucleosomes* give to chromatin its beaded appearance.

Since the same collection of histones binds to nearly all sections of DNA, histones are thought to play essentially structural roles, as opposed to enzymatic or regulatory roles. Only about half the mass of chromatin proteins is histone. The remaining proteins are not nearly as well characterized with their exact functions yet to be determined. Viral DNA chromosomes are also complexed with proteins, both when multiplying within cells and when packaged into virus particles. The SV 40 viral chromosome is at all time compacted into 24 histone-containing nucleosomes identical it structure to those existing in the chromosomes of their host cells. When freed from its protein components, SV 40 DNA is found as a negative supercoil. In contrast, no histones are complexed with the DNA found in adenovirus particles.

The vertebrate viruses encode a unique DNA binding protein to help package their DNA within their protein capsules. When, however, an adenovirus DNA functions within a cell, its own DNA-binding protein is released and replaced with the same histone components that bind to all other cellular DNA.

DNA Supercoiling Around Nucleosome

In forming the nucleosomes of eucaryotic cells, 200-base-pair-long segments of DNA negatively (left-handedly) supercoil twice around the histone cores. The energy lost by supercoiling may be partially compensated by the energy gained from the ionic and hydrogen bonds formed between the histones and the DNA. The number of supercoils in solution need not be the same as found in chromatin. In chromatin, the double helix is slightly more twisted than in solution, with 10 base pairs per helical turn in chromatin versus 10.5 base pairs per helical turn in naked DNA in solution. As a result, there are fewer superhelices found in solution than found in chromatin.

Histone-like Proteins in Prokaryotes

For many decades, it was believed that bacterial DNA, unlike eukaryotic DNA, occurs naked and does not have a compacted chromatin-like arrangement. But recently, using more genetle preparative procedures, condensed *E.coli* chromosomal sections have been seen that are organized into bead-like packets from which small, basic proteins analogous to the histones can be extracted. That bacterial DNA is complexed with proteins that have histone-like properties undoubtedly reflects its need to be regularly compacted in order to function.

Crystallization of these proteins, DNA-binding protein II, reveals that its 9500-dalton chains associate in pairs as dimers containing extended arginine containing arms able to interact with the phosphates of one turn of the DNA backbone. So it is likely that dimers sited adjacently along prokaryotic DNA are oriented into a helical arrangement with the DNA bound on the outside. Interestingly, the average *degree of supercoiling* (i.e., the number of super-helicle twists per ten base pairs) is about 0.05 for all naturally occuring DNA supercoils, regardless of their source.

Topoisomerases of Supercoiled DNAs

That supercoiling not only occurs but is biologically very important is affirmed by the existence of the enzyme topoisomerase II, which specifically generates the negative supercoils that characterize so much of chromosomal DNA. *Topoisomerases* are a group of enzymes that convert (isomerize) one topological version of DNA into another. They do so by changing the linkage number, which is the number of times two DNA chains twist around each other.

Prokaryotic topoisomerase II (sometimes called *gyrase*) uses the energy of ATP to generate negative supercoils by untwisting DNA in the left handed direction. Its eukaryotic counterpart, however, may be capable of using ATP to generate negative supercoils only in the presence of other, yet to be described proteins. The supercoiling action of topoisomerase II is counterbalanced by a second enzyme, topoisomerase I, which converts supercoiled DNA to the unstrained, energetically more favourable relaxed state. The relative amounts of topoisomerase I and topoisomerase II in cells are finely tuned, tending to create just the right amount of negative supercoiling. Thus, mutations that lower the number of topoisomerase I molecules are viable only if the number of topoisomerase II molecules also decrease.

Both topoisomerase I and II work by catalyzing the breakage and rejoining of DNA phosphodiester bonds. Unlike virtually all other enzymes, their action does not lead to new patterns of covalent bonds. Rather, their role is to create temporary gaps in polynucleotide chains.

When acting topoisomerase do not create free ends but instead become themselves covalently attached to one of the two broken ends (depending on the specific enzymes, either the 3′ and 5′ end). Through such catalysis, a tyrosine group on the topoisomerase becomes linked to the terminal phosphate group of the cut polynucleotide chain. The free energy present in the original phosphodiester bond is thus preserved so that it can be used to rejoin the broken chain. When a DNA chain is broken by a topoisomerase, its broken ends do not fall apart but are held together by the topoisomerase. At no time do the broken ends have the capacity to rotate freely. If that happened, topoisomerase would be limited to only completely relaxing double helices. Instead, individual topoisomerases function to make discrete changes of either one positive turn (topoisomerase I) or two negative turns (topoisomerase II) in the linkage number. Such discrete steps result from DNA chains passing through either transient single-stranded breaks (topoisomerase I) or transient double-stranded breaks (topoisomerase

Table 2.8. DNA Topoisomerases

	Procaryotic (E.coli)		Eucaryotic (Hela)	
	Type I	*Type II*	*Type I*	*Type II*
Approximate MW	100,000	400,000	100,000	309,000
Subunits	Monomer	Tetramer (A_2B_2)	?	Dimer (subunit MW=172,000)
Gene	top A	gyr A Polypeptide MW 105,000; activity inhibited by nalidixic acid and oxolinic acid gyr B Polypeptide MW 95,000; activity inhibited by coumermycin and novobiocin	?	?
Covalent intermediate with DNA	At 5′ P	Subunit A Tyrosine covalent bond to 5′ P	At 3′ P	At 5′ P
Relaxation	Mg^{2+} required; ATP-independent	ATP-independent	No Mg^{2+} required; ATP-independent	ATP-dependent
Negative supercoiling	None	ATP-dependent	None	None

II). Topoisomerases are constructed in such a way that they can undergo reversible conformational changes that create cavities through which DNA chains can pass.

The ATP requirement for bacterial topoisomerase II (gyrase)-induced supercoiling most likely reflects the use of ATP in mediating the necessary conformational changes that lead to chain passage. Topoisomerase I, however, relaxes supercoiled DNA without requiring a cyclical input of energy. Complete understanding of how topoisomerase I and II work can occur only when their precise structures are determined through X-ray crystallographic procedures. Even now, however, the prediction can be made that double helices wrap themselves around topoisomerase II in positive supercoils prior to the start of the cutting-rejoining cycle. Without this restraint on the orientation of the DNA, topoisomerase II would break as well as make negative supercoils, and the net result of its action would be the eventual relaxation of the double helical substrate to the uncoiled state.

Looped and Supercoiled Structure

The fact that linear DNA molecules have free ends would seem to preclude their possession of supercoiled segments. Yet, examination of gently isolated linear eukaryotic chromosomes shows that their chromatin is organized into a large number of successive looped domains organized along a scaffold containing two major DNA-binding proteins. Somehow, attachment to the scaffold prevents the rotation of one domain from being transmitted to adjacent sections, since the DNA within each of these loops independently supercoils and may be under different torsional strain.

There is much evidence both from *Drosophila's* giant salivary chromosomes and from the looped lamp brush chromosomes of the salamander that the individual loops represent functional units of chromatin, with given loops usually being transcribed into one or more very long RNA molecules. How the protein scaffold might maintain independently supercoiled loops was a complete mystery until recently. Now it is less so, following the discovery that a major component of the scaffold is none other than topoisomere II.

The data presented illustrate several points:

1. The sixteen possible nearest-neighbour frequencies occur with a large number of frequencies.
2. The sums of the vertical columns show that the amount of A is equal to the amount of T and that G equals C, thus indicating the DNA was probably replicated correctly.

3. The two DNA strands are of opposite polarity. This is borne out by the frequency equivalence of the pairs of sequences: CpT and ApG, GpT and ApC, GpA and TpC, and CpA and TpG, as predicted by antiparallel DNA strands. Were the two strands of the same polarity, different matching sequences would have been predicted, for example, TpA and ApT, GpA and CpT, CpA and GpT, etc.

To show that newly synthesized DNA is made using a DNA template requires two rounds of nearest neighbour analysis. In the first round the originally isolated DNA is used in the reaction and a set of 16 nearest neighbour frequencies are obtained as described. In the second round the enzymatically synthesized DNA is used in the reaction and a second set of frequencies is produced.

The results show good agreement between the two sets of frequencies, thus indicating that DNA plays a template role in DNA replications in vitro. Since the early, comparatively, crude in vitro DNA synthesis experiments, a large number of refinements have been made. For example, the reaction mixtures now duplicate the conditions that are present in growing cells such that it is possible to synthesize DNA in vitro that is identical in all respects with DNA produced in vivo. The reaction mixtures required for efficient DNA synthesis vary, depending on the DNA used as the template. In general, there is a basic set of enzymes and proteins required for the synthesis of all DNA sources. Beyond that, each DNA requires a number of specific proteins for new synthesis to occur, and the actual set of proteins needed depends on the DNA in question.

Types of DNA's

Prokaryotic DNA

Also called the *circular a superhelicular DNA.* The DNA molecules of prokaryotes and most viruses are circular. A circular molecule may be a covelently closed circle, which consists of two unbroken complementary, single strands, or it may be a *nicked circle,* which has one or more interruption (nicks) in one or both strands. With few exceptions, covalently closed circles are twisted. Such a circle is said to be a *superhelix* or a *supercoil.* It has also been discussed already.

Eukaryotic DNA

It is the filamentous type of DNA found in all eukaryotic cells of animals and plants and has already been discussed.

Extranuclear DNA

Also known as non-chromosomal or cytoplasmic DNA. In cytoplasm this DNA is associated with some of cell organelles as given belows:

(a) Mitochondrial DNA

Mitochondrial DNA is usually found in cyclic double stranded, supercoiled molecules, the exceptions being the linear mitochondrial DNA molecules from *Tetrahymena* and *Parameciui*. Marmialian mitochondrial DNA molecules are not packaged into nucleosomes. They are about 15 kbp long and can therefore code for 15 to 20 proteins. However, some yeast ones are considerably larger and *Saccharomyces* mitochondrial DNA is 75 kbp long.

Plant mitochondrial DNA is much longer still. Yeast mitochondrial DNA molecules have been widely studied since they are readily amenable to genetic analysis. Mitochondrial DNA can only codes for a small proportion of mitochondrial proteins. Mitochondrial DNA from several mammals, *Xenopus* and *Drosophila*, has been sequenced and the sequence analysed. There are genes for two ribosomal RNAs, 22 tRNAs and for 13 protein most or all of which are involved in electron transport. The only region of mammalian mitochondrial DNA which is non-coding is the *D-loop region* involved in the initiation of DNA replication. This is also the region at which transcription of both strands is initiated.

Transcription continues uninterrupted around the cyclic molecule and the transcripts are the processed to give individual messenger, ribosomal and tRNAs. The tRNA forms the puncturation between the various protein coding regions and provides the signals for the processing enzymes. Plant mitochondrial DNA is apparently much more complex than animal mitochondrial DNA. It consists of per mutation of basic structure related to each other by recombination.

(b) Chloroplast DNA

Chloroplast DNA is in general much larger than mitochondrial DNA, being in the molecular weight range 100 million (150 kbp). In contrast to mitochondria which have from one to ten molecules of DNA per organelle, chloroplasts tend to have a very large number of copies of the DNA molecules in each organelle, in some cases greater than one hundred. Like mitochondrial DNA, the DNA in chloroplasts carries the coding information for essential membrane components, tRNA and rRNA. All known chloroplast DNA molecules are cyclic and supercoiled.

(c) Kinetoplasts DNA

The kinetoplast is part of highly specialized mitochondrion found in certain groups of flagellated protozoa such as trypanosome. Its DNA (kDNA) consists of an interlinked series of many thousands of cyclic DNA molecules which vary in size from 0.6 kbp to 2.4 kbp depending from the type of trypanosomes from which they are obtained. These components, which are known as *minicircles*, are further interlinked with a much smaller number of larger circular DNA moiecules of above 30 kbp in size known as *maxicircles* in the majority of system analyzed. While maxicircles appear to perform the connectional functions of mitochondrial DNA in trypanosomes, minicircles are microheterogenous in sequence and size and there is no evidence to suggest that they are ever transcribed so that it is possible that they may fulfil some structural rather than coding role.

Satellite DNA

At the other extreme DNA with a $Cot_{1/2}$ value as low as 10^{-3} moles of nucleotide seconds $litre^{-1}$ consists largely of *satellite DNA*. This represents highly repeated sequences of which there may be a million or more copies per haploid genome, which are usually quite short and are arranged in tandem arrays. The origin of the name satellite relates to the method of its isolation on caesium chloride buoyant density gradients of sheared DNA where it will sometimes form a satellite band separate from main DNA band, due to its differing content of adenine and thymine residues.

The simplest known DNA is poly [d(A-T)], which occurs in certain crabs. Other satellites can have any number upto several hundred base pair which are repeated in tandem fashion along the genome. The distribution of satellite DNA among chromosomes varies. Some chromosomes have virtually no satellite sequences while others are largely composed of satellite sequences. In general, satellite DNA appears to be concentrated near the centromere of the chromosomes in the heterochromatin fraction. DNA sequence analysis has shown that the basic repeat unit of satellite DNA is itself made up of sub repeats. For example, the major momse satellite has a repeating structure of 234 base pairs made up of four related 58 and 60 bp segments each in turn made up of 28 and 30 bp sequences. The satellite between and within the related species are themselves related in an evolutionary sense by cyclic rounds of multiplication and divergence of an initial short sequence.

Fold-back DNA

Also known as ***palindromic DNA***. It is a special class of DNA sequences comprising 3-6% eukaryotic DNA. The size range is from 300 to 1200 base pairs and the molecules have a $Cot_{1/2}$ value of less than 10^{-5} moles of nucleotide second $litre^{-1}$. This palindromic DNA is represented in all frequency classes and is widely distributed throughout the metaphase chromosomes. Some palindromic DNA arises when two copies of the ***Alu*** sequences are present close to one another in opposite orientations and it was a result of the ability of such DNA to renative instantaneously.

Tautomeric form of DNA

An important chemical feature of DNA is the position of the hydrogen atoms in the purine and pyrimidine bases. Before 1953, many chemists thought that some of these hydrogen atoms randomly moved from one ring nitrogen or oxygen atoms to another and so could not be assigned a fixed location. Now we realize that although such movements, called ***tautomeric shifts*** do occur, these hydrogens have preferred atomic locations.

The nitrogen atoms attached to the purine and pyrimidine rings are usually in the ***amino*** (NH_2) form and only rarely assume the ***imino*** (NH) configuration. Likewise, the oxygen atoms attached to the C6 atoms of guanine and thymine normally have the ***keto*** (C-O) from and only rarly take up to ***enol*** (COH) configuration. It is essential to the biological functioning of DNA that the hydrogen atoms have relatively stable locations. If the hydrogen atoms had no fixed positions, adenine could often pair with cytosine and guanine with thymine, and sequence of the bases on the two intertwined chains would not necessarily be complementary. If that were the case, the DNA could not function as a genetic molecule, for it is the complementary relationship between the opposing chain sequences that gives DNA its capacity for self-replication.

Replication of the double helix involves strand separation followed by formation of complementary DNA chains using the now free single strands as templates to attract the appropriate partner bases dictated by AT and GC base-pairing rules. The amino forms adenine and cytosine and the enol forms of guanine and thymine must indeed occur only very rarely compared to their amino and keto alternatives. Otherwise, the many errors (mutations) incurred during DNA replication would be incompatible with orderly cell growth and division.

Unique DNA

In human cells it is generally classified as that DNA which has a $Cot_{1/2}$ value of about 1000 moles of nucleotide seconds litre^{-1}. It comprises about half of the total haploid DNA content and is thought to consist of the sequences coding for most enzyme functions for which there is only one or a small number of genes per haploid genome. The genes coding for the various chains of globin or the enzyme glucose 6-phosphates fall into this category.

Repetitive DNA

In this type of DNA certain bases or nucleotides are repeated many times. Among the repetitive DNA is a fraction which reanneals with a Cot value of between 100 and 1000 moles of nucleotides seconds litre^{-1}. The sequences represented in this group are generally thought to be those coding for proteins which form major structural components of the cell such as the histones. The genes for rRNA and tRNA also fall into this category.

Z-DNA or Left Handed DNA

On the basis of the arrangement of pentose sugar forming chains of DNA two forms of DNA have been reported. Usually the sugar chain has the pentose sugar with ascending sequence of carbon atoms. This is right handed DNA or B-DNA. Recently in 1979, another form of DNA have pentose sugar with ascending arrangement of carbon atom has been reported and named Z-DNA or left handed DNA.

(a) Resemblances between Z-DNA and B-DNA

(i) Both are double helical.

(ii) The two strands of double helix are antiparallel in both DNAs.

(iii) Both forms exhibit G≡C pairing.

(b) Differences between Z-DNA and B-DNA

(i) Z-DNA has left handed helical sense as against right handed helical sense of B-DNA.

(ii) Due to a different arrangement of molecules within the Z-DNA polymer, the phosphate backbone follows a zig-zag course, while in B-DNA it is regular.

(iii) In Z-DNA, the sugar residues have alternating orientation so that repeating unit is a dinucleotide as against the B-DNA where repeating unit is a mononucleotide and the orientation of sugar molecules is not alternating.

(iv) In Z-DNA, one complete helix i.e. a twist through 360°, has twelve base pairs or six repeating dinucleotide units (12 base pairs) while in B-DNA, one complete helix has only ten base pairs or ten repeating units.

(v) Because twelve base pairs are accommodated in one helix in Z-DNA, as against ten in B-DNA, the angle of twist per repeating unit (dinucleotide) is 60° as against 36° in B-DNA.

(vi) One complete helix is 45Å in Z-DNA it is 34Å in B-DNA.

(vii) Since bases get more length spread out in Z-DNA and since the angle of tilt is 60°, they are more closer to the axis and hence the diameter of the Z-DNA molecule is 18Å, whereas it is 20Å in B-DNA.

Denaturation and Renaturation of DNA

Single-stranded DNA

Single-stranded DNA molecules rarely occur naturally. The best known example is the DNA molecule from the small spherical bacteriophage ϕX174 first isolated by *Sinsheimer* in 1959 but it is also found in the filamentous bacteriophages such as fd. Among the animal viruses single-stranded DNA is found in the parvo viruses. Some of these naturally occurring single-stranded DNA's are circular and other linear. The former can be distinguished by their insensitivity to exonucleases. In general they are comparatively small molecules with less than 10,000 nucleotides.

The molar proportions of the bases do not show the usual equivalence of A and T, and G and C, required for double helix formation and the bases react readily with formaldehyde since the amino groups are not protected by hydrogen bondings as they are in double stranded DNA. Single-stranded DNA molecules are very much more flexible than double-stranded molecules of the same size since they lack the rigid double helical structure. In general they behave in solution in a manner similar to RNA molecules.

Hypochromic Effect

When double stranded DNA molecules are subjected to extremes of temperature of pH, the hydrogen bonds in the double helix are ruptured and the DNA collapses into two single-stranded molecules. If heat is used as the denaturant, the temperature at which this collapse occurs is known as the T_m or transition temperature. The absorption at 260 nm of any polynucleotide is due to that of its component bases. However, this absorption tends to be suppressed in the double-stranded DNA molecule where the bases are stacked above one another and inhibited from swinging out freely in solution in hydrogen bonds, and consequently it is very much lower than that of equimolar amounts of the component bases or nucleotides free in solution. This inhibition of absorption is to some extent relieved when the DNA molecule goes through the helix-coil transition and the bases are no longer so rigidly stacked although they still interact to some extent.

As a consequence of this absorption of DNA solutions rises by about 20-30 percent as the DNA undergoes the melting process, and the transition process is usually measured by ultraviolet spectroscopy. The increase in absorbance on melting is known as the *hyperchromic effect.*

Such a transition can be characterized by several factors:

1. *The nature of the DNA.* Homogenous DNA, such a viral DNA, melts over a short temperature range but heterogeneous DNA melts over a longer range.
2. *The (G + C) content of the DNA.* The melting temperature of any DNA can be related to its (G+C) content since this base pair confer extra stability on the molecule. In DNA molecules such as those from bacteriophage l, in which some regions are richer in (G+C) than others, the transitions of both regions can be observed.
3. *The nature of the solvent.* In low concentrations of counterion the transition temperature is low and its width broad. At high concentrations of counterion the T_m is raised and with width of the transition becomes sharp. However, if 'denaturing' salts such as sodium perchlorate are present, addition of more salt will lower the T_m since the rupture of apolar bonds caused by the anion will overcome the ionic stabilization of the cation.

The helix-coil transition is also associated with a change in density of the DNA molecule, the single-stranded DNA being more dense than the equivalent double stranded form with the same (G+C) content.

The Renaturation of DNA

When two DNA strands are returned from the extreme conditions which caused them to melt to their original state they may reassociate to form a double helix again. However, whether or not they do so depends on a variety of factors. In principle the process should follow simple second-order kinetics but, because of the high probability of two sequences which were not previously matched forming an imperfect union which is then difficult to dissociate, the degree of renaturation can vary from less than 1 percent according to the nature of the sample and the conditions of renaturation.

1. *The nature of the sample.* Simple sequence DNA molecules such as $d(G)_n$ $d(C)_n$ have no difficulty finding the appropriate sequence with which the reanneal and do so without difficulty. However, if a eukaryotic DNA sample from a large genome is reannealed, clearly some of the sequences will have to encounter many other non-complementary sequences in solution before they find the correct partner. Thus if most of the DNA from eukaryotic cells is quickly cooled to a low temperature, so that the diffusion of the DNA in solution is inhibited, few of the single strands will reassociate.
2. *The temperature of the reassociation process.* At very low temperatures (4°C) not only a diffusion limited but, for a DNA molecule which has become mismatched with a strand having only a few complementary bases, the opportunities to break away and continue its search for the correct complementary strands are reduced. Consequently DNA which is heated beyond the T_m and then quickly cooled to a low temperature will be denatured whereas solutions maintained at high temperatures below the T_m may renature.
3. *The size of the DNA fragments.* Large string-like fragments of single-stranded DNA encounter diffusion problems and frequently cannot reanneal correctly since this effectively reduces their opportunities for finding complementary strands. For this reason DNA is often sheared for reannealing experiments.
4. *The ionic strength of the solution.* Two highly charged DNA molecules are likely to repel one another, and the presence of salt is necessary to mask this repulsion in renaturation.
5. *The concentration of the DNA.* Clearly, at higher concentrations the probability of two complementary strands encountering each other is raised.
6. *The time allowed for reannealing experiments.* If renaturation is allowed under ideal conditions, two DNA samples of identical concentration should take different times to reanneal according to the genome size. For this reason the term *Cot value* has been defined for the study of reannealing of DNA and also for the study of the formations of DNA-RNA hybrids. *Co* represents the DNA concentration and *t* represents time in seconds. This is again monitored by ultraviolet spectroscopy or, more frequently, by hydroxyapatite chromatography since this can be used to separate double- and single-stranded DNA molecules and can be used at higher concentrations and hence for shorter times.

3

Replication of DNA

During cell division the chromosomes which are made up of DNA molecules undergo spletting and the sister chromatids move apart to maintain the usual number of chromosomes. Each chromatid has a single chain of DNA. To make it double and to convert chromatid upto chromosome the duplication of DNA becomes essential.

Replication of the DNA Molecule

Kornberg showed that DNA can replicate in a test tube with no cells present. All that is required is a mixture containing DNA, a specific enzyme (which he called DNA-polymerase), and a mixture of the four precursors: the nucleoside triphosphates deoxy-ATP, deoxy-CTP, deoxy-GTP, and deoxy-TTP. If any one of the four nucleoside triphosphates is omitted from the reaction mixture, DNA does not replicate itself. The intact DNA serves as a template for the reaction–a guide to the exact placement of nucleotides in the new strand. Where there is a T in the template, there must be an A in the new strand, and so forth.

Two other models of how the double helix might replicate were suggested after the original paper by Watson and Crick. In the model of conservative replication, the original double helix would serve somehow as a template but would either be reconstituted or, perhaps, would never unwind at all. Thus the new molecule would contain none of the atoms of the original. According to the model of dispersive replication, fragments of the original molecule would serve as templates, assembling two molecules, each containing old and new parts, perhaps at random. In *semiconservative replication* (the model proposed by *Watson* and *Crick*), the original two strands would separate, and each would function as the template for a new partner. Each molecule produced by semiconservative replication would therefore consist of one old and one new strand. After a short time, experimental work confirmed the model of semiconservative replication.

Synthesis of DNA from monomeric units of deoxyribonucleotides using a pre-existing DNA molecule as a template is DNA replication. Three mechanistic models have been proposed regarding DNA replication. These are:

(i) Dispersive

(ii) Conservative

(iii) Semi-conservative.

(i) Dispersive

According to this model, it was proposed that the original DNA is broken into many fragments, which are more or less uniformly distributed among the progenies. Inside the progeny, these fragments grow to form complete DNA segment. However, according to this model one would expect unequal distribution of genes among progenies leading to great variations among them. It would lead to irregular

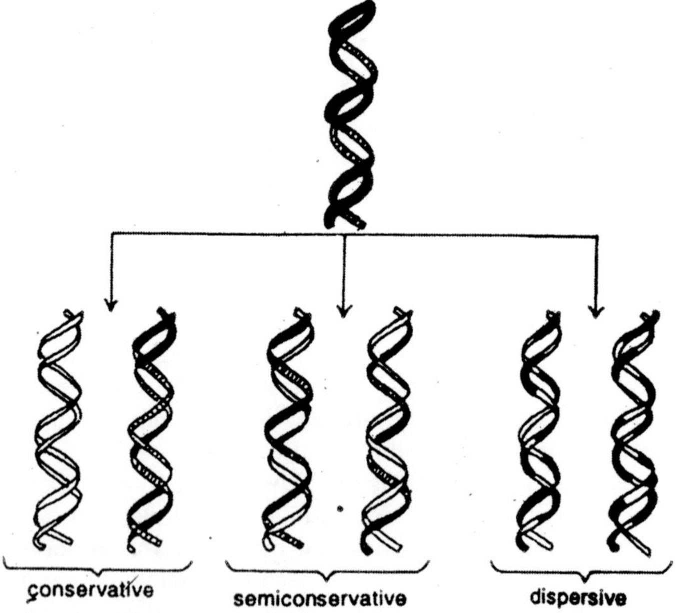

Fig. 3.1. A diagram illustrating three possible modes of DNA replication.

arrangement of nucleotides. However, these expectations are not realized and hence the dispersive mechanism of DNA replication does not find support.

(ii) Conservative

According to this model, the original DNA is copied as such and then one copy of the complete DNA is transmitted to one of the progenies. Thus, according to this model, both primary and secondary structures of the DNA are conserved. There is neither breakage nor unwinding of the parental DNA during replication. The experimental evidence however, do not support this model as well, although such a mechanism of DNA replication is possible structurally and chemically.

(iii) Semi-conservative

This mechanism is based on Watson-Crick's model of DNA structure and is now accepted mechanism of DNA replication. The replication of DNA involves the unwinding of double delical DNA molecule but no breakage of the separated strands. Thus, the primary but not the secondary structures of parental DNA are conserved in progenies.

The idea of semiconservative mechanism of DNA replication came from the careful studies of the double helical model of DNA structure. The complimentary strand structure is an indication of the possible strand separation and formation of complimentary strands on each of the free single strands. During semiconservative replication, the hydrogen bonds between the complimentary base pairs are broken and the two separated strands start replicating as soon as a few bonds are broken. The two strands do not separate completely before the new strands are formed. That is why during the process of DNA replication some 'Y' shaped regions in the DNA molecule are observed. Each strand can act as a template or mould for the formation of new complimentary chains. The nucleotides are put together into short pieces of DNA by one of the DNA polymerases. The short segments of the newly synthesized DNA are known as *Okazaki fragments.* The complimentarity of the strands is maintained because of the selective pairing between A and T and G and C. The small Okazaki fragments are joined together to form complete DNA strands by the enzyme polynucleotide ligase. In experiments, where ligase is selectively inhabited, the accumulation of Okazaki fragments take place.

Experimental Proof

A clever experiment by Matthew Meselson and Franklin Stahl convinced the scientific community that semiconservative replication is the correct model. Working at the California Institute of Technology in 1957, they devised a simple way to distinguish old strands of DNA from new ones. The key was to use a "heavy" isotope of nitrogen. Heavy nitrogen (^{15}N) is a rare, nonradioactive isotope that makes molecules more dense than chemically identical molecules containing the common isotope ^{14}N. To study DNA of different densities (that is, DNA containing ^{15}N versus DNA containing ^{14}N), Meselson, Stahl, and Jerry Vinograd invented a new centrifugation procedure that allowed them to determine the density of DNA from a specific sample. At a certain molarity, a solution of cesium chloride (CsCl) has a density very close to that of DNA; at high gravitational forces produced in an ultracentrifuge, cesium ions sediment to some extent, thus establishing a density gradient. When a DNA sample is dissolved in CsCl and centrifuged at about 100,000 times the force of gravity, the DNA gathers in a layer in the centrifuge tube at a position where the density of the CsCI solution equals that of the DNA. To cluster in this way, DNA that is initially lower in the tube, where the density is greater than its own, must rise; DNA that is in a region of lower density must sink.

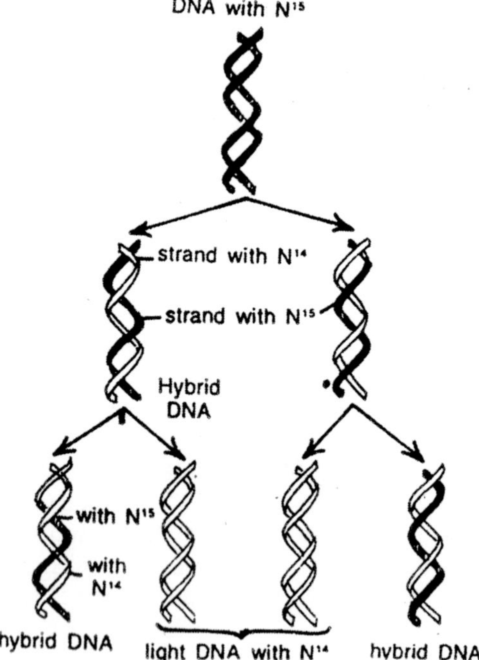

Fig. 3.2. Meselson and Stahl's experiment.

After developing this method of measuring DNA densities, Meselson and Stahl could begin experimenting. They grew a culture of the bacterium *Escherichia coli* for 17 generations on a medium in which the nitrogen source (ammonium chloride, NH_4Cl) was made with ^{15}N instead of ^{14}N. As a result, all the DNA in the bacteria was "heavy." Another culture was grown on medium with ^{14}N. They extracted DNA from both cultures. When these extracts were combined and centrifuged with CsCl, two separate DNA bands formed, showing that this method would work for separating DNA samples of slightly different densities.

Meselson and Stahl then conducted their main experiment. They grew another culture on ^{15}N medium and *transferred* it to normal ^{14}N medium. *E.coli* reproduces every 20 minutes, and Meselson and Stahl collected some of the bacteria from each generation after the transfer. They extracted DNA from the samples. After DNA was duplicated and the cells divided to produce each new generation, the DNA banding in the density gradient was different from the original banding. Initially, the DNA was uniformly labeled with ^{15}N and hence was relatively dense. After one generation, when the DNA had been duplicated once, all the DNA was of an intermediate density. After two generations, there were two equally large DNA bands: one of low density and one of intermediate density. In samples from subsequent generations, the proportion of low-density DNA increased steadily.

These data can be explained by the semiconservative model of DNA replication. The high-density DNA had two ^{15}N strands, the intermediate-density DNA had one ^{15}N and one ^{14}N strand, and the low-density DNA had two ^{14}N strands. In the first round of DNA replication, the strands of the double helix, both heavy with ^{15}N, separated; during the process of separation, each acted as the template for a second strand, which contained only ^{14}N and hence was less dense. Each double helix then consisted of one ^{15}N and one ^{14}N strand and was of intermediate density. In the second replication, the ^{14}N-containing strands directed the synthesis of partners with ^{14}N, creating low-density DNA, and the ^{15}N strands got new ^{14}N partners.

If the DNA had replicated in accord with either of the other models, the results would have been quite different. Under the conservative model, after one generation there would have been two bands–one for heavy DNA (^{15}N–^{15}N) and the other for light (^{14}N–^{14}N); no DNA of intermediate density would have formed at any time. If dispersive replication had taken place, the first round of replication would have produced DNA of intermediate density, but the density of all the DNA would have decreased with each subsequent replication. The crucial observation proving the semiconservative model was that intermediate-density DNA ($^{15}N-^{14}N$) appeared in the first generation and continued to appear in subsequent generations.

Cairn's Autoradiography Experiment

J. Cairns demonstrated the semiconservative type of replication by using a technique of autoradiography as given in the following lines:

In autoradiography technique, the material is first supplied with a suitable radioactive material like tritiated thymidine (H^3-TdR; H^3 is heavy isotope of hydrogen and it replaces normal hydrogen in thymidine to give rise to tritiated thymidine). Tritiated thymidine is used since this will selectively label only DNA and will not label RNA, since thymine base is absent in RNA. The tritiated thymidine gets incorporated into DNA and replaces ordinary thymidine. The material is then sectioned or else the cells may be broken down to release the intact bacterial chromosome on slides. These slides are then covered by photographic emulsion and stored in the dark. During this storage the particles emitted by tritiated thymidine will expose

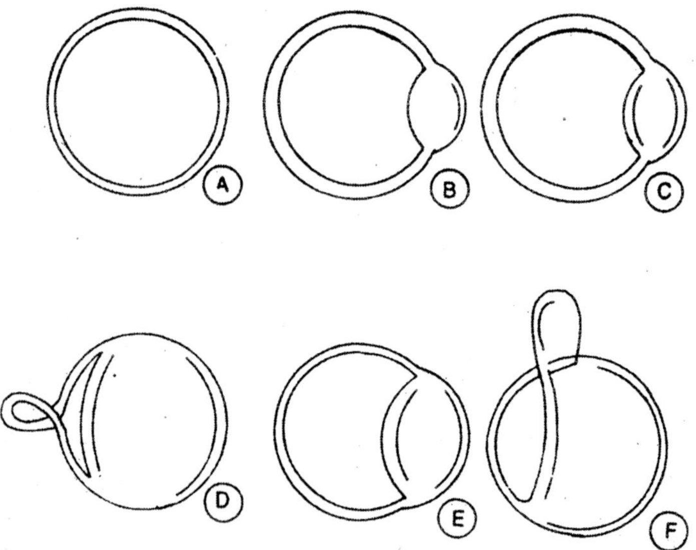

Fig. 3.3. Cairn's model for circular DNA replication.

the film, which can be developed. This photograph will then show the regions of the presence of tritium and thus indirectly the presence of labelled DNA.

Using this above technique, the replication of DNA could be easily followed by *J. Cairns* and the results were reported in 1963. During incorporation of tritiated thymidine, autoradiographs could be prepared at regular known intervals. Autoradiographs from this replicating material prepared at regular known intervals demonstrated semi-conservative mode of replication. It would be expected that in regions, where one cycle of replication of DNA has been completed, the density of dots will be higher than in the region where the replication has not taken place. The lighter density of dots is considered to indicate that only one of the two strands is labelled, while a heavier density of dots will indicate that only one of the two strands is labelled, while a heavier density of dots will indicate that both strands are labelled. Such a situation was actually observed. The rate at which the replication proceeds could also be worked out by measuring the length of DNA undergoing replication in a known interval of time. The generation time was worked out by *Cairns* as 30 minutes in *E.coli.* The length of the chromosome was worked out to be about 1 mm. The rate of replication as obvious would be approximately 30μ to 40μ per minute (1 mm = 1000μ).

That after replication, one of the two strands in the daughter DNA molecules is derived from the parent molecule and the other is synthesized afresh. In θ shaped figure, which is obtained in the second cycle of replication is presence of label, the two areas in the split region would never be equally labelled. For instance, one arc would be twice as heavily labelled as the other arc. This is what was actually observed by *Cairns.* Thus the observation support the semiconservative type of replication of DNA.

Taylor's Experiment on Vicia faba Root Tips

J.H. Taylor (1957) and his coworkers also demonstrated the semiconservative method of DNA duplication in the root tip cells of *Vicia faba* by autoradiography. The roots were grown in a medium containing radioactive thymidine, so that the radioactivity is incorporated in the DNA of these cells. The outline of this labelled chromosome on a photographic film appears in the form of scattered black dots of silver grains. When these root tips with labelled chromosomes were transferred to the unlabelled medium containing colchicine and studied for radioactivity, the following observtions were made:

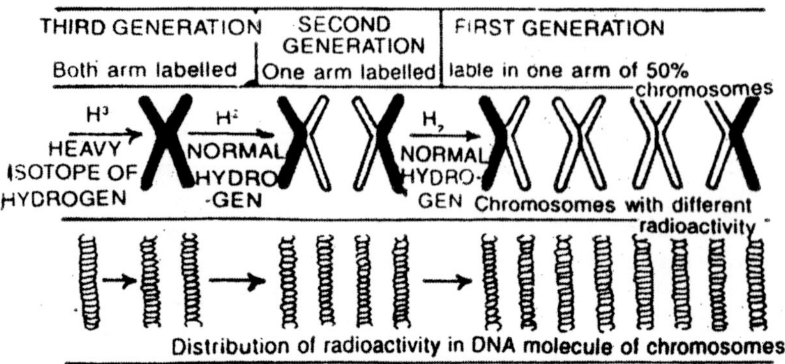

Fig. 3.4. Taylor's experiment on Vicia faba.

1. In the chromosomes of first generation the radioactivity was found to be uniformly distributed in both the chromatids, because in them the original strand of DNA double helix was labelled with radioactivity and the new one was non-labelled.
2. In the chromosomes of second generation only one of the two chromatids in each chromosome was radioactive.

This experiment demonstrates the semiconservative method of chromosome replication. But as it is known that each chromatid is composed of only one double helical molecule of DNA, it is the replication of DNA.

Role of Enzymes and Proteins Replication

The complicated process of DNA replication in *E.coli* requires many different proteins and enzymes, some of which also function in other cellular process such as the repair of DNA damage and genetic recombination. The key enzyme in the synthesis of new DNA is DNA polymerase.

In *E.coli* there are three DNA polymerases. Each catalyzes the DNA template-directed condensation of deoxyribonucleoside 5′-triphosphates. There are significant differences in their activity in DNA synthesis and in the nuclease (DNA or RNA break-down) activities associated with them. All three polymerases catalyze new DNA synthesis in the 5′-3′ direction. The rate of DNA synthesis catalyzed by the three enzymes varies circular prokaryotic DNA molecules replicate bidirectionally (in both directions away from the origin of replication). In these cases there are two replication forks that are mirror images of one another; picture two Ys joined head-to-head by the two pairs of arms. The area of a double-stranded DNA molecule that has denatured for replication is called a ***replication bubble.***

In *E.coli* the unwinding of the DNA is catalyzed by an enzyme called helicase, the product of a gene called *rep*. For each 10 base pairs of DNA unwound, one turn of the helix becomes untwisted. As a result, as the helix unwinds a torsional strain is imposed on the DNA. That is, since the DNA molecule is circular, the pulling apart of the helix at one location will cause increased tightening of the molecule elsewhere, much like what happens when the strands of a piece of rope that is fixed at each end are pulled apart. This torsional strain is relieved by the action of enzymes called topoisomerases, which cause single-stranded breaks in DNA away from the replication forks and then allow one strand to rotate relative to the other strand. The unwinding of the DNA produces single-stranded regions, which are stabilized by DNA *single-stranded binding* (SSB) *proteins.* Each SSB protein is a tetramer of a 74,000 dalton subunit, and this tetramer binds to eight bases of single stranded DNA. Over 250 tetramers bind to each replication fork. Once the strands have started to unwind, the internal bases become available for the formation of bonds with bases in the new chain. Initiation of DNA replication next takes place. In *E.coli*, primases bind to the single-stranded DNA of both arms of the replication fork and synthesize short primers. The primers are lengthened by the

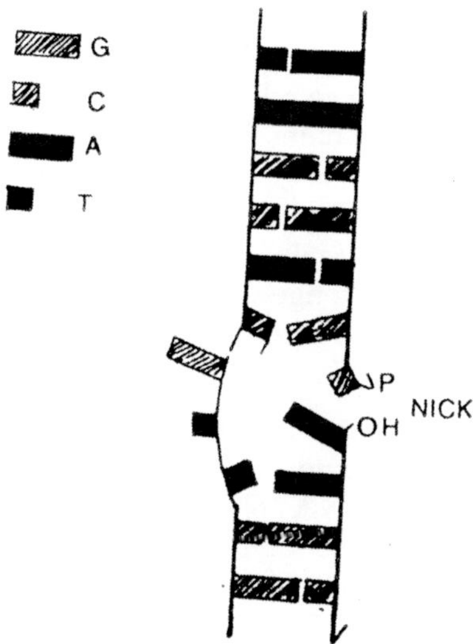

Fig. 3.5. A nicked duplex DNA molecule such as is formed by an endonuclease enzyme.

action of DNA polymerase III, which synthesizes the complementary DNA chains to the template strands. Note that, because of the opposite polarities of the two DNA template strands and the requirement for a 5′- to -3′ direction of new DNA synthesis, the primers are located at different relative positions on the two template strands.

The DNA helix continues to unwind. On the template on the left the new strand continues to be synthesized continuously (in the same direction as the direction of unwinding). However, because DNA polymerases can only synthesize DNA in the 5′-to-3′ direction, the new synthesis on the template on the right has gone as far as it can. Thus, a new initiation of DNA synthesis occurs on the right-hand template on the newly exposed single-stranded template close to the site of helicase activity. As before, a primer is made, which is lengthened by DNA polymerase III action. The overall result is that the new DNA strand being made on the left-hand template is synthesized continuously while the new DNA being made on the right-hand template is synthesizd discontinuously so that Okazaki fragments are produced. Eventually the Okazaki fragments are joined together into a continuous DNA strand. This process requires the activities of the enzymes DNA polymerase I and DNA ligase. If we consider two adjacent Okazaki fragments, the 3′ end of the newer DNA fragment is adjacent to but not joined to the primer 5′ end of the previously synthesized fragment. The DNA polymerase III that has just completed the synthesis of the newer fragment now dissociates from the DNA and DNA polymerase I takes its place. This enzyme continuous the 5′-to-3′ synthesis of the newer DNA fragment, at the same time removing the primer of the older fragment through the action of the 5′-to-3′ exonuclease function of DNA polymerase I. When DNA polymerase I is finished, there is a gap between the two new DNA fragments, and this gap is sealed in a reaction catalyzed by DNA ligase, thereby producing a longer DNA strand. The whole process is completed until all the DNA is replicated.

Replication Fork

How is semiconservative DNA replication accomplished? That is depends upon enzymes should not surprise you. We now know that there are different types of DNA polymerases, with different functions, and that the replication process is intricate. Kornberg's basic observations, including the need for a DNA template and for a mixture of nucleoside triphosphates, still hold. He also showed that nucleotides are always added to the growing chain at the same end: the 3′ end, the

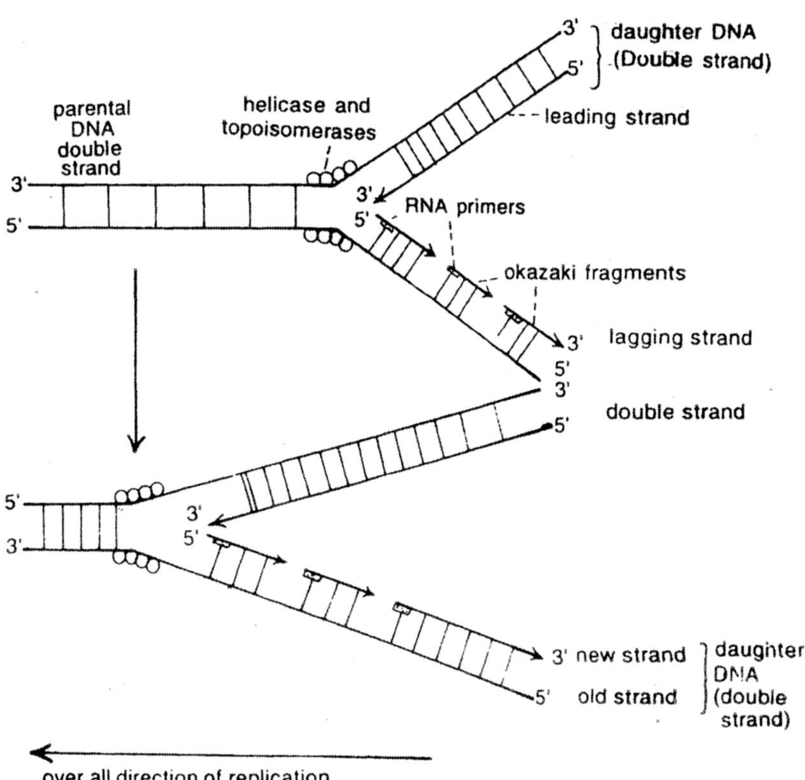

Fig. 3.6. Diagrammatic representation of discontinuous synthesis of DNA.

end at which the DNA strand has a free hydroxyl groups on the 3′ carbon of its terminal deoxyribose. This hydroxyl groups reacts with a phosphate group on the 5′ carbon of the deoxyribose of a deoxyribonucleoside triphosphate, and thus the chain grows. Bonds linking the phosphate groups of the deoxyribonucleoside triphosphate break and thereby release the energy for this reaction. Two of the phosphate groups diffuse away, and one becomes part of the sugar-phosphate backbone of the growing DNA molecule.

For double-stranded DNA to replicate, the strands must unwind and separate from each other. Only then can they function as templates for the synthesis of new, complementary strands. The unwinding results in a *replication fork*–a moving, Y-shaped structure that is the region where new DNA strands are being synthesized. Recall that the two original strands are antiparallel, with the 3′end of one strand paired with the 5′ end of the other. As the replication fork moves along the parent DNA molecule, an enzyme, DNA polymerase III, catalyzes the replication of both strands. How can this be accomplished, given that new nucleotides are added only at the 3′ end of a polynucleotide chain?

One parent strand is being exposed beginning at its 3′ end, which presents no problem–its complementary strand is synthesized continuously as the replication fork proceeds. This daughter strand is called the *leading strand*. The other daughter strand, the *lagging strand*, is produced in discontinuous spurts (100 to 200 nucleotides at a time in eukaryotes; 1,000 to 2,000 at a time in prokaryotes). The discontinuous stretches are synthesized just as the leading strand is, by adding the 5′ end of a nucleotide to the 3′ end of the daughter strand, but the stretches are synthesized in the opposite direction with respect to the replication fork. These stretches of new DNA for the lagging strand are called *Okazaki fragments* after their discoverer, the Japanese biochemist Reiji Okazaki. While the leading strand grows continuously "forward," the lagging strand grows in shorter, "backward," stretches with gaps between them. The gaps between the Okazaki fragments are then filled in by DNA polymerase I, and another enzyme, *DNA ligase*, links the fragments.

Working together, two DNA polymerases, DNA ligase, and several other proteins do the complex job of DNA synthesis with a speed and accuracy that are almost unimaginable. In *E.coli*, the complex makes new DNA at a rate in excess of 1,000 base pairs per second and makes mistakes in fewer than one base in 10^8–10^{12}.

On a bacterial chromosome, DNA synthesis begins at just one point, the *origin of replication*. Each chromosome of a eukaryote, on the other hand, has many origins of replication. DNA synthesis may thus proceed simultaneously in many areas of a single eukaryotic chromosome. Synthesis proceeds in both directions from an origin of replication as two replication forks move away from it.

ENZYMES USED IN REPLICATION

The enzymatic synthesis of DNA is a complex process, primarily because of the need for high fidelity in copying the base sequence and for physical separation of the parental strands. The number of steps that must be completed is far too great to be accomplished by a single enzyme and, in fact, about twenty proteins are known at present to be necessary. Thus, in an effort to provide some understanding with a minimum of confusion, each step in the process will be treated separately. We will consider the basic chemistry of polymerization, the source of the precursors, the problems raised by the chemistry of polymerization, the means of initiating and terminating synthesis, and the mechanisms for eliminating replication errors.

Polymerases

In 1957, Arthur Kornberg showed that in extracts of *E.coli* there exists a DNA polymerase (now called *polymerase I* or *pol I*). This enzyme was able to synthesize DNA from four precursor molecules–namely, the four deoxynucleoside 5′-triphosphates (dNTP), dATP, dGTP, dCTP, and dTTP–as long as a DNA molecule to be copied (a *template* DNA) was provided. Neither 5′-monophosphates nor 5′-diphosphates, nor 3′-(mono-, di-, or tri-) phosphates can be polymerized–only the 5′-triphosphates are substrates for the polymerization reaction; soon we will see why this is the case. Some years later, it was found that pol I, though playing an essential role in the replication process, is not the major polymerase in *E.coli*; instead, the enzyme responsible for advance of the replication fork is polymerase III or pol III. Pol III also exclusively uses 5′-triphosphates as precursors and requires a DNA template before polymerization can occur. Pol I and pol III have many features in common and, in fact, a few types of DNA molecules replicate by using only pol I. The overall chemical reaction catalyzed by both DNA polymerases is:

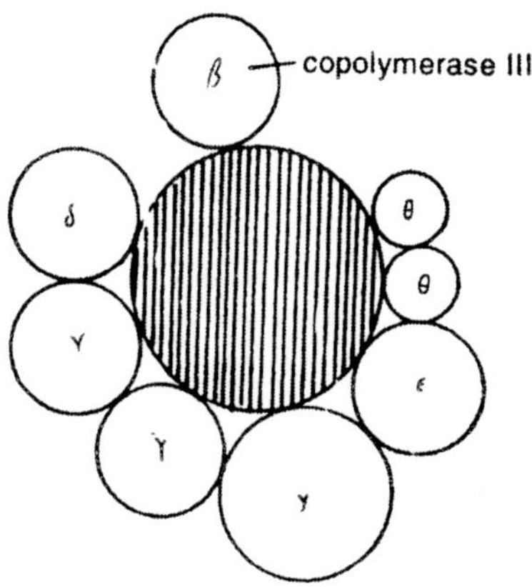

Fig. 3.7. DNA polymerase-III holoenzyme. It consists of subunits α, β, γ, ε, θ, δ and ε.

Poly(nucleotide)$_n$-3′-OH + dNTP → Poly(nucleotide)$_{n+1}$-3′-OH + PP in which PP represents pyrophosphate cleaved from the dNTP.

Pol I and pol III have many features in common. Both enzymes only polymerase deoxynucleoside 5′-triphosphates and can do so only while copying a template DNA. Furthermore, polymerization can only occur by addition to a *primer*–that is, an oligonucleotide hydrogen-bonded to the template strand and whose terminal 3′-OH group is available for reaction (that is, a "free" 3′-OH group). The meaning of a primer is already made clear, which depicts six potential template molecules; of these, only three can be said to be active–(c), (e), and (f)–each of which has a free 3′-OH group. The lack of activity with (d) and the direction of synthesis with (e) and (f) indicate that nucleotides do not add to a free 5′-P group. The lack of any synthesis with (a) or (b) indicates that addition to a 3′-OH group cannot occur if there is nothing to copy. Thus we draw two conclusions:

1. Both a primer with a free 3′-OH group and a template are needed.
2. Polymerization consists of a reaction between a 3′-OH group at the end of the growing strand and an incoming nucleoside 5′-triphosphate. When the nucleotide is added, it supplies another free 3′-OH group. Since each DNA strand has a 5′-P terminus and a 3′-OH terminus, strand growth is said to proceed in the 5′→3′ (5′-to-3′) direction.

Occasionally polymerases add a nucleotide terminus that cannot hydrogen-bond to the corresponding base in the template strand. This may be purely a mistake or may result from the tautomerization of adenine and thymine. In any case, it is important that the unpaired base be removable while its incorrectness is recognizable–namely, as an unpaired base at the 3′-OH terminus of a growing strand.

Pol I responds to an unpaired terminal base by terminating polymerizing activity, because the enzyme requires a primer that is correctly hydrogen-bonded. When such an impasse is encountered, a 3′→5′ exonuclease activity, which may be thought of simply as pol I running backwards or in the 3′→5′ direction, is stimulated, and the unpaired base is removed. After removal of this base, the exonuclease activity stops, polymerizing activity is restored, and chain growth begins again. This exonuclease activity is called the *proofreading* or *editing function* of pol I.

Another function of polymerase I is that of a 5′→3′ exonuclease. This activity has the following features:

1. Nucleotides are removed from the 5′-P terminus only, one by one.
2. More than one nucleotide can be removed by successive cutting.
3. The nucleotide removed must have been base-paired.

4. The nucleotide removed can be either or the deoxy-or the ribo-type.
5. Activity can be at a nick as long as there is a 5´-P groups.

The main function of the 5´→3´ exonuclease activity is to remove ribonucleotide primers. The 5´→3´ exonuclease activity at a single-strand break (nick) can occur simultaneously with polymerization. That is, as a 5´-P nucleotide is removed, a replacement can be made by the polymerizing activity. Since pol I cannot form a bond between a 3´-OH group and a 5´-monophosphate, the nick moves along the DNA molecule in the direction of synthesis. This movement is called *nick translation.*

Experimental conditions can be chosen so that polymerization will occur at a single-strand break without concomitant 5´→3´ exonuclease activity. The growing strand then displaces the parental strand. This is thought to be an important step in the mechanism of genetic recombination. Of all *E.coli* polymerases known to date, polymerase I is the only one capable of carrying out an unaided displacement reaction. In other strand displacement reactions, auxilary proteins are required and ATP is cleaved to fuel the unwinding of the helix; this will be discussed when the events at a replication fork are described.

Pol III is a very complex enzyme. In its most active form it is associated with eight other proteins to form the *pol III holoenzyme,* occasionally termed pol III. The term holoenzyme refers to an enzyme that contains several different subunits and retains some activity even when one or more subunits is missing. The smallest aggregate having enzymatic activity is called the *core enzyme.* The activities of the core enzyme and the holoenzyme are usually very different. Genes encoding five of the subunits have been identified; these are called *dnaE, dnaN, dnaQ, dnaX,* and *dnaz.* The *dnaE* protein possesses the major polymerizing activity but each of the subunits, except for the *dnaQ* protein is essential for replication. Pol III shares wtih pol I a requirement for a template and a primer but its substrate specificity is much more limited. For instance, pol III cannot act at a nick nor is it active with single-stranded DNA primed by either a DNA or RNA nucleotide fragment. The principal activity *in vitro* is on gapped DNA in which the gap is less than 100 nucleotides long. Such a gap is akin to the state of the DNA at a replication fork–that is, the parental strands are separated and bear short single-stranded regions ahead of the growing daughter chain.

Pol III cannot carry out strand displacement either, and another system is needed to unwind the helix in order that a replication fork will be able to proceed. The enzyme, like pol I, possesses a 3′→5′ exonuclease activity which performs the major editing function in DNA replication. This function is carried out by the *dnaE* subunit, which is also the major polymerizing subunit, as we have just mentioned. The *dnaQ* subunit plays an important role in editing, also, but the biochemical basis of the role is not yet known; it probably interacts with the *dnaE* subunit. The principal evidence supporting the view that it is involved in providing fidelity to the replication process is that bacteria containing a mutation in the *dnaQ* gene have a somewhat higher mutation frequency. Pol III also possesses a 5′→3 exonuclease activity; however, the enzyme acts only on single-stranded DNA so that it cannot carry out nick translation. The biological role of the 5′→3′ exonuclease activity of pol III is unknown at present. Although pol III holoenzyme is the major replicating enzyme in *E.coli*, much less is known about it than about pol I, because it is a more complex enzyme. Study of pol III is currently an active field of research.

All known polymerases (for both DNA and RNA) are capable of chain growth in only the 5′→3′ direction; that is, the growing end of the polymer must have a free 3′-OH group. It is possible for the following reasons that the enzymes evolved in this way to facilitate editing. If 3′→5′ growth were to occur, the growing strand would also be terminated with a 5′-triphosphate and the 3′-OH group of the incoming nucleotide would react with it. Chemically this is certainly acceptable but since the bonds formed contain only a single phosphate, an editing function would leave a free 5′-monophosphate. In order for chain growth to proceed, an enzymatic system would be needed to enter the replication fork and convert the monophosphate to a triphosphate. There is already a great deal going on in the replication fork, so that it would seem more economical for the cell to require 5′→3′ growth exclusively. However, the observation that chain growth proceeds in only one direction introduces what is probably the greatest complication in the entire replication process; this will be described shortly.

DNA Ligase

Neither replication from a primed circular single strand nor gap filling results in a continuous daughter strand. Discontinuity results because no known polymerase can join a 3′-OH and a 5′-monophosphate group. The joining of these groups is accomplished by the enzyme *DNA ligase*, which functions in replication and other important processes. *E.coli* DNA ligase can join a 3´-OH group as long as both are termini of adjacent base-paired deoxynucleotides–the enzyme cannot bridge a gap.

In the usual polymerization reaction, the activation energy for phosphodiester bond formation comes from cleaving the triphosphate. Since DNA ligase has only a monophosphate to work with, it needs another source of energy. It obtains this energy by hydrolyzing either ATP or nicotine adenine dinucleotide (NAD); the energy source depends upon the organism from which the DNA ligase is obtained. The *E.coli* DNA ligase uses NAD.

Continuous and Discontinuous DNA Replication

Autoradiographic evidence leads us to believe that replication is occurring simultaneously on both strands. *Continuous* replication is

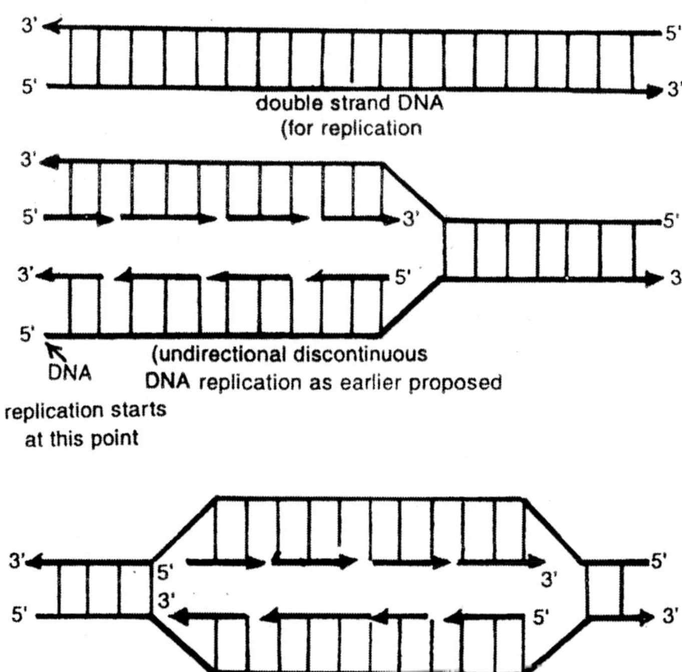

Fig. 3.8. Discontinuous DNA replication.

of course, possible on the 3′→5′ template strand, which begins with the necessary 3′-OH *primer.* (Primer is double-stranded DNA–or, as we shall see, a DNA-RNA hybrid–continuing as single-stranded DNA template. The strand being synthesized has a 3′-OH available). A *discontinuous* form of replication takes place on the complementary strand, where it occurs in short segments, backward, away from the Y-junction. These short segments, called *Okazaki fragments* after *R. Okazaki* who first saw them, average about 1,500 nucleotides in prokaryotes and 150 in eukaroytes. The strand synthesized continuously is referred to as the *leading strand,* and the strand synthesized discontinuously is referred to as the *lagging strand.*

Once initiated, continuous DNA repliction can proceed indefinitely. DNA polymerase III on the leading strand template has what is called high *processivity*: once it attaches, it does not release until the entire strand is replicated. Discontinuous replication, however, requires the repetition of four steps primer synthesis, elongation, primer removal with gap filling, and ligation.

Primer synthesis and elongation

In order for Okazaki fragments to be synthesizd, a primer must be created de novo (Latin, from the beginning). None of the DNA polymerases can create that primer. Instead, one of two enzymes, either *RNA polymerase,* the transcribing enzyme or, more commonly, *primase,* an RNA polymerase coded for by the *dnaG* gene, creates the primer. It is from two to sixty nucleotides, depending on the species, at the site of Okazaki fragment initiation. The result is a short RNA primer that provides the free 3′-OH group that DNA polymerase III needs in order to synthesize the Okazaki fragment. DNA polymerase III continues until it reaches the primer RNA of the previously synthesized Okazaki fragment. At that point it stops and releases from the DNA.

All three prokaryotic polymerases not only can add new nucleotides to a growing strand in the 5′→3′ direction but also can remove nucleotides in the opposite 3′→5′ direction. This property is referred to as *3′→5′ exonuclease activity.* Enzymes that degrade nucleic acids are classified as *exonucleases* if they remove nucleotides from the end of a nucleotide strand or as *endonuclease* if they can break the sugar-phosphate backbone in the middle of a nucleotide strand. At first glance, exonuclease activity seems like an extremely curious property for a polymerase to have–curious unless we think about its ability to check complementarity. If the complementarity is improper, which is to say that the wrong nucleotide has been inserted, the polymerase can remove the incorrect nucleotide, put in the proper one, and continue on its way. This is known as the *proofreading* function of the DNA polymerase. In addition, the RNA primers of Okazaki fragments can be removed by exonuclease activity.

Role of primer

DNA polymerase I is a polymerase when it adds nucleotides, one at a time, and an exonuclease when it removes nucleotide one at a time. To complete the Okazaki fragment, DNA polymerase I acts in both capacities. (Mutants of DNA polymerase I cannot properly connect Okazaki fragments.) DNA polymerase I completes the Okazaki fragment by removing the previous RNA primer and replacing it with DNA nucleotides. When DNA polymerase I has completed its nuclease and polymerase activity, the two previous Okazaki fragments are almost complete. All that remains is a single phosphodiester bond to be made.

Ligation

DNA polymerase I cannot make the final bond to join the Okazaki fragment to the previously synthesized DNA. An enzyme, *DNA ligase*, completes the task by making the final phosphodiester bond in an energy-requiring reaction.

A question of evolutionary interest is why RNA is used for priming of DNA synthesis. Why not use DNA directly and avoid the exonuclease and resynthesis activity? One possible answer is that because priming is inherently more error-prone then regular DNA synthesis it is best for the cell to have the primer nucleotides removed and replaced by DNA synthesized in a less error-prone fashion before DNA synthesis is completed. If the priming nucleotides are RNA, then DNA polymerase I can recognize and remove them in the final stage of Okazaki fragment synthesis, at which time it replaces them with DNA nucleotides inserted with a low error rate.

Another question of evolutionary interest is why DNA synthesis cannot take place in the 3′→5′ direction. Perhaps the answer has to do with proofreading and the exonuclease removal of incorrect nucleotides. When an incorrect nucleotide is found and removed, the next nucleotide brought in, in the 5′→3′ direction, will have a triphosphate end available to provide the energy for its own incorporation. Consider what would happen if the polymerase were capable of adding nucleotides in the opposite direction. The energy for the diester bond would be coming from the triphosphate already attached in the growing 3′→5′ strand. Then, if an error in complementary were detected and the most recently added nucleotide were removed from the 3′→5′ strand by the polymerase, the last nucleotide in the double helix would no longer have a triphosphate available to provide energy for the diester bond with the next nucleotide brought in. Continued polymerization would thus require additional enzymatic steps to provide the energy for the process to continue. This could slow the process down considerably. As it is, the process works at a speed of about four hundred nucleotides incorporated per second with an error rate of about one incorrect pairing per one hundred thousand bases, improved to a rate of only one mistake in ten million by exonuclease proofreading. (Other repair system can improve this error rate another thousandfold, to about one error every 10^{10} times an average base is replicated.

The Initiation of DNA Replication

Each replicon (e.g., the *E.coli* chromosome or a segment of eukaryote chromosome) must have a region in which DNA replication is initiated. In *E.coli* this region is referred to as the genetic locus *oriC*. In order for DNA replication to begin, several steps must occur. First, the specific origin site must be recognized by the appropriate protein. Then the site must be opened and stabilized. And, finally, a replication fork must be initiated in both directions, involving continuous and discontinuous DNA replication. Although many of the proteins involved are known, all the steps at the enzymatic level are not, and hence our understanding is a bit sketchy. Following is a description, most of whose steps are known.

OriC, the origin of replication in *E.coli,* is about 245 base pairs long and is recognized by proteins called *initiator proteins* that open up the double helix. The initiator proteins then take part in the attachment of *primosomes*, a complex of two proteins: a primase, which creates RNA primers, and DNA *helicase*, which unwinds DNA at the Y-junction. As the primosomes move along, they create RNA primers used by DNA polymerase III to initiate Okazaki fragments. At some point, leading-strand synthesis begins and Y-junction activity then proceeds as outlined earlier.

In some phages and plasmids, the initiation of replication is not with an RNA primer but with a protein. This protein provides the primer configuration with an OH group from an amino acid. The generality of this type of priming has not been established. Another interesting protein interaction at the origin of replication involves the reverse of initiation of DNA synthesis, the prevention of the initiation of DNA synthesis. This is accomplished by a newly discovered protein, called an "off switch." That is, this protein binds to the DNA at *oriC* and apparently prevents DNA replication from beginning. It does this by its binding activity that prevents the initiator proteins from opening the DNA. Thus this protein may be a very important component in control of the cell cycle stopping the cell cycle from beginning. Presumably, when the appropriate time comes for the cell cycle to begin, the protein is removed.

Events at the Y-Junction

We now have the image of DNA replication proceeding by a primosome, moving along the lagging strand template, opening up the DNA (helicase activity), and creating RNA primers (primase activity). One DNA polymerase III moves along the leading-strand template generating the leading strand by continuous DNA replication, whereas a second DNA polymerase III moves backward, away from the Y-junction, creating Okazaki fragments. *Single-strand binding proteins* (ssb proteins) keep single-stranded DNA stabilized (open) during this process, and DNA polymerase I and ligase are connecting Okazaki fragments.

This simple picture is slightly complicated by the fact that a single DNA polymerase seems to do the entire lagging strand, rather than dropping off the DNA at the completion of an Okazaki fragment and being replaced by a new one at the newest primer near the Y-junction. In addition, there is evidence that the lagging and leading-strand synthesis is coordinated. The *replisome* model has arisen in which both copies of DNA polymerase III are attached to each other and work in concert with the primosome at the Y-junction. According to this model, a single replisome consisting of two copies of the DNA polymerase III *holoenzyme* (each actually made of seven subunits), a helicase, and a primase, move along the DNA. The leading-strand template is immediately fed to a polymerase, whereas the lagging -strand template is not acted on by the polymerase unit an RNA primer has been placed on the strand, meaning that a long (fifteen hundred base) single strand has been opened up.

As the replisome moves along, another single-stranded length of the lagging-strand template is formed. At about the time that the Okazaki fragment is completed, a new RNA primer has been created. The Okazaki fragment is released and a new Okazaki fragment is begun, starting with the latest primer taking the replisome back to the same configuration, but one Okazaki fragment farther along.

Supercoiling

The simplicity and elegance of the DNA molecule masks an inevitable problem of coiling. Since the DNA molecule is made from two strands that wrap about each other, certain operation, such as DNA replication and its termination, meet topological difficulties. Up to this point, we have seen the circular *E.coli* chromosome in its "relaxed" state. However, there are enzymes in the cell that cause DNA to become overcoiled (positively *supercoiled*) or undercoiled (negatively supercoiled). Positive supercoiling comes about either from too many turns of the DNA in a given length or from the molecule wrapping around itself.

Positive supercoiling comes from having the circular duplex wind about itself in the same direction as the helix twists (right handed), whereas negative supercoiling comes about by having the duplex wind about itself in the opposite direction as the helix twists (left handed). The former state increases the number of turns of one helix around the other side (the *linkage number*, L), whereas the latter decreases it. The three forms of DNA, all have the same sequence yet differ in their linkage number. They are referred to as topological isomers (*topoisomers*). The enzymes that create or alleviate these states are called *topoisomerases.*

Topoisomerases affect supercoiling by either of two methods. Type I topoisomerases break one strand of a double helix and, while binding the broken ends, pass the other strand through the break. The break is then sealed. Type II topoisomerases (e.g., *DNA gyrase* in *E.coli*) do the same sort of thing only instead of breaking one strand of a double helix, they break both and pass another double helix through the temporary gap.

As DNA replication proceeds, positive supercoiling builds up ahead of the Y-junction. This is eliminated by the action of topoisomerases that either create negative supercoiling ahead of the Y-junction in preparation for replication or alleviate positive supercoiling after it has been created. The major components of DNA replication in *E.coli* are summarized in table 3.1.

Termination of Replication

The termination of the replication of a circular chromosome presents no major topological problems. The theta-structure replication finishes with both Y-junctions having proceeded around the molecule. The leading strand on one template closes in on the lagging strand begun in the other direction with the same happening on the other template. The process stops with about twenty-five twists remaining at no particular spot on the chromosome (there is no "termination" locus). A topoisomerase then release the two circles and DNA polymerase I and ligase close them up.

Table 3.1. Summary of the Enzymes Involved in DNA Replication in E.coli.

Enzyme (Protein)	*Genetic Locus*	*Function*
DNA polymerase I	*pol A*	Gap filling and primer removal
DNA polymerase II	*pol B*	?
DNA polymerase III		
α subunit	*dnaE* (*polC*)	DNA replication
β subunit	*dnaN*	DNA replication
γ subunit	*dnaX*	DNA replication
δ subunit	?	DNA replication
ε subunit	*dnaQ*	3′→5′ exonuclease
θ subunit	?	DNA replication
τ subunit	*dnaX*	DNA replication
Initiator protein	*dnaA*	Binds to origin of replication
RNA polymerase subunit	*rpoA, B, C, D*	RNA primer in some system
Primase	*dnaG*	RNA primer in some system
DNA ligase	*lig*	Closes nicked DNA strands
Helicase	*rep*	Unwinds DNA for replication
Ssb proteins	*ssb*	Single-strand stability
DNA topoisomerase I	*topA*	Supercoiling of DNA
DNA topoisomerase II		
α subunit	*gyrA* (*nalA*)	ATPase
β subunit	*gyrB* (*cou*)	Cutting, closing of DNA

Several different mechanisms have been explored for the termination of the linear chromosomes of some viruses and all eukaryotic genomes. Linear molecules have the problem of completing the last Okazaki fragment. An RNA primer on the very tip of the 3′→5′ template cannot be replaced by DNA polymerase I, assuming even that a final primer can be put on the very tip of the molecule. In eukaryotes, an enzyme, telomerase, attaches repeats of a short sequence at each chromosome tip.

Replication Models

The model of DNA replication that we have presented here comes primarily from evidence gathered in *E.coli*, which replicates by way of the *theta*-structure intermediate. However, two other modes of replication occur in circular chromosomes: rolling-circle and D-loop.

Rolling-Circle Model

In the *rolling-circle* mode of replication, a nick (a break in one of the phosphodiester bonds) is made in one of the strands of the circular DNA, resulting in replication of a circle and a tail. This form of replication occurs in the Hfr *E.coli* chromosome, or the F plasmid, during conjugation. The F^+ or Hfr cell retains the circular daughter while passing the linear tail into the F^- cell. This method is also used in several phages, which fill their heads (protein coats) with linear DNA replicated from a circular parent molecule.

In the model for rolling-circle replication, the nick made in one strand creates a free 3´-OH end and a free 5´-PO_4 end. Synthesis of a new circular strand occurs by addition of nucleotides to the 3´ end using the complementary intact strand as a template. No primer is needed because the original break produces a primer configuration (3´-OH). As nucleotides are added to one end of the broken strand in a continuous fashion, the other end is displaced as a 5´-PO_4 tail. As replication of the circular templates occurs, the 5´-PO_4 tail is replicated in a discontinuous manner, and the resulting double helix can be severed from the double-helical circle by a nuclease. DNA ligase closes the replicated circular strand and can join the ends of the replicated tail into a circle in the F^- cell, or can package the linear molecule in a phage head, depending upon which type of circular DNA has been replicated.

D-Loop Model

Chloroplasts and mitochondria have their own circular DNA molecules that appear to replicate by a slightly different mechanism than those described. The origin of replication is at different point on each of the two parental template strands. Replication begins on one strand, displacing the other while forming a displacement loop or *D-loop* structure. Replication continues until the process passes the origin of replication on the other strand. Replication is then initiated on the second strand, in the opposite direction. The result is two circles. Some species have chloroplasts and mitochondria with circular DNAs that have multiple D-loops formed.

Eukaryotic DNA Replication

As we saw earlier, linear eukaryotic chromosome usually have multiple origins of replication resulting in figures referred to as "bubbles" or "eyes." Multiple origins allow eukaryotes to replicate their larger quantities of DNA in a relatively short time, even though eukaryotic DNA replication is considerably showed by the presence of histone proteins associated with the DNA to form chromatin. For example, the *E.coli* replication fork moves about twenty-five thousand base pairs per minute, whereas the eukaryotic Y-junction moves only about two thousand base pairs per minute. The number of replications in eukaryotes varies from about five hundred in yeast to as many as sixty thousand in a diploid mammalian cell.

Much less is understood about eukaryotic DNA replication because of the complexity of eukaryotes and the relatively shorter time during which they have been studied effectively. We presume that eukaryotes have solved the same problems faced by prokaryotes in a similar, but not identical, fashion. For example, eukaryotes have five types of DNA polymerases, named DNA polymerase α, β, γ, δ, and ε. DNA polymerases γ, δ, and ε have exonuclease activity. DNA polymerases α and δ are the major replicating enzymes, with polymerase α replicating the lagging strand and polymerase δ replicating the leading strand. The role of polymerase ε is unclear; it seems capable of regular leading- or lagging-strand replication. DNA polymerase β is the major repair polymerase (like polymerase I in prokaryotes). DNA polymerase γ appears to be concerned primarily with mitochondrial DNA replication.

Table 3.2. Eukaryotic DNA Polymerases

Enzyme	*Function*
DNA polymerase α	Replication of nuclear chromosomes (lagging strand)
DNA polymerase β	Repair of nuclear chromosomes
DNA polymerase γ	Replication of mitochondrial chromosomes
DNA polymerase δ	Replication of nuclear chromosomes (leading strand)
DNA polymerase ε	Probably replication of nuclear chromosomes

4

PROTEIN SYNTHESIS

Proteins are the molecules responsible for catalyzing most intracellular chemical reactions (enzymes), for regulating gene expression (regulatory proteins), and for determining many features of the structues of cells, tissues, and viruses (structural proteins). A protein is composed of one or more chains of amino acids that are covalently joined. The chains of amion acids are called *polypeptides*. The 20 different amino acids commonly found in natural polypeptides can be in any number and any order. Because the number of amino acids in a polypeptide molecule usually ranges from 100 to 1000, the number of different protein molecules that is possible is enormous.

Each amino acid contains a carbon atom (the α carbon) to which is attached one carboxyl group (–COOH), one amino group (–NH_2), and a side chain commonly called an R group. The R group are generally chains or rings of carbon atoms bearing various chemical groups. The simplest side chains are those of glycine (–H) and of alanine (CH_3). Polypeptide chains are formed when the carboxyl group of one amino acid joins with the amion grioup of a second amino acid; the resulting chemical bond is an ordinary covalent bond called a *peptide bond*. Thus, the basic unit of a protein is a polypeptide chain in which α-carbon atoms alternate with peptide units to for a backbone having an ordered array of side chains.

The two ends of every polypeptide molecule are distinct. One end has a free –NH_2 group and is called the *amino terminus*; the other end has a free –COOH group and is the *carboxyl terminus*. Polypeptides are synthesized by adding individual amion acids to the carboxyl end of the growing chain. Conventionally the amion acids of a polypeptide chain are numbered starting with the amino acid at the amion end.

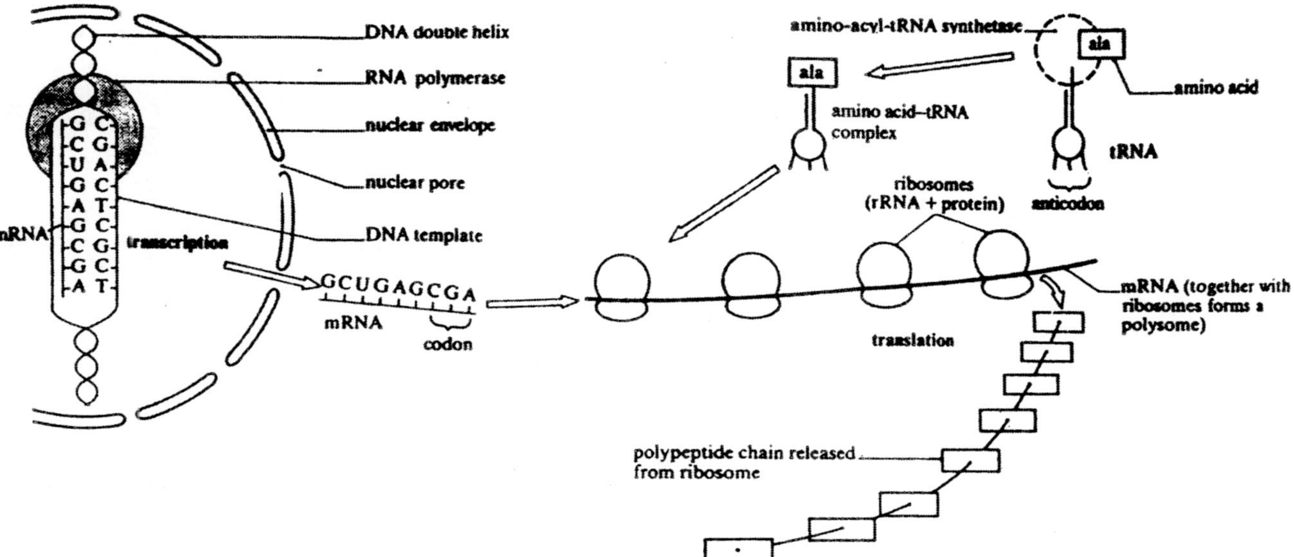

Fig. 4.1. Simplified summary diagram of the major structures and processes involved in protein synthesis in the cell.

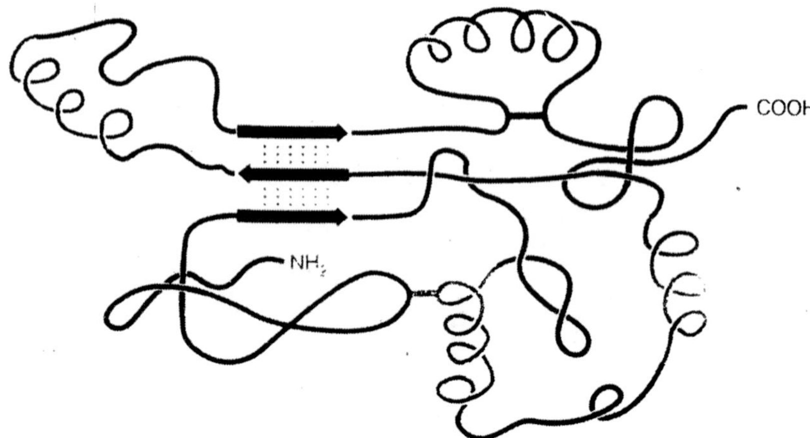

Fig. 4.2. A schematic diagram of the path of the backbone of a polypeptide, showing possible ways in which the polypeptide may be folded. Heavy black arrows represent β sheets; the dots joining the arrows represent hydrogen bonds.

Most polypeptide chains are highly folded, and a variety of three-dimensional shapes have been observed. The manner of folding is determined primarily by the sequence of amino acids –in particular, by noncovalent interactions between the side chains–and so each polypeptide chain tends to fold into a unique three-dimensional shape as it is being synthesized. In some cases, protein folding is interactions with other proteins in the cell. The rules of folding are complex and shape cannot usually be predicted from the amino acid sequence except for the simplest proteins. On the average. the molecules fold so that amino acids with charged side chains tend to be on the surface of the protein (in contact with water) and those with uncharged side chains tend to be internal. Specific folded configurations also result from hydrogen-bonding between prepaid groups. Two fundamental polypeptide structures are the α helix and the β sheet. Covalent bonds may also form between the sulfur atoms of some pairs of cysteines.

Many protein molecules consist of more than one polypeptide chain. When this is the case, the protein is said to contain *subunits.* The subunits may be identical or different. For example, hemoglobin, the oxygen carrier of blood, consists of four subunits, two each of two different types.

Relations Between Genes and Polypeptides

Most genes contain the information for the synthesis of only one polypeptide chain, Furthermore, the *sequence* of nucleotides in a gene determines the *sequence* of amino acids in polypeptide. This point was first proved by studies of the tryptophan synthetase gene *trpA* in *E. coli*, a gene in which many mutations had been obtained and accurately mapped. The effects of numerous mutations on the amino acid sequence of the enzyme were determined by directly analyzing the amino acid sequences of the wildtype and mutant enzymes. Each mutation was found to result in a single amino acid substituting for the wildtype amino acid in the enzyme; more importantly, *the order of the mutations in the genetic map was the same as the order of the affected amino acids in the polypeptide chain*. This attribute of genes and polypeptdes is called *colinearity*, which means that the sequnce of base pairs in DNA determines the sequence of amino in the polypeptide in a colinear or point-to-point manner. Colinearity is universally found in prokaryotes. However, we will see later that in eukaryotes noninformational DNA sequences interrupt the continuity of most genes, the order but not the spacing between the mutations correlates with amino acid substitution.

Transcription

The first step in gene expression is the synthesis of an RNA molecule copied from the segment of DNA that constitutes the gene. The basic features of the production of RNA are described in this section.

General Features of RNA Synthesis

The essential chemical characteristics of the enzymatic synthesis of RNA resemble those of DNA synthesis.

1. The precursors is the synthesis of RNA are the four ribonucleoside 5'-triphosphates– namely, adenosine triphosphate (ATP), guanosine triphosphate (GTP), cytidine triphosphate (CTP. and uridine triphosphate (UTP). They differ from the DNA precursors only in that the sugar is ribose rather than deoxyribose and the base uracil (U) replaces thymine(T).
2. In the formation of RNA, a sugar-phosphate bond is formed between the 3'-hydroxyl group of one nucleotide and the 5'-triphosphate of a second nucleotide. This is the same chemical reaction as that which occurs in the synthesis of DNA, but the enzyme is different.

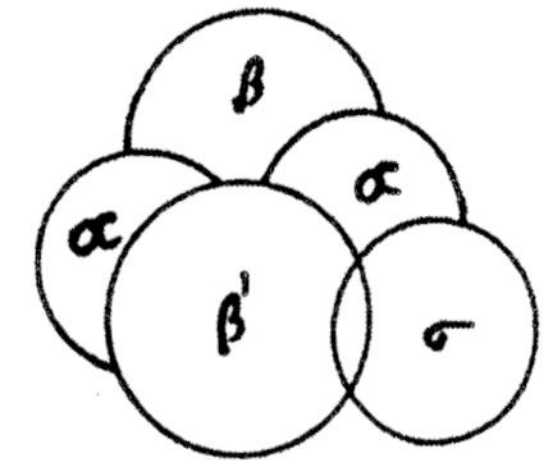

Fig. 4.3. A model of the structure of prokaryotic RNA polymerase showing association of five polypeptides $(\alpha_2\beta\beta\ \sigma)$

3. The sequence of bases in an RNA molecule is determined by the base sequence of the DNA template. Each base added to the growing end of the RNA chain is chosen for its ability to base-pair with the DNA template strand: thus, the bases C, T, G, and A in a DNA strand cause G, A, C, and U, respectively, to be added to the growing end of an RNA molecule.
4. Nucleotides are added only to the 3'-OH end of the growing chain: as a result, the 5' end of a growing RNA molecule bears a triphosphate group. (The 5' → 3' direction of chain growth is the same as that in DNA synthesis.)

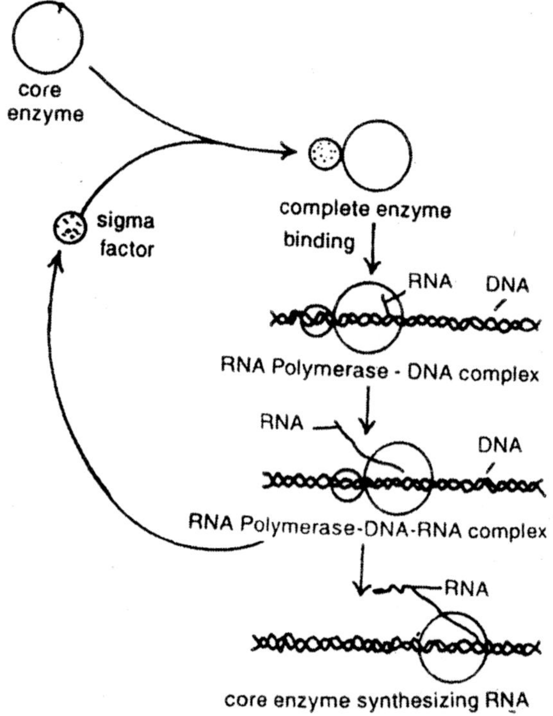

Fig. 4.4. Role of sigma factor and core enzyme of RNA polymerase during transcription.

A significant difference between DNA polymerase and RNA polymerase is that *RNA polymerase is able to initiate chain growth without a primer.*

An important feature of RNA synthesis is the following:

Introns and Exons

In 1977 biologists were surprised to discover that the DNA of a eukaryotic gene is longer than its corresponding mRNA. It should be the same length because the mRNA is a direct copy. It was discovered that immediately after the mRNA is made, certain sections of the molecule are cut out, before it is used in translation. The sections of the gene that code for these unused pieces of RNA are called ***introns***. The remaining sections of the gene are the code for the protein and are called ***exons***. The size and arrangement of introns is very variable and characteristic for a particular gene. In prokaryotes there are no introns.

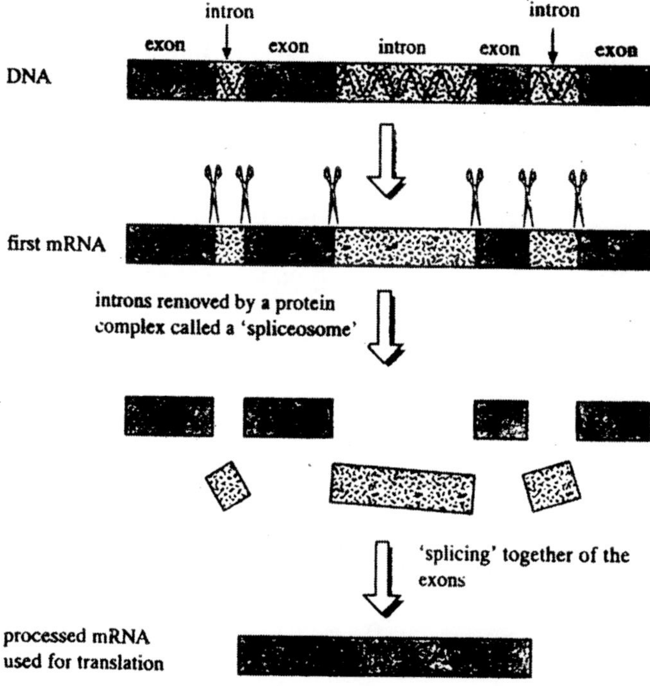

Fig. 4.5. Exons, introns and intron splicing.

One possible function for introns has come with the discovery that the same mRNA may have different introns removed in different cells. The gene therefore has alternative introns and can code for different, though similar, proteins. This increases its potential use.

An example is the calcitonin gene. Two different forms of mRNA can be produced by this gene, depending on which introns are removed.

One is produced in the thyroid gland and codes for the protein calcitonin, which had 32 amino acids. Calcitonin is a hormone which acts to lower calcium levels in the blood. The other is produced in the hypothalamus and codes for a protein with 37 amino acids which is similar to calcitonin and is called CGRP (calcitonin gene-related peptide). This is a powerful vasodilator agent. It is also released from nerve endings in some parts of the peripheral nervous system.

Protein Synthesis in Prokaryotes

There are three major steps in protein synthesis: initiaion, elongation, and termination. These will be discussed in turn.

Initiation

Initiator tRNA. The first amino acid in the syntheis of all bacterial polypeptides is N-formyl-methionine (fmet), which is a modified methionine amino acid in which the α-amino group is "blocked" and therefore cannot participate in peptide bond formation. The formyl group is added to the methionine after the amino acid has become attached to a specific tRNA, called tRNA. fmet. This reaciton is catalyzed by the enzyme *transformylase*. In many cases the fmet that starts a polypeptide chain is subsequently removed by enzymatic action.

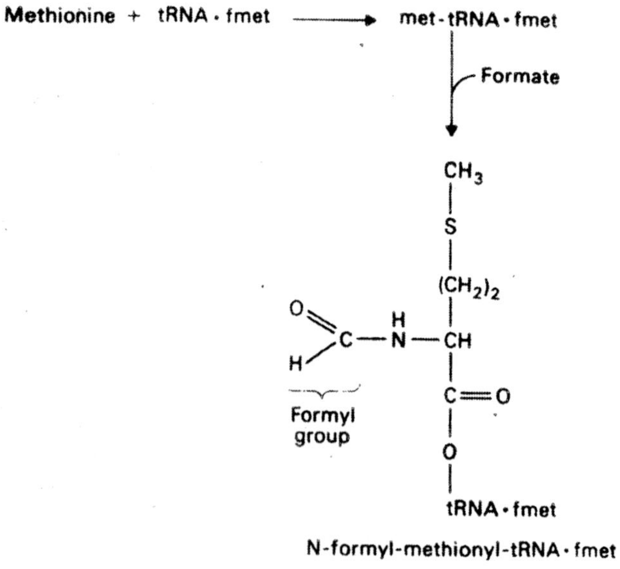

Methionine + tRNA · fmet ⟶ met-tRNA · fmet

↓ Formate

$$\begin{array}{c} CH_3 \\ | \\ S \\ | \\ (CH_2)_2 \\ | \\ \overset{O}{\underset{H}{\gt}}C-\overset{H}{N}-CH \\ | \\ C=O \\ | \\ O \\ | \\ tRNA \cdot fmet \end{array}$$

Formyl group

N-formyl-methionyl-tRNA · fmet

Fig. 4.6. Synthesis of N-formylmethionyl-tRNA.

Biochemical analysis has shown that at least two species of tRNA that can be charged with methionine are present in all prokaryotic organisms: one is that tRNA involved with initiaton, and the other species is responsible for the insertion of methionine elsewhere in the polypeptide chain. The latter rRNA is designed rRNA.met. Both of these tRNAs in bacteria are aminoacylated by the same enzyme, but only tRNA, fmet is a substrate for the transformylase-catalyzed reaction. Both tRNAs read AUG (the only methionine codon), but in addition tRNA.fmet can recognize GUG and UUG codons. RNA-sequencing studies have shown that both molecules havae an anticodon that is complementary to AUG. The two tRNA molecules do differ in some other properties. For example, the binding of fmet-TRNA.fmet to ribosomes is catalyzed by an initiator factor whereas the binding of met-tRNA.met is catalyzed by elongation factors. The two tRNAs apparantly bind to the ribosome at different sites. Thus it is clear that fmet.tRNA.fmet must have a strucutre that is specific for its role in initiation.

Ribosome binding sites

In bacteria, the first step in initiation is the formation of a complex between the 30S ribosomal subunit, fmet-TrNA, and an mRNA molecule. The 50S subunit is added later to form the active 70S ribosome (monosome). The mRNA may contain information for one to several distinct polypeptide chains. For each of the segments coding for a polypeptide, there is a specific nucleotide sequence for orienting the mRNA correctly and in the right reading frame on the ribosome. These sequences are called the ***ribosome binding sites.***

Message origin	Ribosome binding site sequence
E. coli lac Z	UUC ACA CAG GAA ACA GCU AUG ACC AUG AUU
E. coli trp B	AUA UUA AGG AAA GGA ACA AUG ACA ACA UUA
E. coli RNA polymerase β	AGC GAG CUG AGG AAC CCU AUG GUU UAC UCC
Phage λ cro	AUG UAC UAA GGA GGU UGU AUG GAA CAA CGC

Fig. 4.7. Some prokaryotic ribosome binding sites. The initiation codon, AUG, is boxed. The larger boxed regions indicate the regions of contiguous complementarity (including allowable G-U base pairs) to the 3' end of 16S rRNA.

Most of the ribosome binding sites have apurine-rich sequence about 8 to 12 bases upstream from the AUG start codon. This equence and other bases in this region are complementary to a pyrimidine-rich region, including at least CCUCC at the 3' end of 16S rRNA. The mRNA region that binds in this way is called the *Shine-Dalgarno sequence* after the discoverers of this relationship. Thus it appears that the formation of complementary base pairs between mRNA and 15 S rRNA in the 30S ribosomal subunit allows the ribosomes to locate and bind to the initator regions in the mRNA.

Initiation factors and initiation

In addition to mRNa, fmet-tRNA, and ribosomal subunits, three protein initiaion factors (IF-1, IF-2, and IF-3) and GTP are required for the initiation process to occur. First the properties of the initiation factors are discussed and then the scheme proposed for the initiation processes in protein synthesis is presented.

1. *IF-3.* The IF-3 factor weighs 23,000 daltons and functions in binding mRNA to the 30S subunit. It also acts as a dissociation factor for separating the 30S and 50S subunits after polypeptide synthesis is complete. Like all the IFs, IF-3 is found bound to free 30S subunits and can be released by washing the subunits in 0.5 M ammonium chloride.

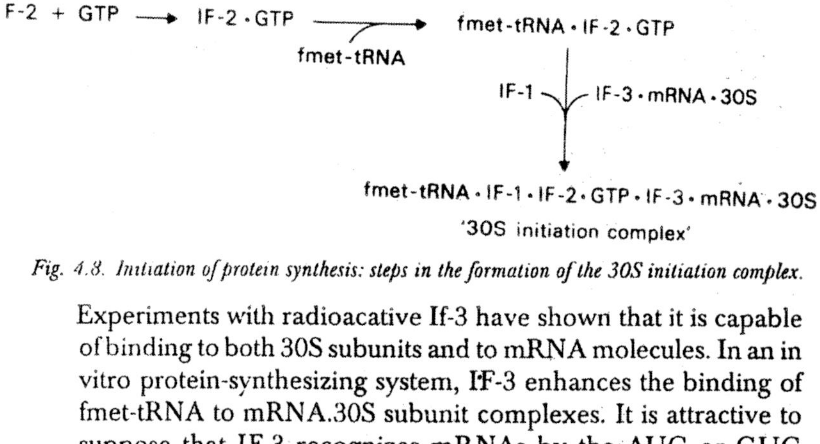

F-2 + GTP → IF-2 · GTP →(fmet-tRNA) fmet-tRNA · IF-2 · GTP

↓ IF-1, IF-3 · mRNA · 30S

fmet-tRNA · IF-1 · IF-2 · GTP · IF-3 · mRNA · 30S
'30S initiation complex'

Fig. 4.8. Initiation of protein synthesis: steps in the formation of the 30S initiation complex.

Experiments with radioacative If-3 have shown that it is capable of binding to both 30S subunits and to mRNA molecules. In an in vitro protein-synthesizing system, IF-3 enhances the binding of fmet-tRNA to mRNA.30S subunit complexes. It is attractive to suppose that IF-3 recognizes mRNAs by the AUG or GUG initiation codons, but there is no solid evidence on this point.

In summary, the initiation reaction in which IF-3 is involved is:

IF-3 + mRNA + 30S subunit → (IF-3.mRNA.30S) complex

2. *IF-2.* The 80,000 dalton IF-2 protein in involved with the binding of the initiator tRNA to the IF-3.mRNA.30S complex. The high-energy molecule GTP is used in this reaction. In vitro experiments have shown that IF-2 and GTP will bind to form a complex that is stabilized when it is turn forms a complex with fmet-tRNA. This latter complex then binds with the IF-3.mRNA.30S complex and

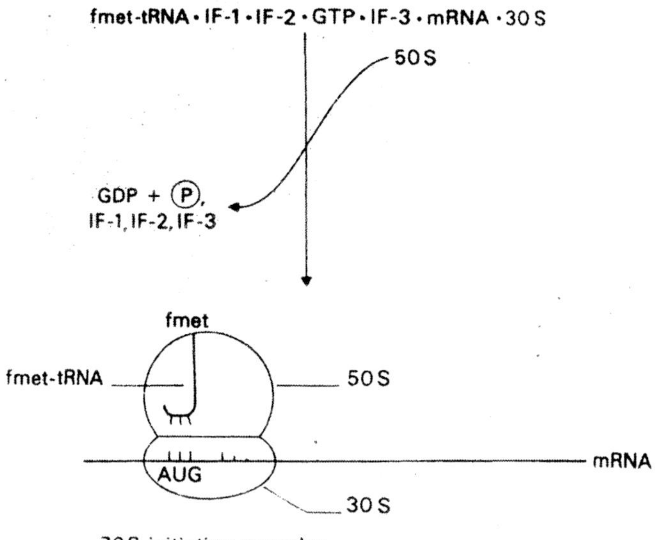

Fig. 4.9. Initiation of protein synthesis: addition of 50S ribosomal subunit to 30S initiation complex leads to formation of 70S ribosome in frame on the mRNA.

the IF-1 protein factor (9000 daltons) to forms the 30S initiation complex.

3. *Dissociation of initiation factors from the initiation complex.* The initiation factors function to bring fmet-tRNA, mRNA and 30S subunits into a stable association. The next step is the addition of a 50S subunit to form a 70S initiation complex. This leads to the hydrolysis of GTP to GDP + P and the release of three intiation factors. The factors can then be used for further initiaton reactions on the same or different mRNA.

Elongation

The 70S ribosome has two sites for binding aminoacyl-tRNA. In protein synthesis, charged tRNA binds first to a site called the A (aminoacyl) site. Then the amino acid it carries becomes joined to the growing polypeptide chain carried by the tRNA at the site called the P (peptidyl) site by the formation of a peptide bond.

It is not known whether the fmet-tRNA enters the A site and then moves to the P site or whether it enters the P site directly. Before further protein synthesis can occur, however, the fmet-tRNA must become located in the P site hydrogen-bonded to the start codon on the mRNA. Once this has occurred a cyclic sequence of events commences in which one amino acid at a time is added to the growing polypeptide chain. This is called *elongation.*

Binding of aminoacyl-tRNA

The charged tRNA with the complementary anticodon to the codon in the reading frame of the A site becomes bound to that site of the

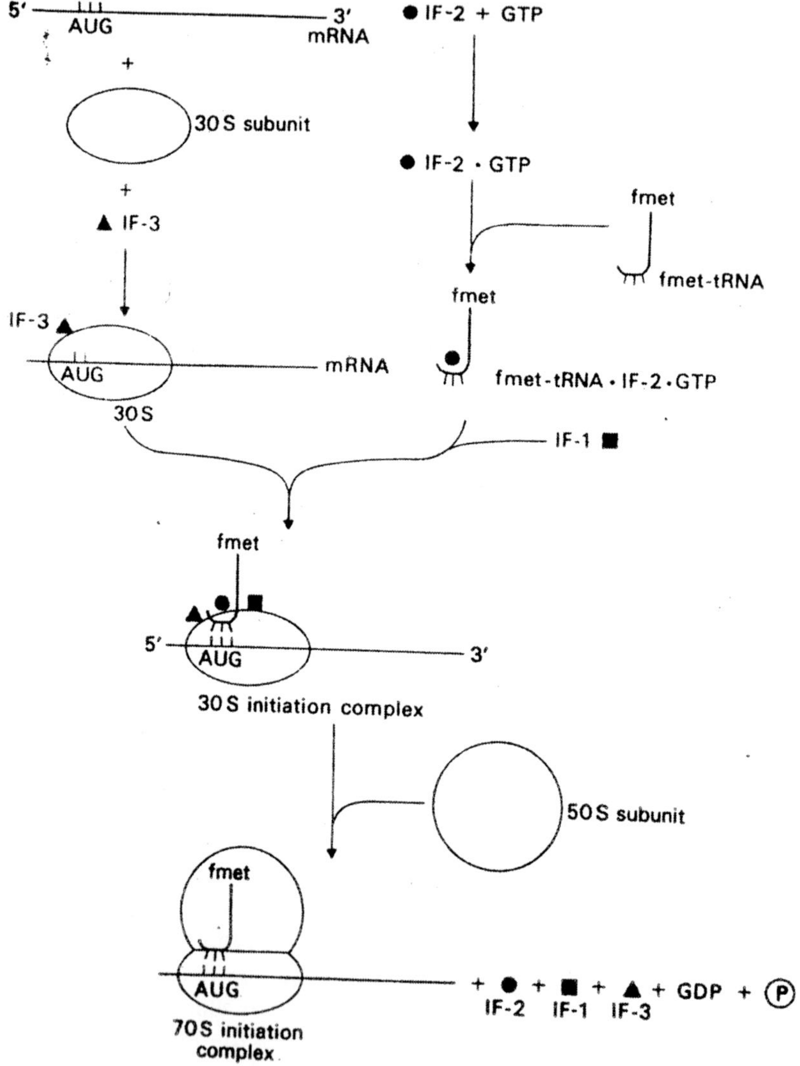

Fig. 4.10. Summary of the steps in the initiation of protein synthesis in prokaryotes.

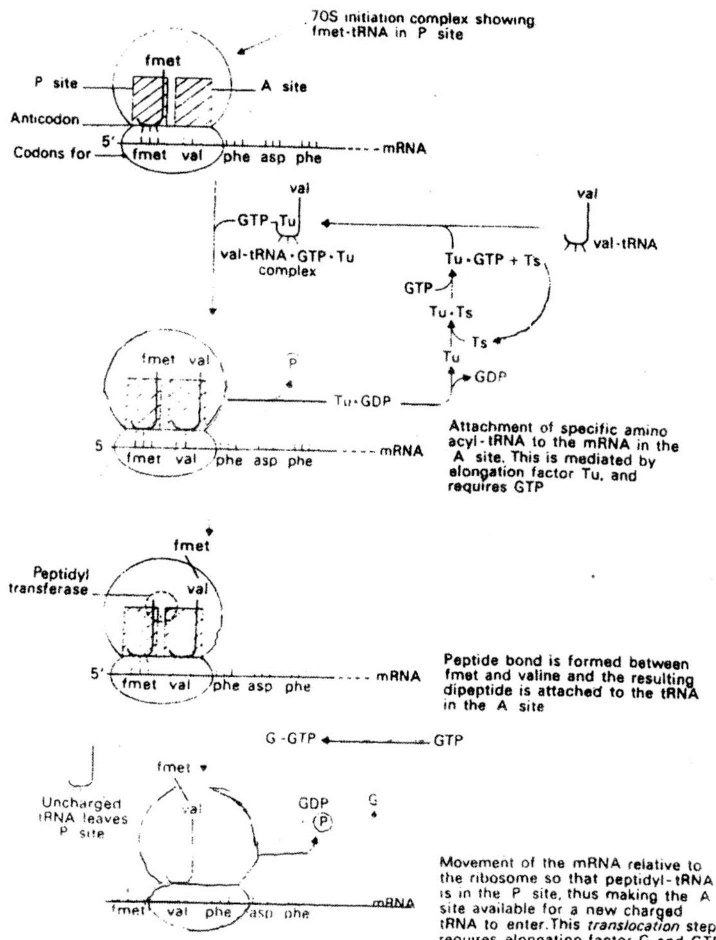

Fig. 4.11. Summary of the elongation (peptide bond formation) and translocation steps in protein synthesis.

ribosome in a reaction requiring ***elongation factor*** T (EE-T) and GTP. This factor can be isolated from the soluble proteins of *E. coli*, and by column chromatography it can be separated into two polypeptides: Ts, which is stable and weighs about 30,000 daltons, and Tu, which is unstable and weights 42,000 daltons.

EF-T has been shown to bind with GTP and this is postulated to bring about the dissociation of the factor into the two polypeptides, resulting in the formation of an EF-Tu.GTP complex and releasing free EF-Ts. The next step in the elongation process is the binding of aminoacyl-tRNA to the complex to produce an aminoacyl-tRNA.Tu.GTP complex. There is evidence that this complex is an intermediate in aminoacyl-tRNA binding to ribosomes. Once the charged tRNA is bound in the A site, GTP is hydrolyzed as a result of the enzymatic action of one or more 50S ribosomal proteins. This hydrolysis cause the release of EF-Tu in a complex with GTP. the latter is released and the elongation factor can reassociate with EF-Ts. The process can then be repeated with another aminoacyl-tRNA.

(a) Tu + Ts (Elongation factor T) + GTP → Tu + GTP + Ts

(b) Tu + GTP + aa-tRNA → aa-tRNA · Tu · GTP
(amino acyl-tRNA) complex

(c) aa-tRNA · Tu · GTP + active 70S ribosome → aa-tRNA·70S (charged tRNA) enters A site) + Tu·GDP + P_i (released from ribosome)

(d) Tu · GDP + Ts → Tu · Ts

Experiments with an analog of GTP that cannot be hydrolyzed have shown that GTP hydrolysis is required for release of EF-Tu from the ribosome but it is not needed for aminoacyl-tRNA binding to the ribosome. Other experiments have shown that binding of fmet-tRNA to the ribosome does not require EF-Tu.

Peptide bond formation

At the beginning of the this stage, a tRNA carrying the growing polypeptide chain is located in the P site, and an aminoacyl-tRNA is located in the A site. Thee tRNAs are maintained in positions conducive for peptide bond formation of the hydrogen bonds between the

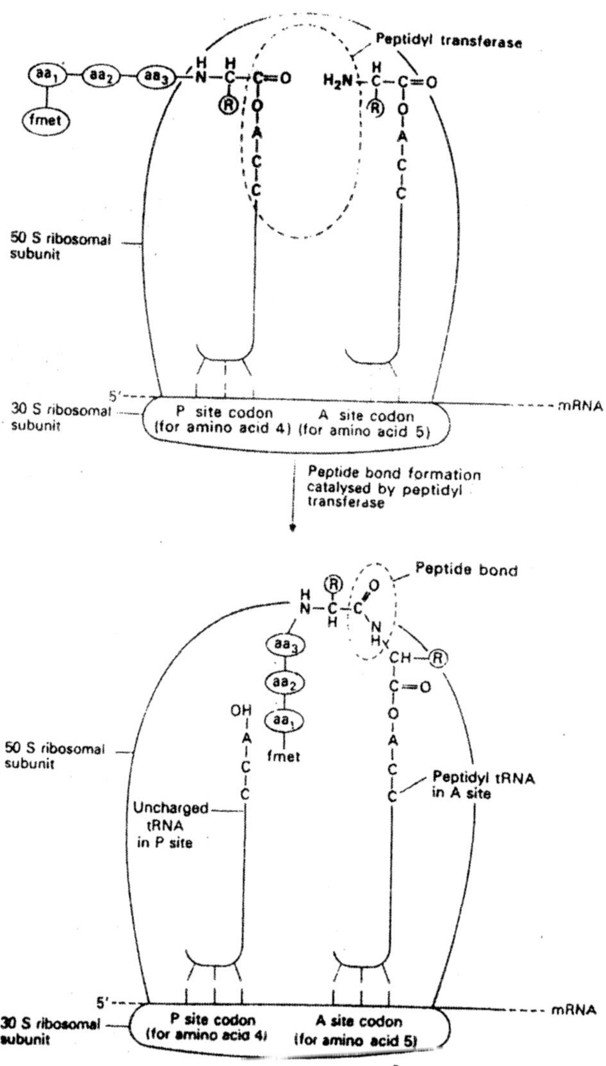

Fig. 4.12. Diagrammatic representation of peptide bond formation on ribosomes catalyzed by peptidyl transferase.

respective codons and anticodons and by the tertiary structure of the ribosome. The peptide bond is formed with the aid of the enzyme, *peptidyl transferase*, which is a ribosomal protein of the 50S subunit. The end result of the reaction is that the polypeptide chain is one amino acid longer, and the growing polypeptide chain has been transferred from the tRNA in the P site to the tRNA in the A site. The tRNA in the P site, which now has no amino acid bound to it, is called an *uncharged tRNA*.

Translocation

Once the peptide bond has been formed and the polypeptide chain is on the tRNA in the A site, the next step is advancement of the ribosome precisely one codon (three nucleotides) down the mRNA, a process called *translocation*. During this translocation event, the peptidyl-tRNA remains attached to the mRNA by codon-anticodon–pairing properties and thus becomes located in the P site. The A site is then vacant, and the aminoacyl-tRNA specified by the new codon there becomes bound by the process already described. The uncharged tRNA left in the P site after peptide bond formation is also released from the ribosome during translocation.

Elongation factor G (EF-F), a 72,000–84,000 dalton protein, and GTP hydrolysis are needed for translocation to occur, but it is not yet known how the translocation mechanism works. One GTP molecule is hydrolyzed for each translocation event. It appears that EF-G molecule is hydrolyzed for each translocation event. It appears that EF-G leaves the ribosome after translocation, since EF-Tu and EF-G cannot interact with the ribosome at the same time.

Termination

The end of the polypeptide chain is indicated on a mRNA molecule by a specific *chain-terminationg* (stop) *codon*. Three such codons are known: UAA, UAG, and UGA. No naturally occurring tRNA has an anticodon for any of these stop codons, and therefore no amino acid can be put into the polypeptide. Three specific termination factors have been shown to be involved in regarding the stop codon. they differ in their codon specificity and GTP requirement (Table 4.1).

RF1 and RF2 have overlapping specificities for the stop codons. They have been shown to interact with the termination codons by interaction at the A site. The RF3 factor apparently plays a stimulatory role in RF1 and RF2 activity. There is some evidence for a GTP requirement in the RF3 factor's activty. In any event, chain termination, as mediated by these factors, involves the cleavage of the carboxyl group of the C-terminal end of the polypeptide chain from the tRNa in the P site. This results in the release of the polypeptide and the now uncharged tRNA. The ribosome will then move alorg the mRNA until a new initiation sequence is encountered (as it may be in polycistronic mRNAs), or it will dissociate from the mRNA. If none is found, when the ribosome is released from the mRNA, IF-3 functions to keep the two subunits apart. Thus, when a new 70S initiation complex is formed, the two subunits are drawn randomly from the free pools of 30S and 50S subunits.

Table 4.1. Properties of prokaryotic termination factors.

Termination factor	*Molecular weight (daltons)*	*Stop codons recognized*	*GTP requirement*
RF1	44,000	UAA and UAG	No
RF2	47,000	UAA and UGA	No
RF3	46,000	None	Yes

While the polypeptide chain is being synthesized, the primary sequence of amino acids directs the three-dimensional shape. In other words, the elongating chain begins to assume its final shape as it is being made. Indeed, some enzyme activity can be detected on ribosomes that have not yet completed the synthesis of an enzymatic polypeptide.

Polysomes

Efficient translation of an mRNA molecule cannot be achieved by a single ribosome moving along it. The amount of space a ribosome

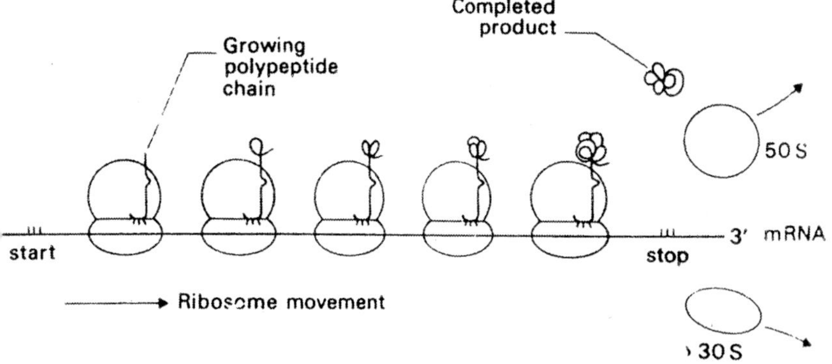

Fig. 4.13. Diagrammdtic representation of a polysome engaged in protein synthesis.

takes up on a mRNA is relatively small, and thus several ribosomes can work on the mRNA at once. The association of a number of ribosomes on a single mRNA chain is called a *polyribosome* or *polysome*, and this allows several polypeptide chains to be made from each mRNA. The length of the polypeptide chain on a given ribosome will be directly proportional to how far the ribosome has moved along the mRNA from the 5' end of the molecule. The existence of polysomes explains why a cell needs so little mRNA, while at the same time it contains so much more protein.

Relationship of Transcription and Translation

In bacteria the mRNA typically becomes associated wth ribosomes while synthesis of the mRNA molecule is continuing . This is possible owing to the lack of a nuclear membranes so that as the 5' end of the growing mRNA molecule is displaced form the DNA as the double helix reforms, the ribosome binding site becomes available. Ribosomes then load on to the mRNA in rapid sequence, the first being close behind the RNA polymerase.

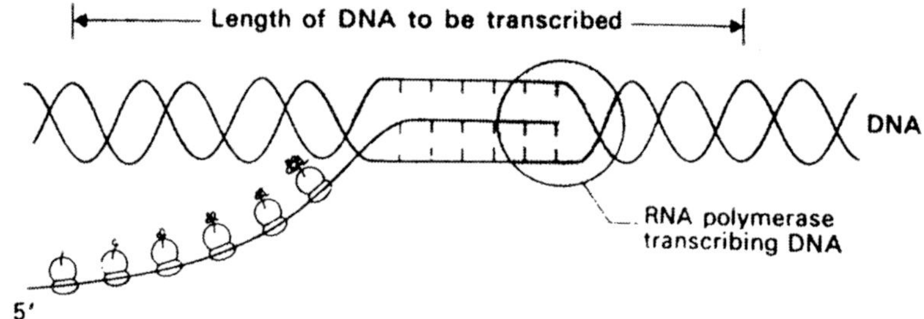

Fig. 4.14. Schematic of the possible translation of an mRNA while it is still being transcribed in prokaryotes.

To give some idea of the rates of these processes, the mRNA for the tryptophan biosynthetic operon is transcribed at a rate of about 1000 nucleotides per minute and the translation process proceeds at about the same rate. Thus approximately 350 amino acids can be polymerized into polypeptide chains each minute.

Another important aspect of translation is the lability of the mRNA. As mentioned in an earlier chapter, prokaryotic mRNAs are considered to be short-lived in that degradation of the molecule by 5'–exonuclease action competes wth the ribosome-mediated initiation of protein synthesis. Continued mRNA synthesis, then, is necessary for continued polypeptide synthesis.

PROTEIN SYNTHESIS IN EUKARYOTES

The steps and mechanisms of protein synthesis are similar in both eukaryotes and prokaryotes. As previously discussed, the ribosomes are different, and this is also the case with the soluble protein factors.

Initiation

Initiation of protein synthesis involves the binding of mRNA to the ribosomes. In eukaryotes it has now been established that the 5'-cap structure is necessary to get efficient binding but the 3' poly(A) sequence apparently is not needed. There is good evidence, however, that the poly(A) sequence stablizes the mRNA during the translation process. The precise mechanism whereby the eukaryotic message binds to the ribosome is not known, although it is most likely that RNA–RNA and RNA–protein interactions are involved. To date the nucleotide sequence has been determined to the 5'-side of the AUG start codon for a number of eukaryotic mRNAs and there are very few common features. Strikingly, while the last 50 nucleotides of *E. coli* 16S rRNA and eukaryotic 18S rRNA are highly homologous and similiar secondary structure models can be built, the eukaryotic rRNA does *not* have the CCUCC sequence characteristic of prokaryotic rRNA. Also, there is no Shine-Dalgarno sequence in the eukaryotic mRNa. Thus it appears that, unlike the case in prokaryotes, there is not a fixed nucleotide sequence that plays a role in ribosome binding to the message.

As in prokaryotes the initiation codon is AUG, and a special initiator methionyl-tRNa recognizes that signal in the message. Unlike its prokaryotic counterpart the methonine carried by the tRNA does not become formylated, since the appropriate enzyme system does not exist in eukaryotic cells. The initiator tRNAcan be distinguished from the met-tRNA that reads AUG codons elsewhere in the message, however, by the fact that it can be formylated in vitro in the presence of an *E. coli* extract. Hence in eukaryotes it is also appropriate to define the two methionine-accepting tRNAs as tRNA.fmet and tRNA.met.

At least in mammals, there are many more initiation factors (labeled eIFs for eukaryotic initiation factors) than in prokaryotes. In many cases the proteins have not been purified to homogeneity, and thus their absolute roles in protein synthesis are uncertain. Between them all, they carry out the initiation events performed by the three prokaryotic IFs. It remains to be seen to what extent the situation in mammals is generalizable throughout the eukaryotes.

Elongation

As in prokaryotes, there are two elongation factors: eEF-1 (equivalent to prokaryotic EF-T) and eEF-2 (equivalent to prokaryotic

EF-G). eEF-1 has been studied in a number of systems, and in general it exists in multiple forms. Purified eEF-1 from rabbit reticulocytes, for example, has a molecular weight of 186,000 and consists of three subunits weighing 62,000 daltons each. In some systems the subunits aggregate to produce molecules of greater than 1 million daltons. Regardng the function of eEF-1, much less is known than in the prokaryotes. The eEF-1 from rabbit reticulocytes, for example, has been shown to bind to aminoacyl-tRNA and to GTP, and thus to facilitate bindng of the aminoactyl-tRNA to the A site in ribosomes. During this step, GTP is hydrolyzed to DGP as a result of GTPase activity of the elongation factor, and an eEF-1.DGP complex is released from the ribosome.

The eukaryotic eEF-2 is similar to prokaryotic EE-G, although the two are not interchangeable in *in vitro* systems. The factor from rabbit reticulocytes has a molecular weight of 96,500–110,000 daltons and, after binding with GTP, binds to the ribosome. This event results in hydrolysis of GTP to GDP, translocation of the ribosome one codon down the message, and release of an eFF-2.GDP complex. All these steps are similar to the events that take place in prokaryotes, one exception being that the eukaryotic factor forms a stable complex with GTP whereas the prokarotic factor does not.

Termination

The same chain termination codons are functional in eukaryotes as in prokaryotes. One release factor has been identified in rabbit reticulocytes, and this has a molecular weight of 115,000 daltons and may be a dimer. This factor recognizes all three chain termination codons and it requires GTP to carry out its function. No stimulatory factor analogous to the porkaryotic RF-3 has been found in eukaryotes.

Protein Synthesis and Cellular Compartmentation in Eukaryotes

In prokaryotes there is no nuclear membrane to separate the transcription process from the translation process. The process of a nuclear membrane and the various modification process peculiar to mRNAs in eukaryotes present many levels at which the regulation of gene expression can be affected. These include transcription itself, the processing of the primary transcript to produce mature mRNA, RNA-RNA splicing, and the movement of mRNA from the nucleus to the cytoplasm.

In eukaryotic cells, the cytoplasm contains a network of inter-connecting channels bounded by membranes called the *endoplasmic reticulum* (ER). The membranes involved are continuous and may in fact connect to the nuclear membrane and the cell membrane. Close examination of the ER reveals that it is differentiated into two types, *smooth* (SER) and *rough* (RER), which are distinguished by the fact that the latter has ribosomes bound to it (hence the rough appearance) whereas the former does not. Thus, ribosomes in the cytoplasm are either membrane bound or free. The membrane-bound ribosomes synthesize proteins that are either secreted from the cell or that are packaged in lysosomes (where the proteins degrade other proteins). Thus, for example, pancreatic cells that secrete enzymes into the intestine are extremely rich in RER. The free ribosomes synthesize all other proteins found in the cell, that is, those in the cytoplasm, nucleus, mitochondria and chloroplasts (if present).

The proteins is to be secreted are made on the ribosomes of the RER and then transferred across the membrane into the channel system. The mRNAs for the secreted proteins must somewhow become associated specifically with the ribosomes of the RER. In 1975, G. Blobel and B. Dobberstein proposed a *signal hypothesis* to explain this. They suggested that there is a unique sequence of codons located to the 3' side of the initiator AUG codon which is present only in mRNAs for proteins that must be transferred across membranes. Translation of these codons results in a specific amino acid sequence at the N-terminal end of the protein. Then they postulated that the special end of the protein facilitates the attachment of the ribosome to the membrane so that the protein can be transferred across it. Once the protein has been completed, they proposed that the ribosome dissociates from the ER.

5

Gene Expression

Jacob and Monod proposed the operon model in 1961 for the co-ordinate regulation of transcription of genes involved in specific metabolic pathways. The operon is a unit of gene expression and regulation which typically includes.

- The *structural genes* (any gene other than a regulator) for enzymes involved in a specific biosynthetic pathway whose expression is co-ordinately controlled.
- Control elements such as an *operator sequence*, which is a DNA sequence that regulates transcription of the structural genes.
- *Regulator gene(s)* whose products recognize the control elements, for example a repressor which binds to and regulates an operator sequence.

The Lactose Operon

Escherichia coli can use lactose as a source of carbon. The enzymes required for the use of lactose as a carbon source are only synthesized when lactose is available as the sole carbon source. The lactose operon (or lac operon) consists of three structural genes: lacZ, which codes for b-balactosidase, an enzyme responsible for hydrolysis of lactose to balactose and glucose; lacY which encodes a galactoside permease which is responsible for lactose transport across the bacterial cell wall; and lacA, which encodes a thiogalactoside transacetylase. The three structural genes are encoded in a single transcription unit, lacZYA, which has a single promoter P_{lac}. This organization means that the three lactose operon structural proteins are expressed together as a polycistronic mRNA containing more than one coding region under

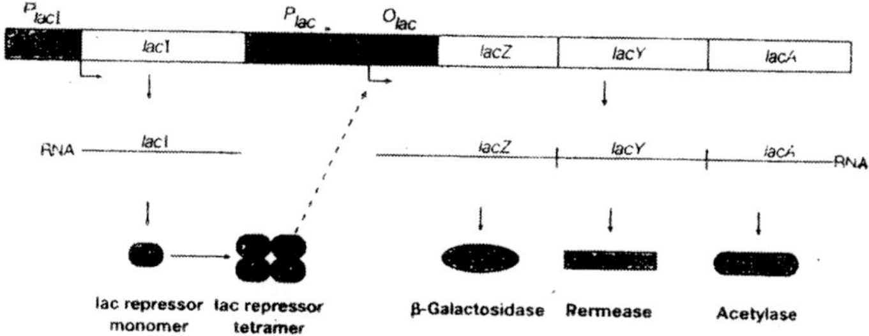

Fig. 5.1. Structure of the lactose operon.

the same regulatory control The lacZYA transcription unit contains an operator site o_{lac} which is positioned between bases –5 and +21 and the 5'-end of the P_{lac} promoter region. This site binds a protein called the lac repressor which is a potent inhibitor of transcription when it is bound to the operator. The lac repressor is encoded by a separate regulatory gene lacI which is also a part of the lactose operon lacI is situated just upstream from P_{lac}.

The Lac Repressor

The lacI gene encodes the lac repressor which is active as a tetramer of identical subunits. It has a very strong affinity for the lac operator-binding site, O_{lac}, and also has a generally high affinity for DNA. The lac operator site consists of 28 bp which is palindromic. (A palindrome has the same DNA sequence when one strand is read left to right in a

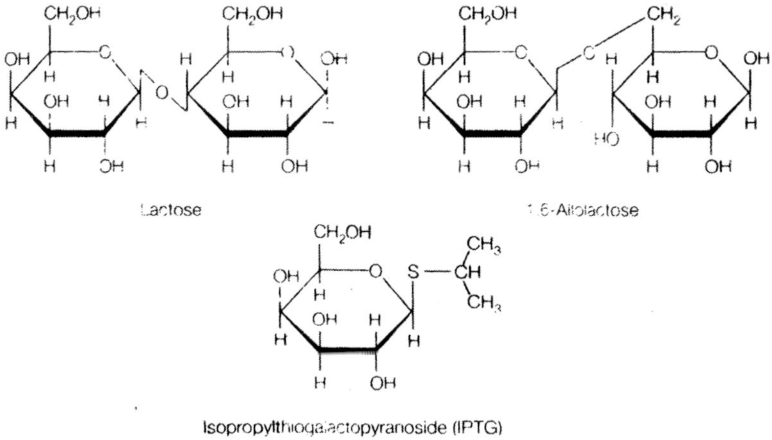

Lactose

1,6-Allolactose

Isopropylthiogalactopyranoside (IPTG)

Fig. 5.2. Structure of lactose, allolactose and IPTG.

5' to 3' direction and the complementary strand is read right to left in a 5' to 3' direction. This inverted repeat symmetry of the operator matches the inherent symmetry of the lac repressor which is made up of four identical subunits. In the absence of lactose, the repressor occupies the operator-binding site. It seems that both the lac repressor and the RNA polymerase can bind simultaneously to the lac promoter and operator sites. The lac repressor actually increases the binding of the polymerase to the lac promoter by two orders of magnitude. This means that when lac repressor is bound to the O_{lac} operator DNA sequence, polymerase is also likely to be bound to the adjacent P_{lac} promoter sequence.

Induction

In the absence of an inducer, the lac repressor blocks all but a very low level of transcription of lacZYA. When lactose is added to cells, the low basal level of the permease allows its uptake, and b-galactosidase catalyzes the conversion of some lactose to allolactose.

Allolactose acts as an inducer and binds to the lac repressor This causes a change in the conformation of the repressor tetramer, reducing its affinity for the lac operator. The removal of the lac repressor from the operator site allows the polymerase (which is already sited at the adjacent promoter) to rapidly begin transcription of the lazZYA genes. Thus, the addition of lactose, or a synthetic inducer such as isopropyl-β-D-thiogalactopyranoside (IPTG) very rapidly stimulates transcription of the lactose operon structural genes. The subsequent removal of the inducer leads to an almost immediate inhibition of this induced

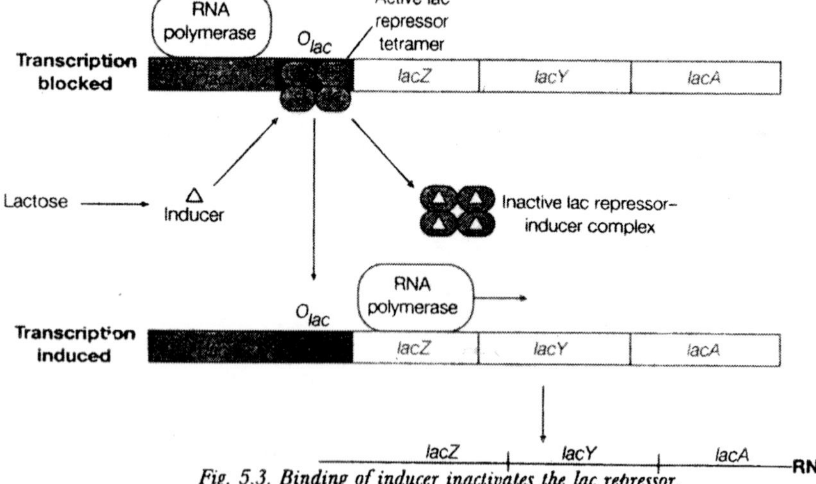

Fig. 5.3. Binding of inducer inactivates the lac repressor.

transcription, since the free lac repressor rapidly re-occupies the operator site and the lacZYA RNA transcript is extremely unstable.

cAMP Receptor Protein

The P_{lac} promoter is not a strong promoter. P_{lac} and related promoters do not have strong-35 sequences and some even have weak–10 consensus sequences. For high level transcription, they require the activity of a specific activator protein called cAMP receptor protein (CRP). CRP may also be called catabolite activator protein or CAP. When glucose is present, *E. coli* does not require alternative carbon sources such as lactose. Therefore, catabolic operons, such as the lactose operon, are not normally activated. This regulation is mediated by CRP which exists as a dimer which can not bind to DNA on its own, nor regulate transcription. Glucose reduces the level of cAMP in the cell. When glucose is absent, the levels of cAMP in E. coli increase and CRP binds to cAMP. The CARP-cAMP complex binds to the lactose operon promoter P_{lac} just upstream from the site for RNA polymerase. CRP binding induces a 90° bend in DNA, and this is believed to enhance RNA polymerase binding to the promoter, enhancing transcription by 50-fold.

The CRP-binding site is an inverted repeat and may be adjacent to the promoter (As in the lactose operon), may lie within the promoter itself, or may be much further upstream from the promoter. Differences in the he CRP-binding sites of the promoters of different catabolic operons may mediate different levels of response of these operons to cAMP *in vivo.*

THE TRP OPERON

The Trypotophan Operon

The trp operon encodes five structural genes whose activity is required for trypotophan synthesis. The operon encodes a single transcription unit which produces a 7 kb transcript which is synthesized downstream from the trp promoter and trp operator sites P_{trp} and O_{trp} Like many of the operons involved in amino acid biosynthesis, the trp operon has evolved systems for co-ordinated expression of these genes when the product of the biosynthetic pathway, trytophan, is in short supply in the cell. As with the lac operon, the RNA product of this transcription unit is very unstable, enabling bacteria to respond to changing needs for tryptophan.

The Trp Repressor

A gene product of the separate trpR operon, the trp repressor, specifically interacts with the operator site of the trp operon. The

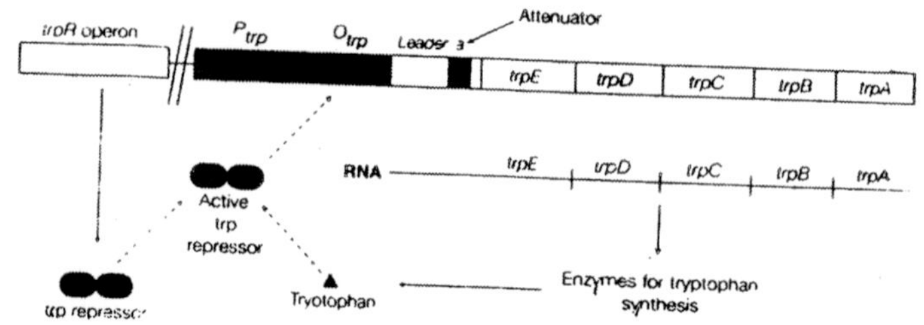

Fig. 5.4. Structure of the trp operon and function of the trp repressor.

symmetrical operator sequence, which forms the trp repressor-binding site, overlaps with the trp promoter sequence between bases -21 and +3 and +3. The core binding site is a palindrome of 18 bp. The trp repressor binds tryptophan and can only bind to the operator when it is complexed with tryptophan. The repressor is a dimer of two subunits which have structural similarity to the CRP protein and lac repressor. The repressor dimer has a structural with a central core and two flexible DNA-reading heads each formed from the carboxyl-terminal half of one subunit. Only when tryptophan is bound to the repressor are the reading heads the correct distance apart, and the side chains in the correct conformation, to interact with successive major grooves of the DNA at the trp operator sequence. Tryptophan, the end-product of the enzymes encoded by the trp operon, therefore acts as a co-repressor and inhibits its own synthesis through end-product inhibition. The repressor reduces transcription initiation by around 70-fold. This is a much smaller transcriptional effect than that mediated by the binding of the lac repressor.

The Attenuator

At first, it was thought that the repressor was responsible for all of the transcriptional regulation of the trp operon. However, it was observed that the deletion of a sequence between the operator and the trpE gene coding region resulted in an increase in both the basal and the activated (depressed) levels of transcription. This site is termed the attenuator and it lies towards the end of the transcribed leader sequence of 162 not that precedes the trpE initiator codon. The attenuator is a rho-independent terminator site which has a short GC-rich palindrome followed by eight successive U residues. If this sequence is able to form a hairpin structure in the RNA transcript, then it acts as a highly efficient transcription terminator and only a 140 bp transcript is synthesized.

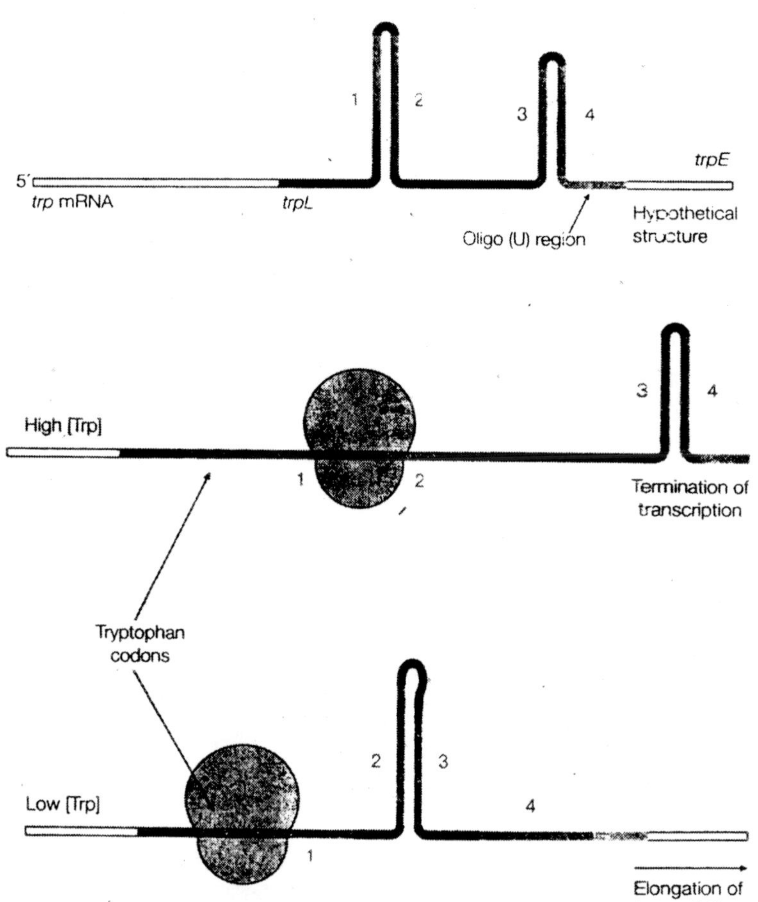

Fig. 5.5. Transcriptional attenuation in the trp operon.

Leader RNA Structure

The leader sequence of trp operon RNA contains for regions of complementary sequence which can form different base-paired RNA structures. These are termed sequences 1, 2, 3, and 4. The attenuator hairpin is the product of the base pairing of sequences 3 and 4. The attenuator hairpin is the product of the base pairing of sequences 3 and 4 (3:4 structure). Sequences 1 and 2 are also complementary and can form a second 1:2 hairpin. However, sequence 2 is also complementary and can form a second 1:2 hairpin. However, sequence 2 is also complementary to sequence 3. If sequences 2 and 3 form a

2:3 hairpin structure, the 3:$ attenuator hairpin cannot be formed and transcription termination will not occur. Under normal conditions, the formation of the 1:2 and 3:4 hairpins is energetically favourable.

The Leader Peptide

The leader RNA sequence contains an efficient ribosome-binding site and can form a 14-amino-acid leader peptide encoded by bases 27–68 of the leader RNA. The 10th and 11th codons of this leader peptide encode successive tryptophan residues, the end-product of the synthetic enzymes of the trp operon. This leader has no obvious function as a polypeptide, and tryptophan is a rare amino acid; therefore, the chances of two tryptophan codons is succession is low and, under conditions of low tryptophan availability, the ribosome would be expected to pause at this site. The function of this leader peptide is to determine tryptophan availability and to regulate transcription termination.

Attenuation

Attenuation depends on the fact that transcription and translation are tightly coupled in E. coli; translation can occur as an mRNA is being transcribed. The 3'-end of the trp leader peptide coding sequence overlaps complementary sequence the two trp codons are within sequence overlaps complementary sequences the two trp codons are within sequence 1 and the stop codon is between sequences 1 and 2. The availability of tryptophan (the ultimate product of the enzymes synthesized by the trp operon) is sensed through its being required in translation, and determines whether or not the terminator (3:4) hairpin forms in the mRNA.

As transcription of the trp operon proceeds, the RNA polymerase pauses at the end of sequence 2 until a ribosome begins to translate the leader peptide. Under conditions of high tryptophan availability, the ribosome rapidly incorporates tryptophan at the two trp codons and thus translates to the end of the leader message. The ribosome is then occluding sequence 2 and, as the RNA polymerase reaches the terminator sequence, the 3:4 hairpin can form, and transcription may be terminated. This is the process of attenuation.

Alternatively, if tryptophan is in scarce supply, it will not be available as an aminoacyl tRNA for translation, and the ribosome will tend to pause at the two trp codons, occluding sequence 1. This leaves sequence 2 free to form a hairpin with sequence 3, known as the anti-terminator. The terminator (3:4) hairpin cannot form, and transcription continues into trpE and beyond. Thus the level of the end product, tryptophan, determines the probability that transcription will terminate early (attenuation), rather than proceeding through the whole operon.

Importance of Attenuation

The presence of tryptophan gives rise to a 10-fold repression of trp operon transcription through the process of attenuation alone. Combined with control by the trp repressor (70fold), this means that tryptophan levels exert a 700-fold regulatory effect on expression from the trp operon. Attenuation occurs in at least six operons that encode enzymes concerned with amino acid biosynthesis. For example, the His operon has a leader which encodes a peptide with seven successive histidine codons. Not all of these other operons have the same combination of regulatory controls that are found in the trp operon. The His operon has no repressor-operator regulation, and attenuation forms the only mechanism of feedback control.

The Arabinose Operon of E. coli

The arabinose operon is another example of glucose-sensitive operon. As with lactose, when arabinose is absent, only a few molecules of the enzymes needed for arabinose catabolism are present in the cell. When arabinose is added (provided glucose is absent), there is a vary rapid increase in the number of arabinose catabolic enzymes. This is controlled by a different mechanism from that described for the lactose operon.

The gene governing the metabolism of arabinose comprise what is called a regulation, which is composed of at least three operons. The *araBAD* operon contains the gene for the enzymes involved with the conversion of L-arabinose to D-xylulose 5-phosphate. The controlling sites for this operon are located adjacent to it. Two operons control the transport of arabinose into the cell: *areE*, which is the structural gene for the L-arabinose binding protein, and *araF*. The regulator gene for the system, *araC*, is located between the *araBAD* controlling site region and the leu operon. The *araC* gene controls the expression of *araBAD* by its positive and negative action in the controlling site region. Since *araBAD*, *araE*, and *araF* are inducible by L-arabinose and controlled coordinately by *araC*, it is assumed that the structures of the three controlling site regions are similar. The following discussion will focus on the *araBAD* operon.

The operon is thought to be controlled as follows. The *araC* gene codes for a P1 protein, which has repressor function and exerts its

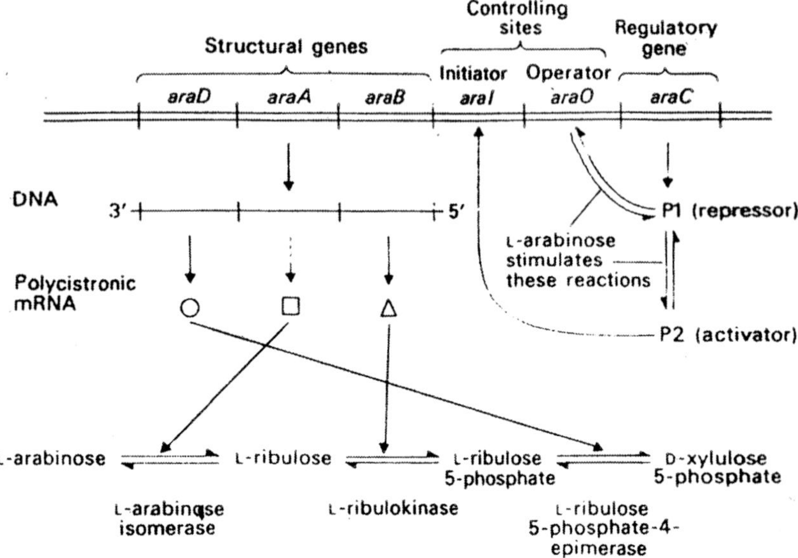

Fig. 5.6. The araBAD (arabinose) operon E. coli and the associated controlling sites and regulatory gene.

effect by binding to that adjacent *araO* (operator) controlling site and preventing the RNA polymerase from binding. Thus the operon is under negative control by P1. When L-arabinose is present, it stimulates the release of P1 from the DNA and the conversion of P1 to P2, which is an activator of the operon. The P2 binds to the *araI* (initiator) site, facilitating RNA polymerase binding and the initiation of transcription. (All this, of course, requires the prior binding of cAMP-CGA complex, and this is thought to occur in the same vicinity.) The three structural genes are transcribed on a single polycistronic mRNA. The operon, then, is under positive control by P2.

Much of this is hypothesis, but there is good evidence for some parts of it. There is genetic evidence for the existence of the aral site, and it is thought that part of this has promoter activity and another part is involved with cAMP-CGA binding. The function of the *araC* gene has been demonstrated by the study of genetic mutants. Mapping experiments with these mutants have shown that *araC* consists of only one cistron. Moreover, *araC* nonsense mutants have been shown to have no cis effect on the *araBAD* operon, thus indicating that *araC* is not in the BAD operon.

Three classes of *araC* alleles are known:

1. *araC*$^+$. The wild type allele renders the operon inducible; the three enzymes are induced by L-arabinose.
2. *araD*$^+$. These occur with high frequency and result in a pleiotropoically L-arabinose-negative phenotype. In other words, the enzymes are not inducible by L-arabinose.
3. *araC*^{c}. These are quite rare and give a pleiotropically constitutive phenotype in that the enzymes are produced even in the absence of the inducer.

As with the regulatory mutants of the lactose operon, diploid studies have been used to obtain an understanding of the function of the *araC* gene. From these studies it was sown there *araC*$^-$ is recessive to C^+ in either the cis or trans arrangement. The C^- alleles are complemented by A^-, B^-, and D^- alleles, and thus the pleiotropic negative phenotype is not the result of a polarity effect on the araBAD operon. The conclusion from studies of C^- and C^+C^- strains was the C^+ produces a protein that, in the presence of L-arabinose, is necessary for the expression of the region. This suggested some positive control in the system and contrasts with the lactose operon C^- mutants, which are constitutive because of the loss of negative control

The Cc alleles are cis- and trans-dominant to C^-, suggesting that they produce the activator P2 in the absence of L-arabinose; this activator is able to turn on an operon that is either cis or trans to the Cc allele. On the other hand, C^+ is dominant to C^c. This suggests that there is some negative control of the operon. Based on the model for regulation of the operon that was presented (which was, of course, proposed on the basis of the data now being discussed), in the absence of arabinose, P1 acts as a repressor, preventing expression of the operon by araCc activator. That is, when P1 is on the operator, transcription ceases even if P2 is present.

There is some biochemical evidence to support the regulatory model. Studies of heat-sensitive *araC*– mutants have shown that both the repressor and activator functions are heat-labile, thereby indicating that arac produces a protein that can serve both functions. The *araC* protein has been purified and it has been shown to have both repressor and activator activity. Indeed there is some evidence that the P2 form of the protein is a dimer of the P1 form. There is also evidence that L-arabinose interacts directly with the *araC* protein to bring about the necessary conversion. Also, there is direct evidence that the activator form of *araC* protein, P2, is required for transcription of the operon. In an *in vitro* system, synthesis of ara mRNA shows an absolute requirement for *araC* protein.

In conclusion, the L-arabinose regulation is under both positive and negative control, with the *araC* protein playing a pivotal role in the regulatory process. The exact nature of the controlling sites remains to be worked out. In contrast to the lactose operon where inhibition must be removed for the genes to be expressed, the arabinose operon requires activation for transcription to begin.

6

Nucleic Acids as the Genetic Material

Earth originated about 5 billion years ago and life originated about 3 billion years ago and that too in water on this earth. All life that has appeared on earth is descended from organisms that were originally very simple. The physical and chemical evolution on earth that preceded the appearance of organisms (prebiotic evolution) was followed by an organismal evolution over the last 4 billion years which culminated in organisms of extraordinary complexity, such as ourselves. Note that present-day organisms–those in existence the past few thousand years–would occupy much less horizontal space in figure than the vertical line marking the interval at 0, for even if 1000 vertical lines could be drawn between 0 and –1, each thin line would represent 1000 thousand (that is, 1 million) years!

It has been reported in this latest thin line of time a great variety of organisms–plant and animal, microscopic to mammoth, incredibly diverse–making up some 2 million kinds, or species. These organisms range from viruses and bacteria through protozoans, sponges, corals and jellyfish, flat, round, and segmented worms, shellfish and starfish, spiders and insects, finned fishes, amphibians, reptiles, birds, mammals, algae and fungi, mosses, ferns, and seed plants in a bewildering tangle of forms of life. By collecting information about these various organisms and by organizing such information–that is, by developing the science of biology–we can group facts about organisms, and establish principles and generalities that apply to many species. Such common threads among species are revealed by studying their structue, form, function and composition.

All organisms have a common origin, and all of them require one or more *cells* and many cell products for their structure and function Cells vary in size and complexity from the tiny and relatively simple cell of a bacterium, about 100 of which can fit across the dot of an *i*, to the giant and relatively complex yolk, which is the single cell of a chicken or an ostrich egg.

Features of a typical cell can be suggested by imagining a plastic bag containing a very porous sponge which is saturated with a thick vegetable soap. This sponge, in turn, surrounds a smaller plastic bag containing noodle soup. The outer plastic bag represents the *cell membrane*, or the outside limit of the cell. The inner plastic bag represents the *nuclear membrane*, the outside limit o the *nucleus*. The sponge represents the *endoplasmic reticulum*, which is a network of membranes (forming channels often interconnected) that also connect the other two membranes. The soups and membranes make up *protoplasm*, which is called *cytoplasm* outside the nucleus. The vegetables in the cytoplasm include various types of bodies (membrane-bound bodies are called *organelles*), such as *ribosomes*, *Mitochondria* and, in green cells, *chloroplasts*. The noodles of the nucleus represent the *chromosomes*. One or a few chromosomes are also present in each mitochondrion and chloroplast. Nucleus-containing cells are *eukaryotic* cells, and organisms composed of such cells are *eukaryotes*. The cells of bacteria and blue-green algae, on the other hand, can be likened to a single plastic bag, the size of a mitochondrion or smaller, containing soup with relatively few vegetables and only a noodle or two. Such cells consists largely of a small mass of protoplasm bounded by a cell membrane, containing ribosomes and one or a few chromosomes. Lacking a nucleus the cells are said to be *prokaryotic*; the cells also lack an endoplasmic reticulum, mitochondria, and chloroplasts. Although all *prokaryotes* are single-celled organisms, eukaryotes may also be composed of single cells, as are protozoans such as Amoebae or of as many as trillions of cells, as in each human being.

The common features of cell structure and form are accompanied by common features of cell function. The thousands of chemical reactions and physical changes that occur in protoplasm comprise the cell's *metabolism*. Metabolism occurs primarily at sites and surfaces provided by membranes, organelles, and large molecules (including chromosomes). The reactions that synthesize more energy-containing substances or protoplasm are *anabolic;* the reactions which degrade energy-providing substances or protoplasm are *catabolic*. By drawing raw materials and energy from the environment and processing them through metabolism, cells are able to maintain themselves, grow, and divide to produce more cells. All of today's organisms are characterized functionally, therefore, by their capacity for two basic functions; (1) *self-maintenance*, which involves growth, replacement, and/or repair of parts of an organism; and (2) *self-reproduction*, which involves making more of the same kind of organisms.

Requirements for the Genetic Material

The genetic material is of central importance to cell function and therefore must fulfil a number of basic requirements:

1. It must contain the information for cell structure, function, and reproduction in a stable form. This information is encoded in the sequence of basic building blocks of the genetic material.
2. It must be possible to replicate the genetic material accurately such that the same genetic information is present in descendant cells and in successive generations.
3. The information coded in the genetic material must be able to be decoded to produce the molecules essential for the structure and function of cells.
4. The genetic material must be capable of (infrequent) variation. Specifically, mutation and recombination of the genetic material are the foundations for the evolutionary process.

The nucleic acids, *deoxyribonucleic acid* (DNA) and ribonucleic acid (RNA), meet all these requirements.

Presence of Genetic Material

Organisms have unique requirements at the structural level, at the functional level and at the macromolecular level. Each organism must possess a facility or factor that (1) persists during the entire existence of the organism, (2) is repeated in each of its progeny, (3) is different in different organisms, and (4) contains the information needed for synthesizing characteristic proteins and nucleic acids. For purposes of scientific investigation, we assume that we are dealing with a material factor rather than a spiritual one. Since this material factor must contain the instructions for the creation, or genesis, of an organism, we can call it *genetic material*. All the different organisms that have appeared on earth are likely to have had a single ancestor, (1) there is only one basic tye of genetic material, and (2) the genetic material can be changed yet still be reproduced. The single most important feature of all organism is, with this view, genetic material. This material is characterized by being preserved, replicated, capable of replicating its modifications, and contains the information for unique protein and nucleic acid. The study of the properties and functions of genetic material thus comprises the core of the study of all organisms, that is, of the science of biology.

Some mature viruses, or *virions*, are composed only of nucleic acid and protein in combination. Clearly, their genetic material must be one or the other or some combination of these two types of macromolecule. A determination can be made through the infection experiments described in the next section.

A Genetic View

The concept of the gene has been the focus of some hundred years of work to establish the basis of heredity. A gene is a sequence of DNA that carries the information representing a protein. Until very recently this would have been an adequate (if incomplete) bio-chemical description. The sequence of DNA could be identified as a continuous stretch of nucleotides, related to the protein sequence by a colinear read-out. But now it is clear that the sequence representing protein is not always continuous; it may be interrupted by sequences not concerned with specifying the protein. So genes may be in pieces that are put together during the process of gene expression. The large number or genes that make up the genome of any species are organizd into a comparatively small number of chromosomes. The genetic material of each chromosome consists of an extremely long stretch of DNA, containing many genes in a linear order. How many genes are present in total has been a puzzle for a long time.

Recently the view that each gene may reside by itself as a unique entity has been superseded by the realization that, in many cases, there may be clusters of related genes that constitute small families. The genome generally has been viewed as rather stable, subject to changes in overall constitution and organization only on an extraordinarily slow evolutionary time scale. This contrasts with recent evidence that in some instances there may be rearrangements that occur regularly; and there may be components of the genome that are relatively mobile. The concept of the gene has therefore undergone an evolution in which, although many of its traditional properties remain, exceptions have been found to show that none constitutes an absolute rule. Starting from the discovery of the gene as a fixed unit of inheritance, its properties were defined in terms of its residence at a definite position on the chromosome, this in turn leading to the view that the genetic material of the chromosome is a continuous length of DNA representing many genes.

A gene may exist in alternative forms that result in the expression of a different characteristic (such as red versus white flower colour). These forms are called *alleles*. The law of independent segregation states that these alleles do not affect each other when present in the same plant, but segregate unchanged by passing into different gametes when the next generation forms. In a true-breeding organism, a *homozygote*, both alleles are the same. But a mating between two parents each of which is homozygous for a different allele generates a hybrid or *heterozygote*. If one allele is *dominant* and the other is *recessive*, the organism will have the appearance or *phenotype* only of the dominant type (so the heterozygote is indistinguishable from the true-breeding dominant parent). But Mendel's first law recognizes that the genetic constitution or *genotype* of the hybrid comprises the presence of both alleles. This is revealed, when the hybrid is crossed with another hybrid to form the second generation. The critical point is that the alleles do not mix, but are physical entities whose interaction is at the level of expression.

Alleles may exhibit *incomplete* (*partial*) *dominance* or no dominance (sometimes known as *codominance*). In the latter case the heterozygote is distinguished from the homozygotes by properties that are intermediate between them. In the snap-dragon, for example, a cross between red and white generates hybrids with pink flowers. Although there is no dominance, the same rule is observed that the first hybrid (F_1) generation is uniform; and the same ratios are generated in a further hybrid cross, although three phenotypes can be distinguished instead of two.

The Independence of Different Genes

Mendel's second law which is the independent assortment summarizes of different genes. When a plant that is dominant for two different characters is crossed with a parent that is recessive for both, as before the F_1 consists of plants uniformly of the dominant type. But in the next hybrid cross, two classes of plants are seen. One consists of the two *parental types*. The other consists of new phenotypes, representing plants with the dominant feature of one parent and the recessive feature of the other. These are called *recombinant types*; and they occur in both possible (*reciprocal*) combinations. The ratios in which these occur can be explained by supposing that gamete formation involves an entirely random association of one of the alleles for the first character with one of the alleles for the other.

All four possible types of gamete are formed in equal proportion, and associate at random in forming the zygotes of the next generation. Once again, the characteristic ratio of phenotypes conceals a greater variety of genotypes, which can be confirmed by the *backcross* to the recessive parent. Essentially the backcross provides a method for looking directly at the genotype of the organism that is being examined.

The law of independent assortment therefore says that the behaviour of any pair (or greater number) of genes can be predicted overall by the rules of mathematical combination. Thus the assortment of one gene does not influence the assortment of the other. Implicit in this concept is the view that assortment is a matter of statistical probability and not an exact result. The ratios of progeny types will approximate increasingly closely to the predicted numbers as a greater number of crosses are performed.

Chromosomes in Heredity

When the chromosomal theory of inheritance was simultaneously proposed by Sutton and by Boveri, it resolved the discussion that had been continuing for some time on the role of chromosomes. They had already been implicated in heredity, although in a somewhat hazy manner, in 1903 it was realized that their properties corresponded exactly with those ascribed to Mendel's particulates units of inheritance. The cell theory established in the middle of the nineteenth century proposed that all organisms are composed of cells, and that these can arise only from preexisting cells.

Early cytology showed that a "typical" cell consists of a dense nucleus separated by a membrane from the less-dense surrounding cytoplasm. Within the nucleus, the granular region of *chromatin* could be recognized by its reaction with certain stains. Not long after Mendel's work, it was found that the chromatin consists of a discrete number of thread-like particles, the *chromosomes*. Chromosomes can be visualized in most cells only during the process of cell division.

The two types of division in sexually reproducing organisms explain both the perpetuation of the genetic material and the process of inheritance as predicted by Mendel's laws. During growth of the organism, the *cell cycle* falls into two parts. The long period of *interphase* represents the time during which the cell engages in its synthetic activities and reproduces its components. Then the short period of *mitosis* is an interlude during which the actual process of division into two daughter cells is accomplished. The products of the series of mitotic divisions that generate the entire organism are called the *somatic cells*. At the end of mitosis, each daughter cell can be seen to start its life with two copies of each chromosome. These are called *homologous*.

The total number of chromosomes in what is known as the *diploid set* can therefore be described as *2n*. The typical somatic cell exists in the diploid state. During interphase, a growing cell duplicates its chromosomal material. This is not evident at the time and becomes apparent only during the subsequent mitosis. During mitosis, each chromosome appears to split longitudinally to generate two copies. These are called *sister chromatids*. At this point, the cell contains *4n* chromosomes, organized as *2n* pairs of sister chromatids. In other words, there are two (homologous) copies of each sister chromatid pair. The series of events, by which mitotic division is accomplished. Essentially the sister chromatids are pulled toward opposite poles of the cell, so that each daughter cell receives one member of each sister chromatid pair. Now these are chromosoems in their own right. The *4n* chromosomes present at the start of division have been divided into two sets of *2n* chromosomes. This process is repeated in the next cell cycle. Thus mitotic division ensures the constancy of the chromosomal complement in the somatic cells.

Meiosis generates cells that contain the *haploid* chromosome number, *n*. This involves two, successive divisions. Once again, the chromosomes have been duplicated previously, so that the cell enters division with *4n*. At the time of first division, the homologous pairs of sister chromatids *synapse* or *pair* to form *bivalents*. Each bivalent contains all four of the cell's copies of one homologue. The first division causes each bivalent to segregate into its two component sister chromatid pairs. This generates two sets of *2n* chromosomes, each set consisting of *n* sister chromatid pairs.

Now the second meiotic division follows, in which both of the sets of *2n* divide again. This division resembles the mitotic division, since one member of each sister chromatid pair segregates to a different daughter cell. The overall result of meiosis is to divide the starting number of *4n* chromosomes into four haploid cells. These may then give rise to mature eggs or sperm. In forming these gametes, homologoues of paternal and maternal origin are separated, so that each gametes gains only one of the two homologues of its parent.

A critical feature in relating this process to the predictions of Mendel's laws was the realization that nonhomologous chromosomes undergo segregation independently, so that either member of one homologous pair enters the gamete at random with either member of a different homologous pair.

Genes occur in allelic pairs, one member of each pair having been contributed by each parent; the diploid set of chromosomes results from the contribution of a haploid set by each parent. The assortment of nonallelic genes into gametes should be independent of origin, nonhomologous chromosomes undergo independent segregation. The critical provision is that each gamete obtains a complete haploid set, and this is fulfilled whether viewed in terms of factors or chromosomes.

ARRANGEMENT OF GENES

Linear

The proof that genes reside on chromosomes required a demonstration that a particular gene is always present on a particular chromosome. This was provided by the properties displayed by a variant of the fruit fly *Drosophila melanogaster* obtained by Morgan in 1910. This white-eyed male appeared spontaneously in a line of flies of the usual red eye colour.

The red eye colour is the *wild type* of character. The white eye is the *mutant* phenotype. The event responsible for generating the mutant phenotype is a genetic change called a *mutation*. Most mutations prevent the gene in which they occur from functioning properly or at all; and so the study of mutation is for the most part the study of inactive alleles.

Sometimes it is possible for a mutation to be reversed by a genetic change that restores the original state of the genetic material. This is called a *reversion* to the wild type. The white mutation could be located on a particular chromosome because of its association with sexual type.

In many sexually reproducing organisms there is an exception to the rule that chromosomes occur in homologous pairs whose separation (disjunction) at meiosis produces identical haploid sets. Male and female sets may differ visibly in chromosome constitution, the most common form of difference being the replacement of one member of a homologue pair with a different chromosome in one of the sexes. This pair is referred to as the *sex chromosomes*, and the remaining homologous pairs are called the *autosomes*.

The chromosome complements of the two sexes can be described as 2A + XX and 2A + XY, where the haploid set of autosomes is denoted A and the two sex chromosomes are X and Y. The sex with the complement of 2A + XX is called *homogametic*; if forms gametes only of the type A + X. The sex with the complement of 2A + XY is called *heterogametic*; it forms equal proportions of gametes of the types A + X and A + Y. The random union of gametes from one sex with gametes from the other sex perpetuates the equal sex ratio at zygote formation. In *Drosophila* the females are homogametic.

A critical prediction of Mendel's laws is that the results of a genetic cross should be the same regardless of orientation– that is, irrespective of which parent introduces which allele. But the reciprocal crosses with white eye in *Drosophila* give different results.

The cross of white male × red female gives the entirely red F_1 expected if red is dominant and white is recessive. But in the F_2 all the white-eye flies that reappear are males. In the reciprocal cross of red male × white female, all the F_1 males are white-eyed and all the females are red eyed. Crossing these gives an F_2 with equal proportions of white and red eyes in each sex. This pattern of inheritance exactly follows that of the sex chromosomes. If the alleles for red and white eyes are carried on the X chromosome, with no locus for eye colour present on the Y chromosome, the phenotype of a male will be determined by the single allele present on its X chromosome. This allele will be transmitted to all of its daughters and none of its sons. This is the typical pattern of *sex linkage.*

The separation of chromosomes seen at meiosis explains the independent assortment of genes that are carried on different chromosomes. But the number of genetic factors is much greater than the number of chromosomes. Each chromosome may bear as many as 1000 to 10000 genes. The quickening pace of genetics after the turn of the century was shown by the passage of only three years between the report of the first eye-colour mutant and Stretevant's study in 1913 of the pattern of inheritance of six sex-linked mutations. Morgan had shown in 1911 that each of these factors shows the same sex-linked pattern as white/red eye colour. By the same logic each must therefore be carried by the X chromosome.

Morgan proposed that the cause of genetic linkage is the simple mechanical result of the location of the factors in the chromosomes. He suggested that the production of genetic recombinant classes can be equated with the process of *crossing over* that is visible during meiosis. Early in meiosis, at a stage when there are four copies of each chromosome organized in a bivalent, pairwise exchanges of material occur between the closely associated (synapsed) homologue pairs. This exchange is called a *chiasma.*

In maize and in *Drosophila*, suitable *translocations* had occurred in which a part of one chromosome had broken off and become attached to another. This allows the translocation chromosome to be distinguished by its appearance from the normal chromosome. In suitable crosses, it is then possible to show that the formation of genetic recombinants occurs only when there has been a physical crossing-over between the appropriate chromosome regions. As the distance between the genes increases, the probability of crossing-over between them will increase. Thus if crossing-over is responsible for recombination, genes lying near each other will be tightly linked, and genetic linkage will decrease with physical distance apart. Reversing the argument, genetic linkage can be taken to be measure of physical distance.

The concept that genes on the same chromosome are linked was extended by Sturtevant with the proposal that the extent of recombination between them can be used as a *map distance* to measure their relative locations. This is expressed as the percent recombination; that is,

$$\text{Map distance} = \frac{\text{Number of recombinants} \times 100}{\text{Total number of progeny}}$$

Distance is given in terms of *map units*, defined by 1 map unit (or centiMorgan) equal 1% recombination.

Thus if two genes A and B are 10 units apart, and it is 5 units from B to a further gene C, the direct measure of distance between A and C will be close to 15. The genes can therefore be placed in *linear order.* A crucial point in the construction of maps is that the distance between genes does not depend on the *alleles* that are used, but only on the gene *loci.* The locus defines the *position* on the chromosome at which the gene for a particular trait resides; the various alternative forms of the gene–that is, the alleles used in mapping–all reside at the *same* location.

The genetic mapping is concerned with identifying the locations of gene loci, which are fixed and in a linear order. In a mapping experiment the same result is obtained irrespective of the particular combination of alleles. Sturtevant concluded that his results "form a new argument in favour of the chromosome view of inheritance, since they strongly indicate that the factors investigated are arranged in a linear series, at least mathematically."

This last qualification is interesting. Although a genetic map can be constructed that represents a chromosome as a linear array of genes, this does not prove that the genetic content of the chromosome is *physically* a continuous array of genes. Many other models were considered before it became clear that a chromosome contains a single thread of genetic material. Linkage is not displayed between all pairs of genes located on a single chromosome.

The maximum recombination that occurs in the 50% predicted by Mendel's second law. (Although there is a high probability that recombination will occur between two genes lying far apart on a chromosome, each individual recombination event involves only two of the four associated chromosomes, thus generating 50% crossover between the genetic factors).

The factors that are well separated may show no direct linkage, in spite of their presence on the same chromosome. However, each can be linked via intermediate factors, and so a genetic map can be extended beyond the limit of 50 map units directly measurable between any pair of genes.

In fact, because of the distortion of linkage measured directly between markers that are not closely linked, map distances usually are based on measurements between fairly close factors, and may be subject to corrections from the simple percent recombination. The *linkage group* includes all those genes that can be connected either directly or indirectly by linkage relationships. Those close together show direct linkage; those more than 50 map units apart in practice assort independently.

As linkage relationships are extended, each gene identified in an organism can be placed into linkage with a group of previously mapped genes. Thus the genes fall into a discrete number of linkage groups. Genes in one linkage group always show independent assortment with regard to genes located in other linkage groups. The number of linkage groups is the same as the number of chromosomes.

The relative lengths of the linkage groups are similar to the actual relative sizes of the chromosomes. The example of *Drosophila* (where it happens to be particularly easy to measure the chromosome lengths). Mendel's concept of the gene as a discrete particulate factor can therefore be extended into the concept that the chromosome constitutes a linear unit divided into many genes, whose physical arrangement may underlie their genetic behaviour.

Identification of DNA as the Genetic Material

Two classic experiments led to the identification of DNA as the genetic material and, in so doing, laid the foundation for molecular genetics. These experiments are described in the following section.

The Transformation Experiments

The development of the current idea that DNA is the genetic material began with an observation in 1928 by Fred Griffith, who was studying the bacterium responsible for human pneumonia–i.e., *Streptococcus pneumoniae* or *Pneumonoccus*. The virulence of this bacterium

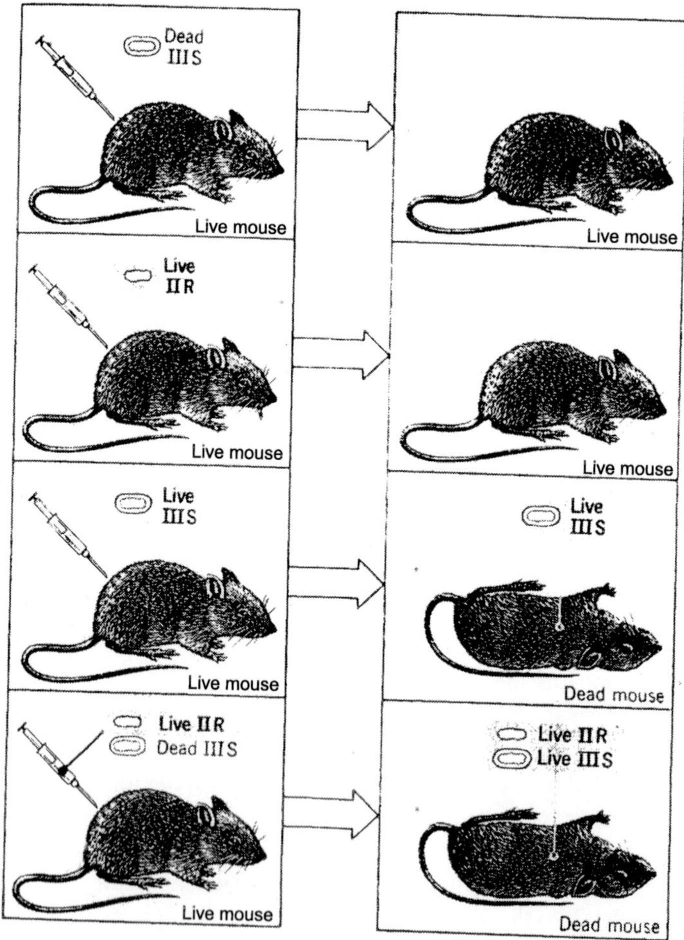

Fig. 6.1. Griffith's demonstration of transformation in pneumococcus.

was known to be dependent on a surrounding polysaccharide capsule that protects it from the dense systems of the body. This capsule also causes the bacterium to produce smooth-edged (S) colonies on an agar surface. It was known that mice were normally killed by S bacteria. Griffith then isolated a rough-edged (R) colony mutant, which proved to be both nonencapsulated and nonlethal. He subsequently made a significant observation–namely, whereas both R and heat-killed S were nonlethal, a mixture of live R and heat-killed S was lethal. Furthermore, the bacteria isolated from a mouse that had died from such a mixed infection were only S–i.e., the live R had somehow been replaced by or *transformed* to S bacteria. Several years later it was shown that the mouse itself was not needed to mediate this transformation because when a mixture of R and heat-killed S was grown in a culture fluid, living S cells were produced. A possible explanation for this surprising phenomenon was that the R cells restored the viability of the dead S cells; but this idea was eliminated by the observation that living S cells grew even when the heat-killed S culture in the mixture was replaced by a cell extract prepared from broken S cells, which had been freed from both intact cells and the capsular polysaccharide by centrifugation. Hence, it was concluded that the cell extract contained a *transforming principle*, the nature of which was unknown.

15 years later when Oswald Avery, Colin MacLeod, and Maclyn McCarty partially purified the transforming principle from the cell extract and demonstrated that it was DNA. These workers modified known schemes for isolating DNA and prepared samples of DNA from S bacteria. They added this DNA to a live R bacterial culture; after a period of time they placed a sample of the S-containing R bacterial culture on an agar surface and allowed it to grow to form colonies. Some of the colonies (about 1 in 10^4) that grew were S type.

To show that this was a permanent genetic change, they dispersed many of the newly formed S colonies and placed them on a second agar surface. The resulting colonies were again S type. If an R colony arising from the original mixture was dispersed, only R bacteria grew in subsequent generations. Hence the R colonies retained the R character, whereas the transformed S colonies bred true as S. Because S and R colonies differed by a polysaccharide coat around each S bacterium, the ability of purified polysaccharide to transform was also tested, but no transformation was observed. Since the procedures for isolating DNA then in use produced DNA containing many impurities, it was necessary to provide evidence that the transformation was actually caused by the DNA alone.

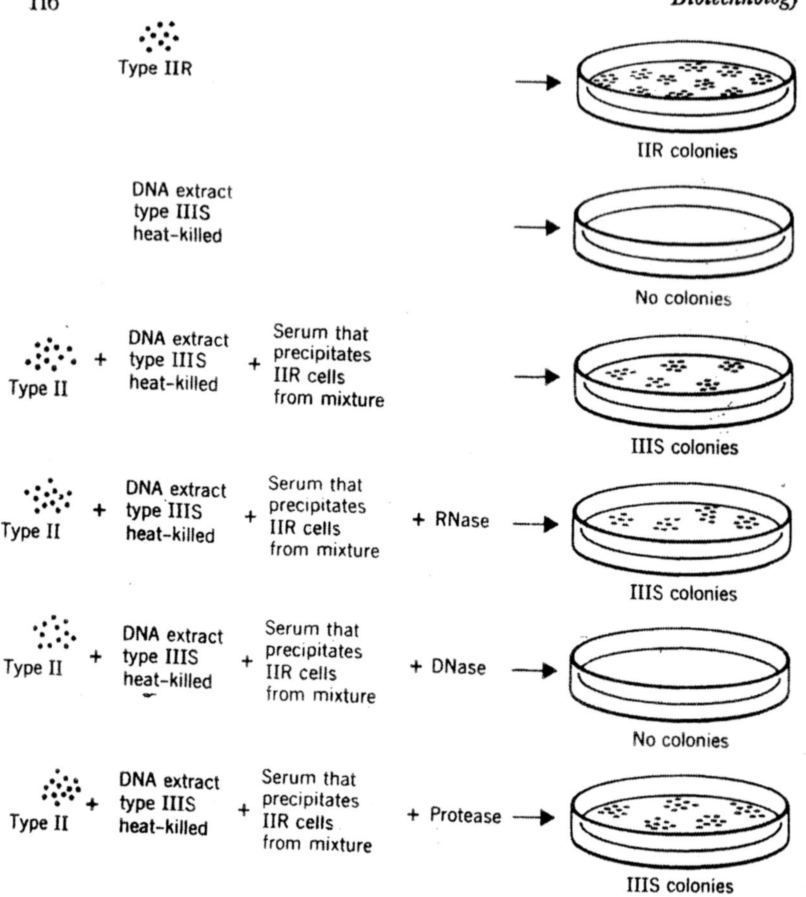

Fig. 6.2. Avery, MacLeod, and McCarty's proof that the "transforming principle" is DNA.

This evidence was provided by the following four procedures.

1. Chemical analysis showed that the major component was a deoxyribose-containing nucleic acid.
2. Physical measurements showed that the sample contained a highly viscous substance having the properties of DNA.
3. Experiments demonstrated that transforming activity is not lost by reaction with either (a) purified proteolytic (protein-hydrolyzing) enzymes–trypsin, chymotrypsin, or a mixture of both–or (b) ribonuclease (an enzyme that depolymerizes RNA).
4. It was demonstrated that treatment with materials known to contain DNA-depolymerizing activity (DNAse) inactivated the transforming principle.

The transformation experiment was not accepted by the scientific community as proof that DNA is the genetic material, because it was widely believed that DNA was a simple tetranucleotide incapable of carrying the information required of a genetic substance.

The tetranucleotide hypothesis was based on chemical analyses that indicated that DNA consists of equimolar amounts of the four bases; this conclusion was based on inadequate chemical procedures for analyzing base composition and on the use of higher organisms as

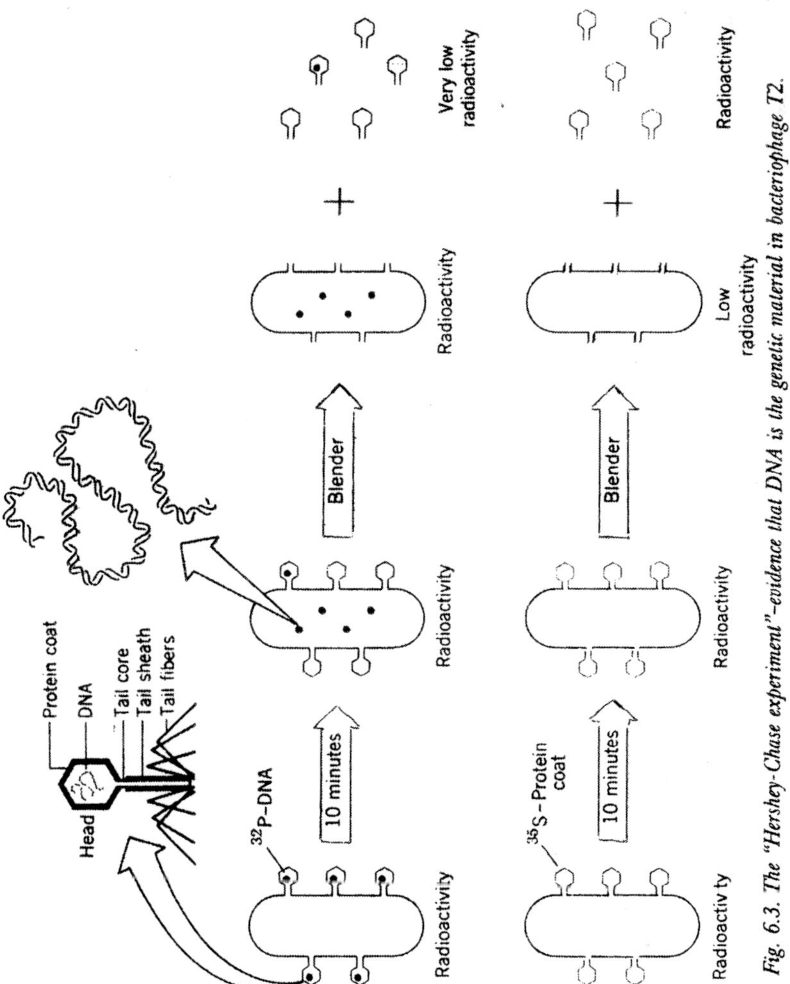

Fig. 6.3. The "Hershey-Chase experiment"–evidence that DNA is the genetic material in bacteriophage T2.

sources of DNA; in these organisms the base composition is not far from being equimolar. However, as techniques improved and a greater range of organisms was examined, it was found that the base composition varies widely from one species to the next and that DNA is not a simple small molecule. With these results, the idea that DNA is the genetic material became acceptable. The following experiment provided the final proof.

The Blendor Experiment

An elegant confirmation of the genetic nature of DNA came from an experiment with *E.coli* phage T2. This experiment, known as the *blendor experiment* because a kitchen blendor was used as a major piece of apparatus, was performed by Alfred Hershey and Martha Chase, who demonstrated that the DNA injected by a phage particle into a bacterium contains all of the information required to synthesize progeny phage particles. A single particle of phage T2 consists of DNA (now known to be a single molecule) encased in a protein shell.

The DNA is the only phosphorus-containing substance in the phage particle; the proteins of the shell, which contain the amino acids methionine and cysteine, have the only sulfur atoms. Thus, by growing phage in a nutrient medium in which radioactive phosphate ($^{32}PO_4^{3-}$) is the sole source of phosphorus, phage containing radioactive DNA can be prepared. If instead the growth medium contains radioactive sulfur as $^{35}SO_4^{3-}$, phage containing radioactive proteins are obtained. If these two kinds of labeled phage are used in an infection of a bacterial host, the phage DNA and the protein molecules can always be located by their radioactivity. Hershey and Chase used these phage to show that ^{32}P but not ^{35}S is injected into the bacterium. Each phage T2 particle has a long tail by which it attaches to sensitive bacteria.

Hershey and Chase showed that an attached phage can be torn from a bacterial cell wall by violent agitation of the infected cells in a kitchen blendor. Thus, it was possible to separate an adsorbed phage from a bacterium and determine the component(s) of the phage that could not be shaken free by agitation–presumably, those components had been injected into the bacterium.

In the first experiment ^{35}S-labeled phage particles were adsorbed to bacteria for a few minutes. The bacteria were separated from unadsorbed phage and phage fragments by centrifuging the mixture and collecting the sediment (the pellet), which consisted of the phage-bacterium complexes. These complexes were resuspended in liquid

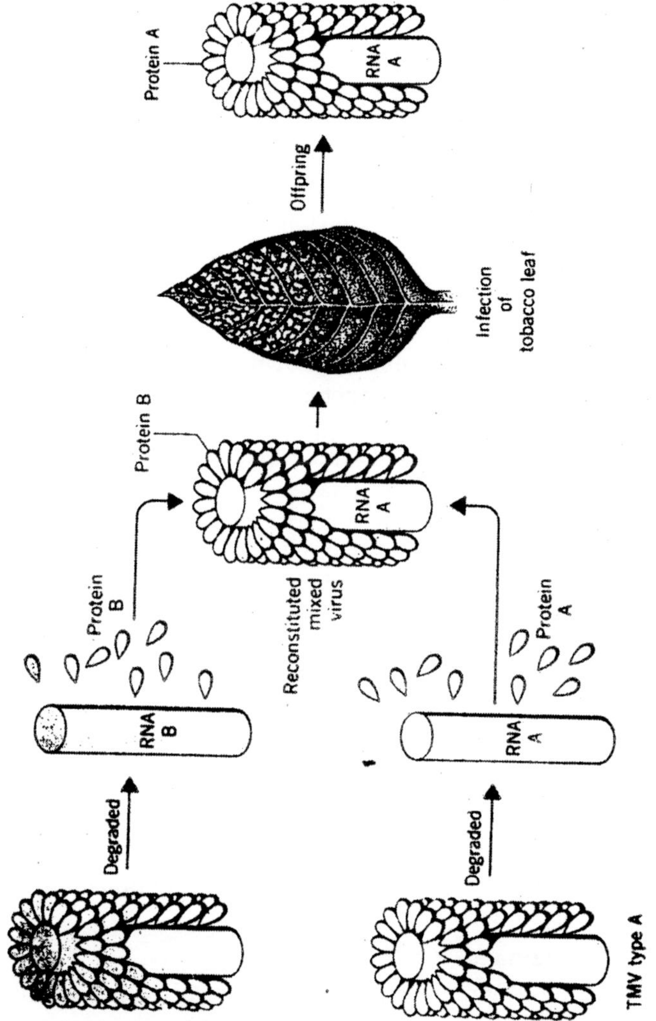

Fig. 6.4. Proof that the genetic material of tobacco mosaic virus (TMV) is RNA, not protein.

and blended. The suspension was again centrifuged, and the pellet, which now consisted almost entirely of bacteria, and the supernatant were collected. It was found that 80 percent of the ^{35}S label was in the supernatant and 20 percent was in the pellet. The 20 percent of the ^{35}S that remains associated with the bacteria was shown some years later to consist mostly of phage tail fragments that adhered too tightly to the bacterial surface to be removed by the blending. A very different result was observed when the phage population was labeled with ^{32}P.

In this case 70 percent of the ^{32}P remained associated with the bacteria in the pellet after blending and only 30 percent was in the supernatant.

Of the radioactivity in the supernatant roughly one-third could be accounted for by breakage of the bacteria during the blending. (The remainder was shown some years later to be a result of defective phage particles that could not inject their DNA.) when the pellet material was resuspended in growth medium and reincubated, it was found to be capable of phage production. Thus, the ability of a bacterium to synthesize progeny phage is associated with transfer of ^{32}P, and hence of DNA, from parental phage to the bacteria.

Transfer Experiments

Another series of experiments, known as ***transfer experiments***, supported the interpretation that genetic material contains ^{32}P but not ^{35}S. In these experiments progeny phage were isolated, after blending, from cells that had been infected with either ^{35}S- or ^{35}P-containing phage and the progeny were then assayed for radioactivity, the idea being that some parental genetic material should be found in the progeny. It was found that no ^{35}S but about half of the injected ^{32}P was transferred to the progeny. This result indicated that though ^{35}S might be residually associated with the phage-infected bacteria, it was not part of the phage genetic material. The interpretation (now known to be correct) of the transfer of only half of the ^{32}P was that progeny DNA is selected at random for packaging into protein coats and that all progeny DNA is not successfully packaged.

Properties of a Genetic Material

To comprise the genes, DNA must carry the information to control the synthesis of the enzymes and proteins within a cell or organism, self-replicate with high fidelity yet show a low level of mutation, and be located in the chromosomes.

Control of the Enzymes

The growth, development, and functioning of a cell are controlled by the proteins within it, primarily its enzymes. Thus the nature of a cell's phenotype is controlled by the protein synthesis within that cell. The genetic material must determine the presence and effective amounts of the enzymes in a cell. For example, given inorganic salts and glucose, an *E.coli* cell can synthesize, through its enzyme-controlled biochemical pathways, all of the compounds it needs for growth, survival, and reproduction.

An enzyme is a protein that acts as a catalyst of a specific metabolic process without itself being markedly altered by the reaction.

Most reactions that are catalyzed by enzymes could occur anyway, but only under conditions too extreme for them to take place within living systems. For example, many oxidations occur naturally at high temperatures. Enzymes allow these reactions to occur within the cell by lowering what is called the *activation energy* (ΔG) of a particular reaction. Most metabolic processes, such as the biosynthesis or degradation of molecules, occur in pathways in which an enzyme facilitates each step in the pathway.

Each reaction product in the pathway is altered by an enzyme that converts it to the next product. The enzyme threonine dehydratase, for example, converts threonine into α-ketobutyric acid. Enzymes are composed of folded polymers of amino acids; the sequence of amino acids determines the final structure of an enzyme. The genetic material determines the sequence of the amino acids.

The three-dimensional structure of enzymes permits them to perform their function. An enzyme combines with its substrate or substrates (the molecules it works on) at a part of the enzyme called the *active site*. The substrates "fit" into the active site, which has a shape that allows only the specific substrates to enter. This view of the way an enzyme interacts with its substrates is called the *lock-and-key model* of enzyme functioning. When the substrates are in their proper position in the active site of the enzyme, the particular reaction that the enzyme catalyzes takes place. The reaction products then separate from the enzyme and leave it free to repeat the process.

Enzymes can work at phenomenal speeds. Some can catalyze as many as a million reactions per minute. Not all of the cell's proteins function as catalysts. Some are structural proteins, such as keratin, the main component of hair. Other proteins are regulatory–they control the rate of production of other enzymes. Still others are involved in different functions; albumins, for example, help regulate the osmotic pressure of blood.

Replication

The genetic material must replicate itself precisely so that every daughter cell receives an exact copy. Some *mutability*, or the ability to change, is also required because we know that the genetic material has changed, or evolved, in the history of life on earth. In their 1953 paper, Watson and Crick had already worked out the replication process based on the structure of DNA. Mutability is also a natural derivative of this process.

Location

It has been known since the turn of the century that genes, the discrete functional units of genetic material, are located in chromosomes within the nuclei of eukaryotic cells: The behaviour of chromosomes during the cellular division stages of mitosis and meiosis mimics the behaviour of genes. Thus the genetic material in eukaryotes must be a part of the chromosomes.

For a long time, proteins were considered the most probable cell components to be genetic material because they have the necessary molecular complexity. Amino acids can be combined in an almost unlimited variety, creating thousands and thousands of different proteins. The first proof that the genetic material is deoxyribonucleic acid (DNA) was provided in 1944 by Oswald Avery and his colleagues. The Watson and Crick model in 1953 ended a period when DNA was thought to be the genetic material, but its structure was unknown.

Storage and Transmission of Genetic Information by DNA

The information possessed and conveyed by the DNA of a cell is of several types:

1. The sequence of amino acids in every protein synthesized by the cell.
2. A start and a stop signal for the synthesis of each protein.
3. A set of signals that determine whether a particular protein is to be made and how many molecules are to be made per unit time.

This information is contained in the sequence of DNA bases. The amino acid sequence and the start and stop signals are not obtained directly from the DNA base sequence but via RNA intermediates. That is, DNA serves as a template for synthesis of specific RNA molecules called messenger RNA. The base sequence of the mRNA is complementary to that of one of the DNA strands and it is form the mRNA sequence that the amino acid sequence is translated. In particular, the base sequence is "read" in order in groups of three bases called *triplets* or *codons* and each group corresponds to a particular amino acid or to a start or stop codon.

The two-stage process has the advantage that the DNA molecule neither has to be used very often nor has to enter the protein-synthetic mechanism. Since protein synthesis occurs continually, one can appreciate why DNA evolved with bases having groups capable of forming hydrogen bonds. First, by specific patterns of hydrogen-bonding–that is, base-pairing–the base sequence of DNA is easily transcribed into an RNA molecule by means of an enzymatic system that adds a base to the polymerizing end of an RNA molecule only if that base can hydrogen-bond with the base being copied. The use of hydrogen-bonding enables the cell to use less genetic material to hold information than if van der Waals forces had been selected in the course of evolution; this is because van der Waals forces are so weak that the error frequency in transmitting information would be much greater than with hydrogen bonds unless more bases (or other molecules) were used for complementary binding.

Transmission from Parent to Progeny

When a cell divides, each daughter cell must contain identical genetic information; that is, each DNA molecule must become two identical molecules, each carrying the information that was contained in the parent molecule. This duplication process is called *replication.* Once again, the value of bases capable of hydrogen-bonding is apparent. That is, to make a replica, the replication system need only require that the base being added to the growing end of a new chain be capable of hydrogen-bonding to the base being copied.

Chemical Stability of DNA

In long-lived organisms a single DNA molecule may have to last one hundred years or more. Furthermore, the information contained in the molecule is passed on to successive generations for millions of years with only small changes. Thus, DNA molecules must have great stability. The sugar-phosphate backbone of DNA is extremely stable. The C–C bonds in the sugar are resistant to chemical attack under all conditions other than strong acid at very high temperatures.

The phosphodiester bond is a little less stable and can be hydrolyzed at room temperature at *p*H 2, but this is not a physiological condition. In considering the stability of the phosphodiester bond, one can see immediately why 2′-deoxyribose rather than ribose is the constituent sugar in DNA. The phosphodiester bond in RNA is rapidly broken in alkali. The chemical mechanism for this breakage requires the OH group on the 2′ carbon of the sugar. In DNA, because deoxyribose is used, there is no 2′-OH group, so the molecule is exceedingly resistant to alkaline hydrolysis. The N-glycosylic bond is also very stable, though it would not be so if the bases were not rings. An alteration in the chemical structure of a base definitely means loss of genetic information.

In a cell there are certainly a large number of chemical compounds that can attack a free base. In considering this problem one can see immediately the value of the double helix. One aspect of this is that the molecule isredundant in the sense that identical information is contained in both strands; that is, the base sequence of one strand is complementary to that of the other strand. In fact, there exist in cells elegant repair systems that can remove an altered base and then by reading the sequence on the complementary strand, can replace the correct base. The more important aspect of this double-helical structure is that the nature of the bases and the duplex structure of DNA provides extreme protection against chemical attack.

The bases are hydrophobic rings having charged groups that contain the genetic information. It is therefore the charged groups that need protection. The hydrophobic nature of the base causes the bases to stack so extensively that water is almost completely excluded from the stacked array. This has the effect that water-soluble compounds are often unable to come into close contact with the "dry" stack of bases. The bases themselves, with the exception of cytosine, are very stable. At a very low rate, however, cytosine is deaminated to form uracil.

Deamination is a disastrous change because the deamination product, uracil, pairs with adenine rather than with guanine. This has two effects: (1) an incorrect base will appear in mRNA and (2) an adenine instead of a guanine will occur in newly replicated DNA strands. There exists an intracellular system, for removing uracil from DNA and replacing it with thymine. The necessity for eliminating a uracil formed by deamination explains why DNA utilizes thymine and not uracil as the base complementary to adenine. If uracil were the normal DNA base, there would be no way to distinguish a correct uracil from an incorrect uracil produced by deamination of cytosine. By using thymine, the cell follows the rule: Always remove a uracil from DNA because it is unwanted.

It is not obvious why RNA uses uracil and not thymine nor why DNA evolved with cytosine rather than with some base that would not deaminate. This may have been an evolutionary accident. It is possible, though, that the original DNA and RNA molecules both contained cytosine and uracil because these were the only pyrimidines available in the primordial sea. Cells then gained the ability to methylate uracil to form thymine because this provides a cell with a criterion for eliminating the result of a cytosine→uracil conversion. It is significant that the final step in the synthesis of thymidylic acid is the methylation of deoxyuridylic acid. The thymidylic acid is then converted to the triphosphate needed for incorporation into DNA.

Mutation

The process by which a base sequence changes is called *mutation*. There are two main mechanisms of mutation:

1. A *chemical alteration* of the base that gives it new hydrogen-bonding properties and causes a new base to be present in a newly replicated daughter molecule.
2. A *replication error* by which an incorrect base or an extra base is accidentally inserted in the daughter molecules.

On the average, mutational changes are deleterious and lead to cell death. Therefore, it is important that not too many mutations occur in a single DNA molecule because otherwise the rare advantageous alteration would always occur in a cell destined to die by virtue of lethal mutations. Thus, the mutation rate must be low or controlled. This is accomplished in two ways. First, the hydrophobic, water-free core of the DNA molecule reduces the accessibility of the DNA to attacking molecules, as we discussed earlier. Second, the cell has evolved several repair mechanisms for correcting alterations and replication errors. These repair systems are not entirely efficient though and allow mutations to occur at a rate that is very low but useful in the long run.

The mutations are usually deleterious, so that it is important that the parental information is not lost. Such loss is prevented in two ways: (1) The other members of the species retain the parental base sequence and (2) a double-stranded molecule is redundant. Normally only one strand is altered, and DNA replicates in such a way that after cell division the DNA molecule in each daughter cell receives only one of the parental single strands. Thus, it is possible for one of the daughter DNA molecules to be normal and the other to be mutant; the mutant may not be able to survive but the cell with the parental base sequence will. If the mutant is better equipped to survive and multiply than the parent organisms or any other member of the species, then after a great many generations, Darwin's principle of survival of the fitest will lead to ultimate replacement of the parental genotype by the mutant phenotype in nature.

RNA as Genetic Material in Small Viruses

As more and more viruses were identified and studied, it became clear that many of them contain RNA and proteins, but no DNA. In all cases so far studied, it is clear that these "RNA viruses" store their genetic information in nucleic acids rather than in proteins just like all other organisms, although in these viruses the nucleic acid in RNA. One of the first experiments that established RNA as the genetic material in RNA viruses was the so-called reconstitution experiment of H. Fraenkel-Conrat and B. Singer, published in 1957. Fraenkel-Conrat and Singer's simple, but definitive, experiment was done with tobacco mosaic virus (TMV), a small virus composed of a single molecule of RNA encapsulated in a protein coat. Different strains of TMV can be identified on the basis of differences in the chemical composition of their protein coats.

By using the appropriate chemical treatments, one can separate the protein coats of TMV from the RNA. Moreover, this process is reversible; by mixing the proteins and the RNA under appropriate conditions, "reconstitution" wall occur, yielding complete, infective TMV particles. Raenkel-Contrat and Singer took two different strains of TMV, separated the RNAs from the protein coats, and reconstituted "mixed" viruses by mixing the proteins of one strain with the RNA of the second strain, and vice versa. When these mixed viruses were used to infect tobacco leaves, the progeny viruses produced were always found to be phenotypically and genotypically identical to the parent strain from which the RNA had been obtained. Thus, the genetic information of TMV is stored in RNA, not protein.

7

GENETIC ENGINEERING

The manipulation of gene is the called the genetic engineering. Genetics is also important in manipulating biological systems for scientific or economic reasons, an endeavor that has made possible the creation of organisms having new phenotypes or genotypes. The fundamental techniques for accomplishing this have been mutagenesis or recombination followed by selection for desired characteristics. When using such techniques, geneticists have been forced to work with the random nature of mutagenic and recombination events, which required selective procedures, often quite complex, to find an organism with the required genotype among the many types of organisms produced. Since the 1970s techniques have been developed by which the genotype of an organism can instead be modified in a *directed* and *predetermined* way. This is alternately called *recombinant DNA technology*, or *genetic engineering*.

Genetic engineering involves isolation of DNA fragments and recombination outside of a cell. Selection of a desired genotype is still necessary but the probability of success is usually many orders of magnitude greater than that with traditional procedures. The basic technique is quite simple: two DNA molecules are isolated and cut into fragments by one or more specialized enzymes and then the fragments are joined together *in a desired combination* and restored to a cell for replication and reproduction.

Current interest in genetic engineering centers on its many practical applications, a few examples of which are the following:

1. Isolation of a particular gene, part of a gene, or region of a genome.
2. Production of particular RNA and protein molecules in quantities formerly thought to be unobtainable.

3. Improvement in the production of biochemicals (such as enzymes and drugs) and commercially important organic chemicals.
4. Production of varieties of plants having particular desirable characteristics (for example, requiring less fertilizer or resistance to disease).
5. Correction of genetic defects in higher organisms, and
6. Creation of organisms with economically important features (for example, plants capable of maturing faster or having greater yield).

Some of these examples will be considered later in the chapter.

Separation of Particular DNA Fragments

In genetic engineering the immediate goal of an experiment is usually to insert a ***particular*** fragment of chromosomal DNA into a plasmid or a viral DNA molecule. This is accomplished by techniques for breaking DNA molecules at specific sites and for isolating particular DNA fragments. DNA fragments are obtained by treatment of DNA samples with a specific class of nuclease. Many nucleases have been isolated from a variety of organisms, and most produce breaks at random sites within a DNA sequence. However, the class of nucleases called *restriction endonucleases*, or, more simply, *restriction enzymes*, consists of sequence-specific enzymes. Most restriction enzymes recognize only one short base sequence in a DNA molecule and make two single-strand breaks, one in each strand, generating 3´-OH and 5´-P groups at each position. Several hundred of these enzymes have been isolated from hundreds of species of microorganisms.

The sequences recognized by restriction enzymes are often *palindromes*–that is, the sequence has symmetry of the form:

A B C	C´B´A´		A B x	B´A´		A B	B´A´
A´B´C´	C B A	or	A´B´x´	B A	or	A´B´	B A

in which the capital letter represent bases, a´ indicates a complementary base, X is any base, and the vertical line is the axis of symmetry. Most of these sequence have 4-6 bases.

One of the most exciting events in the study of restriction enzymes was the observation by electron microscopy that fragments produced by many restriction enzymes spontaneously circularize. These circles could be relinearized by heating, but if after circularization they were also treated with *E.coli* DNA ligase, which joins 3´-OH and 5´-P groups, circularization became permanent. This observation was the first evidence for three important features of restriction enzymes:

1. Restriction enzymes make breaks in palindromic sequences.
2. The breaks are usually not directly opposite one another.
3. The enzymes generate DNA fragments with complementary ends.

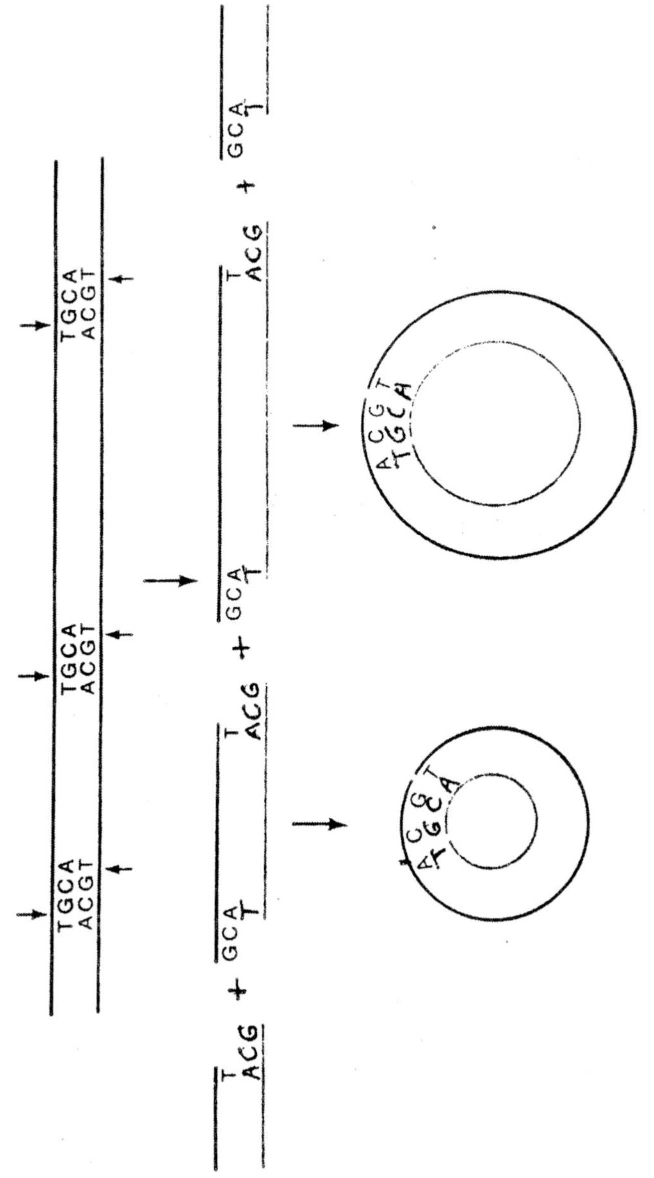

Fig. 7.1. Circularization of DNA fragments produced by a restriction enzyme Arrow indicate cleavage sites.

Examination of a very large number of restriction enzymes showed that the breaks are usually in one of two distinct arrangements: (1) staggered, but symmetric around the line of symmetry (forming *cohesive ends*) or (2) both at the center of symmetry (forming *blunt ends*). Two types of enzymes produce cohesive ends–those yielding a single-stranded extension with a 5′-P terminus and those yielding a 3′-OH extension. Table 7.1. lists the sequences and cleavage sites for several restriction enzymes, some of which generate cohesive sites and others of which yield blunt ends.

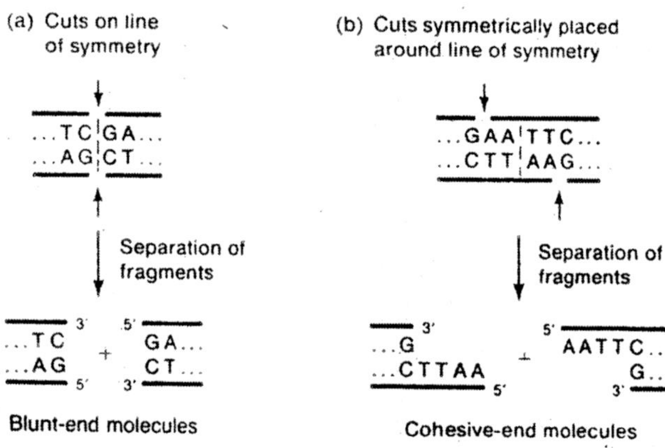

Fig. 7.2. Two types of cuts made by restriction enzymes. The arrow indicate the cleavage sites. The dashed line is the center of symmetry of the sequence.

Most restriction enzymes recognize one base sequence without regard to the source of the DNA. Thus,

> Fragments obtained from a DNA molecule from one organism have the same cohesive ends as the fragments produced by the same enzyme acting on DNA molecules from another organism.

Since most restriction enzymes recognize a unique sequence, *the number of cuts made in the DNA from an organism by a particular enzyme is limited.* A typical bacterial DNA molecule, which contains roughly 3×10^6 base pairs, is cut into several hundred to several thousand fragments, and nuclear DNA of mammals is cut into more than a million fragments. These numbers are large but still small compared to the number of sugar-phosphate bonds in an organism. Of special interest are the smaller DNA molecules, such as viral or plasmid DNA, which may have only 1-10 sites of cutting (or even none) for particular enzymes.

Table 7.1. Some restriction endonucleases, their sources, and their cleavage sites.

Name of enzyme	*Microorganism*	*Target sequence and cleavage sites.*
Generates cohesive ends		
EcoRI	*E.coli*	G↓A A ¦ T T C C T T ¦ A A↑G
BamHI	*Bacillus amyloliquefaciens* H	G↓G A ¦ T C C C C T ¦ A G↑G
HaeII	*Haemophilus aegyptius*	Pu G C ¦ G C↓Py Py↑C G ¦ C G Pu
HindIII	*Haemophilus influenza*	A↓A G ¦ C T T T T C ¦ G A↑A
PstI	*Providencia stuartii*	C T G ¦ C A↓G G↑A C ¦ G T C
TaqI	*Thermus aquaticus*	T↓C ¦ G A A G ¦ C↑T
Generates blunt ends		
BalI	*Brevibacterium albium*	T G G↓¦ C C A A C C↑¦ G G T
SmaI	*Serratia marcescens*	C C C↓¦ G G G G G G↑¦ C C C

Note : The vertical dashed line indicates the axis of symmetry in each sequence. Arrows indicate the sites of cutting. The enzyme TaqI yields cohesive ends consisting of two nucleotides, whereas the cohesive ends produced by the other enzymes contain four nucleotides. Pu and Py refer to any purine and pyrimidine, respectively.

Plasmids having a single site for a particular enzyme are especially valuable. Because of the sequence specificity, ***a particular restriction enzyme generates a unique set of fragments for a particular DNA molecule.*** Another enzyme will generate a different set of fragments from the same DNA molecule. Figure (given below) shows the sites of cutting of *E.coli* phage λ DNA by the enzymes EcoRI and BamHI. A map showing the unique sites of cutting of the DNA of a particular organism by a single enzyme is called a ***restriction map***.

The family of fragments generated by a single enzyme can be detected easily by gel electrophoresis of enzyme-treated DNA, and particular DNA fragments can be isolated by cutting out the portion of the gel containing the fragment and removing the DNA from the gel. Several techniques enable one to locate particular genes on fragments of a restriction map. One of the most generally applicable procedures is *Southern blotting*. In this procedure a gel in which DNA molecules have been separated by electrophoresis is treated with alkali to render the DNA single-stranded (denature the DNA) and then the DNA is transferred to a sheet of nitrocellulose in a way that the relative positions of the DNA bands are maintained. The nitrocellulose, to which the single-stranded DNA tightly binds, is then exposed to radioactive RNA or DNA in a way that leads to renaturation.

Radioactivity becomes stably bound (resistant to removal by washing) to the DNA only at positions at which base sequences

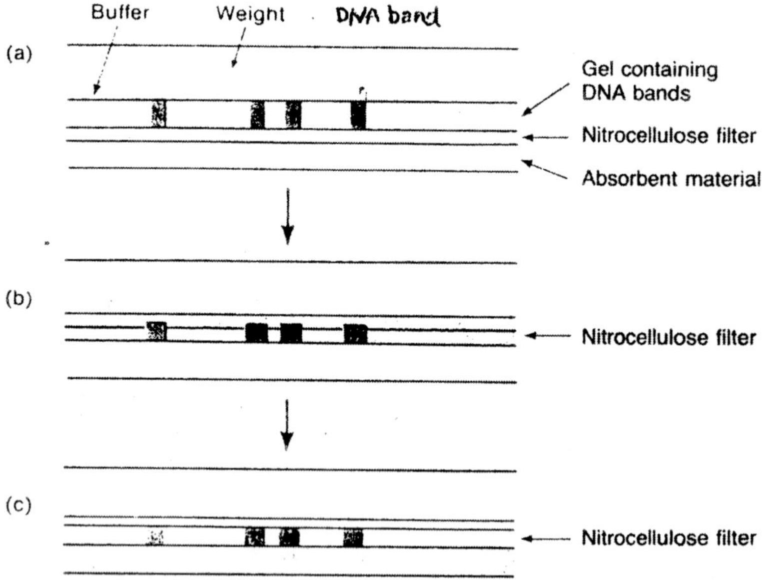

Fig. 7.3. Southern blotting. (a) A weight is placed on a stack consisting of the gel, a nitrocellulose filter, and absorbent material. (b) At a later time the weight has forced the buffer, which carries the DNA, into the nitrocellulose. (c) The lowest layer has absorbed the buffer but not the DNA, which remains bound to the nitrocellulose.

complementary to the radioactive molecules are present. The radioactivity is located by placing the paper in contact with x-ray film; after development of the film, blackened regions indicate positions of radioactivity. If a radioactive mRNA species transcribed from a particular gene is used (for example, mRNA isolated from a specialized cell that predominantly makes one type of mRNA), it will hybridize only with the restriction fragment containing that gene.

Restriction Endonucleases

Restriction enzymes are nucleases and as they cut at an internal position of DNA strand (and not at end) they are known as endonucleases. Restriction enzymes fall into following categories:

Type I : These are most complex, bi-functional (i.e. same enzyme possesses both restriction and modification activity) and are made up of 3 subunits. Recognition site is bipartite and asymmetrical. Cleavage site is nonspecific and is atleast 1000 bp from recognition site.

Type II : These are simplest, separate endonuclease and methylase. Recognition site is short sequence (4-6 bp) often palindromic (with symmetry). Cleavage site is same as or close to recognition site.

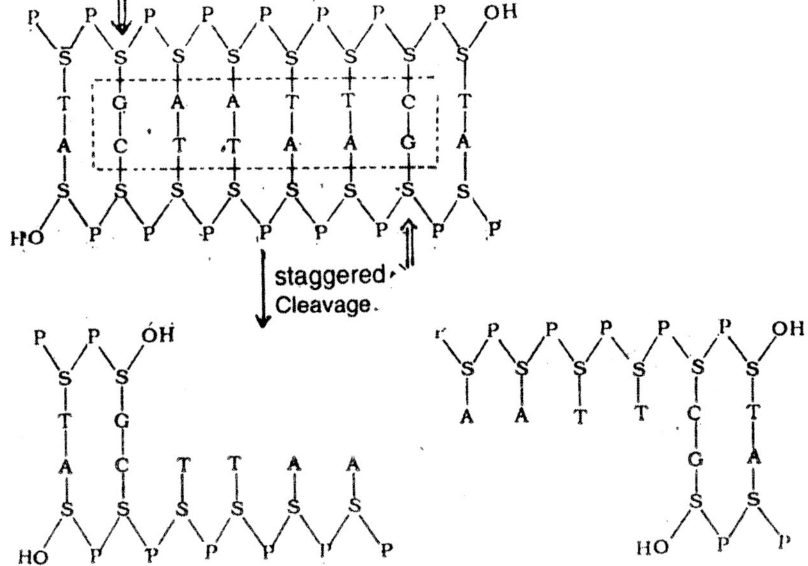

Fig. 7.4. Symmetrical staggered cleavage by Eco RI arrows indicate sites of cleavage.

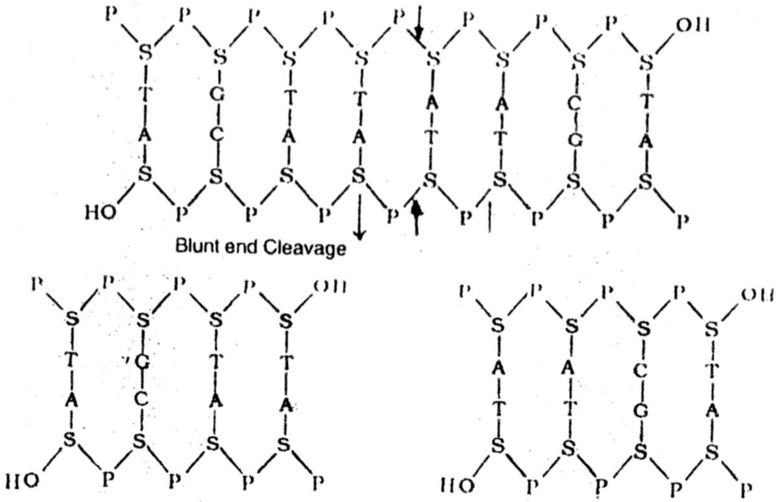

Fig. 7.5. Blunt end cleavage by hind III.

Type III : These are moderate complex, bi-functional, have two subunits. Recognition site is asymmetrical sequence of 5-7 bp. Cleavage site is 24-26 bp downstream from recognition site.

Type II restriction endonucleases are most important tools in gene manipulation techniques. Most of the restriction enzymes cleave DNA at unmethylated target site. Discovery of restriction enzymes (restriction endonucleases) has been useful in most of the modern genetic manipulations. Restriction endonucleases are indispensable tools for recombinant DNA research. They recognize specific oligonucleotide sequence and make double stranded cleavage to generate unique fragment of DNA molecule. Restriction endonucleases are used for dissecting, analysing and re-configuring genetic information at molecular level. Restriction endonucleases are used in: (1) cleavage mapping, (2) preparation of DNA probes, (3) gene cloning etc.

Nomenclature

There are over 2400 type II restriction endonucleases known so far (since 1968) and they recognize more than 188 different restriction sites. These enzymes have been well characterized.

The nomenclature of restriction enzyme is based on various conventions: (1) The generic and specific name of organism in which the enzyme is found are used to provide first part of the designation.

Next number in designation indicates serial number or enzyme reported from same organism. Three letter abbrevation is in italic. (2) Strain or type identification is written as a subscript e.g. Eco_k. If restriction and modification system is genetically specified virus or plasmid, the abbreviated species name of host is given and extrachromosomal element as subscript. (3) When a particular host strain has several different restriction and modification system, these are identified by oman numericals I, II, III etc. following 3 letter abbreviation of host strain. (4) The history and incompleteness of the discovery are reflected in nomenclature of restriction endonucleases.

Unidentified bacteria found with these systems receive the genus-species designate *Uba*. As on 1993 hundreds of enzymes were temporarily named as *Uba* (number). In 1973, nomenclature of restriction enzyme was developed and was based on proposals of Smith and Nathan.

Recognition Sites

Restriction enzymes usually recognize a specific DNA sequence of 4, 5 or 6 nucleotides in length and cleave the DNA within this restriction site. There are 4 bases in DNA, randomly distributed. The expected frequency of any particular sequence can be calculated as 4^n where n is the length of recognition sequence. Thus, tetranucleotide sites will occur every 256 basepair, pentanucleotide sites will occur every 1025 basepairs and hexanucleotide sites will occur every 4096 basepairs. Restriction enzymes either cut (1) Straight across the DNA to give blunt ends, or (2) Straight single strand cuts producing short, single stranded projections at each end of the cleaved DNA to produce cohesive or sticky ends.

Table 7.2. Classification of restriction endonucleases based on recognition site.

(I)	Tetranucleotides recognizing	–	*Alu I, Hae III, Hpa II, Mbo I, Taq I, Msp I* etc.
(II)	Pentanucleotides recognizing	–	*Ava II, Dde I, Eco R II, Hinf I* etc.
(III)	Hexanucleotides recognizing	–	*Kpn I, Xma I, Pst I, Hpal, Bam HI, Hind III* etc.
(IV)	Heptanucleotides recognizing	–	*Me II* etc.

One unit of enzyme (Restriction endonuclease) is defined as the amount of enzyme required to produce a complete digest of 1 μg of substrate DNA in a reaction volume of 0.05 ml in 1 hour under optimal conditions of salt, pH and temperature. All digestions are performed at 37°C and substrate used is lambda DNA. Optimum temperature for most of the restriction enzymes is 37°C. However, in some cases like *Tha I, Taq I, Bcl I, Bst NI* optimum temperature is 60-65°C. Recognition sequences of *Eco RI* are *palindromic.* Sometimes within a same strand sequence may be palindromic.

Table 7.3. Restriction endonucleases.

	Enzyme	*Source*	*5' Recognition site 3'*
1.	*Kpn II*	*Klebsiella pneumoniae*	GGTAC↓C
2.	*Bam HI*	*Bacillus amyloliquifaciens*	G↓GATCC
3.	*Sma I*	*Serratia marcesecens*	CCC↓GGG
4.	*Stu I*	*Streptomyces tubercidicus*	AGG↓CCT
5.	*AVA II*	*Anabenia variabilis*	G↓G(A/T) CC
6.	*Pva I*	*Proteus vulgaris*	CGAT↓CG
7.	*Hind III*	*Hemophilus influenzae Rd*	A↓AGCTT
8.	*Rsa I*	*Rhodopseudomonas sphaeroids*	GT↓AC
9.	*Hin C II*	*Hemophilus influenzae C1161*	GTPyPvAC
10.	*ECORI*	*E.coli RY 13*	G↓AATTC
11.	*Xma I*	*Xanthomonas malvacearum*	C↓CCGGG
12.	*Pst I*	*Providencia stuartii*	CTGCA↓G
13.	*Bgl II*	*Bacillus globigii*	A↓GATCT
14.	*Hpa II*	*Hemophilus parainfluenzae*	C↓CGG
15.	*Alu I*	*Arthrobacter luteus*	AG↓CT
16.	*Hae III*	*Hemophilus aegyptius*	GG↓CC
17.	*Hpa I*	*Hemophilus parainfluenzae*	GTT↓AAC
18.	*Mbo I*	*Moraxella bovis*	N↓GATC
19.	*NCo I*	*Nocardia corallina*	G↓CATGG
20.	*Sac I*	*Streptomyces achromogenes*	GAGCT↓C
21.	*Sal I*	*Streptomyces albus G*	G↓TCGAC
22.	*Sau 3AI*	*Staphylococcus aureus 3AI*	N↓GATC
23.	*Taq I*	*Thermus squatius YTI*	T↓CGA
24.	*Hinc II*	*Haemphilus influenzae Rc*	GTPyPuAC
25.	*Acc I*	*Acinetobacter calcoaceticus*	GT↓(AC)(GT)AC
26.	*Dde I*	*Desulfovibrio desulfuricans*	C↓GNAG
27.	*Eco RII*	*E.coli R-245*	↓CC(A/T)GG
28.	*Hinf I*	*Haemophilus influenzae Rf*	G↓ANTC
29.	*Mst II*	*Microcoleus species*	CC↓TNAGG

If cut is not shown it means exact cutting size is not known. Parentheses indicate that either bracketed base will suffice recognition.

Note : Py, Pu indicate non-specific pyrimidines or purines. N indicates any nucleotide base.

Complementary DNA (cDNA)

A second method for obtaining a gene of interest is to create a *complementary DNA (cDNA)* strand from a strand of messenger RNA (mRNA). This procedure is used when specific mRNA molecules can

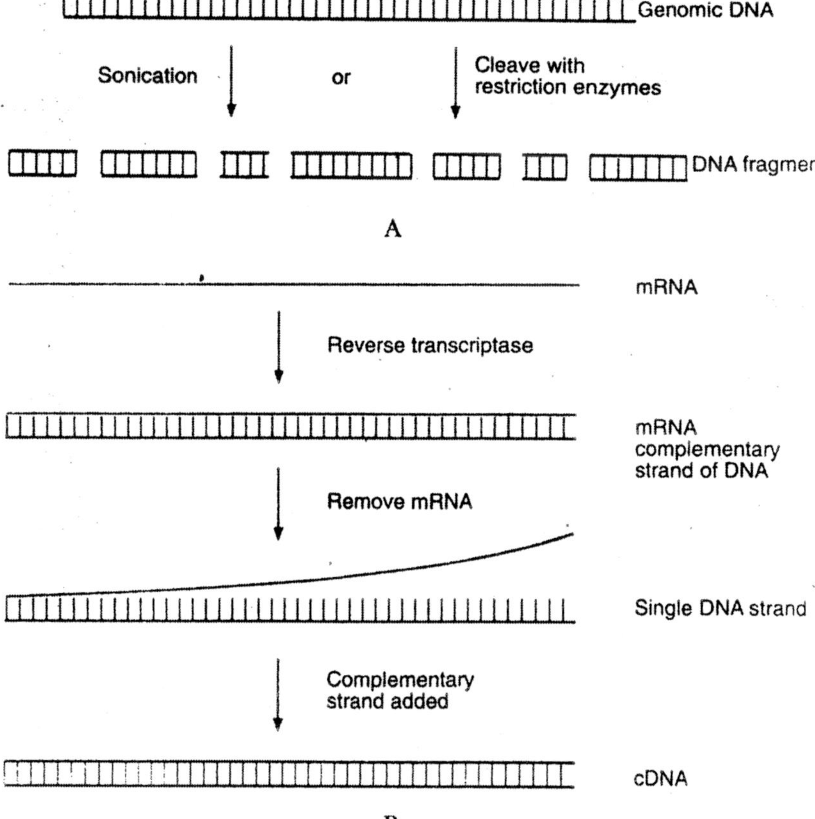

Fig. 7.6. A–To obtain DNA for cloning, large molecules are first broken into smaller segments by sonication or by cleaving them with restriction enzymes. The DNA fragments can then be separated and analyzed to determine which contain a gene of interest. B–If the mRNA encoding a gene of interest can be isolated, a DNA strand that is complementary to this mRNA can be synthesized.

be isolated from cells that produce high levels of a certain protein. For example, a protein called *factor VIII* is necessary for proper blood clotting. It is the protein that most hemophiliacs are missing because of a mutation in the relevant encoding gene. From several cleaver studies, high levels of this protein were found to be produced in the liver along with the key mRNA molecules directing its synthesis

By separating the factor VIII protein and the affiliated mRNA from other proteins and other mRNAs that were also present, it was possible to identify the gene encoding factor VIII. The major advantage of this method is that a "pure" DNA sequence can be produced that does not contain the extraneous DNA sequences (introns) that occur in genomic DNA. Once the mRNA is isolated and purified, it is used as a template to construct a cDNA strand using the unique enzyme *reverse transcriptase,* which synthesizes DNA from RNA. When the cDNA has been formed, the mRNA strand is removed, and a DNA strand that is complementary to the cDNA is then added, resulting in a typical double-stranded molecule. Once DNA has been obtained, the next step in gene cloning is to create an rDNA molecule that can be used for cloning cDNA or any of the thousands of fragments that can be obtained from genomic DNA.

VECTORS

Vectors are the carrier DNAs into which 'foreign' DNAs or genes of interest are spliced to make a recombinant DNA. Vectors along with this 'foreign' DNAs (i.e., recombinant DNA) are then introduced into appropriate host cell and are maintained for study or expression.

Vectors are essentially expected to replicate inside the host cell along with the inserted DNA. The complete technique of isolating a gene of interest and inserting it into a vector and then replicating and maintaining it into a host cell is called as *Gene cloning.*

Types of Vectors

There are two types of vectors:

(a) Cloning vectors.

(b) Expression Vectors.

(a) Cloning vectors

Cloning vectors are used for obtaining millions of copies of cloned DNA segment. The cloned genes in these vectors are not expected to express themselves, at transcription or translation level. Cloning vectors are used for creating genomic library or preparing, the probes or genetic engineering experiments or other basic studies.

(b) Expression vectors

Expression vectors allow the expression of cloned gene, to give the product (protein). This can be achieved through the use of promoters and expression cassettes and regulatory genes (sequences). Expression vectors are used for transformation to generate transgenic plant, animal or microbe where cloned gene expresses to give the product (protein). Commercial production of product of cloned gene may also be achieved by high level expression using the expression vectors.

Promoters in Vectors

Routine manipulations in gene cloning experiments do not require expression of the cloned DNA. But when required, selection of host/ vector system plays important role. If cDNA molecule is used for cloning eukaryotic gene in prokaryotic system then problem of post transcriptional modification is obviated. For given host cell, there may be several types of expression vectors, then in addition of other consideration, a key feature of selection of expression vector is the type of promoter used to direct expression of the cloned sequence.

A vector with very efficient promoter is chosen to maximise the expression. Promoters are referred to as strong or weak depending on their efficiency of expression. Weak promoters are used if overexpression and excess product formation is likely to be toxic to host cell. Promoters are regions with a specific base sequence, to which RNA polymerase will bind. In addition to the strength of promoter, it is also desirable to regulate expression. This is done by using promoters that are either inducible or repressible. This exerts some control.

Arrangement of restriction sites immediately downstream from the promoter is critical. Thus expression vector will have following essential features:

(a) Origin of replication that is functional in target host cell.

(b) Antibiotic resistance genes or other genetic selection mechanism.

(c) Promoter (strong or weak as per suitability) along with regulatory control.

(d) Restriction sites immediately downstream form the promoter.

(e) Unique restriction site for cloning located in a position where the inserted cDNA sequence can be expressed effectively.

For expression of cloned gene in plants or animals only plant specific or animal specific promoters work. Promoters can confer the transferred gene a specific pattern of expression.

Organism	*Gene promoter*	*Induction by*
1. *E.coli*	lac operon, trp operon, λ P_L	IPTG, β-indolylacetic acid, temperature sensitive λ cl protein resp.
2. *A. nidulans*	Glucoamylase	Starch.
3. *S. cerevisiae*	Acid phosphatase Alcohol dehydrogenase galactose utilization Metallothionein	Phosphate depletion glucose depletion galactose heavy metals.
4. *T. reesei*	Cellobiohydrolase	Cellulose
5. Mouse	Metallothionein	heavy metals
6. Human	Heat shock protein	Temperature 740°C
7. Plants (in basal region)	Nopaline synthase (Nos) Mannopine synthase (Mas)	
8. Plants (in leaf tissue)	35 S promoter of Ca MV [high expression (10 fold) at distance of 2 kbp]	
9.	Polyhedrin promoter from baculovirus (used for developing biopesticides or for production of specific chemicals in industry)	

Expression Cassettes

These are gene constructs which allow the insertion of foreign genes, either as transcriptional or transnational fusions, behind specific promoters. PRT series of plasmids which have been derived from pUC 18/19 are the examples.

Expression Vectors

Expression vectors are simply designed to express detectable levels of foreign proteins usually at the bench-scale level (i.e. in shake flask cultures).

Production Vectors

Production level vectors are designed for stable large scale production of gene products at economically significant levels. This means production level vectors have additional genetic elements such as *par* sequences, transcription terminators downstream from the cloned genes etc. This is only artificial distinction. Efficients expression vectors are used successfully to express wide variety of proteins. Plasmids like *pASI* are attractive for production purposes as they have strong *pL* transcriptional unit and also a strong transnational unit.

Cloning in *E.coli*

Plasmid DNA used as vector can be cleaved at a site with restriction endonuclease enzyme and foreign DNA segments can be inserted here. Generally, plasmids with two genes for different antibiotic resistance are used–one to identify bacteria that carry plasmid and second to distinguish chimeric plasmid from the parental vector. Insertion of foreign DNA segment is done in site at one antibiotic resistance gene, therefore resistance is lost to one antibiotic.

The chimeric plasmid is resistant to only one antibiotic while the parental plasmid is resistant to two antibiotics. Multicopy plasmids are more suitable since 10-30 copies of one plasmid in a cell along with the inserted gene will give an amplified response. Addition of chloramphenicol often affects new rounds of chromosome replication while plasmid replication is unaffected. This makes isolation of plasmid easier. A common method for cloning DNA is to cleave both plasmid and insert DNA with the same restriction endonuclease. Most of the plasmids which are used as cloning vectors are plasmids derived from natural plasmids so as to posses desirable features.

pBR 322

This is derived from *E.coli* plasmid Col E1. Bolivar and Rodriguez prepared this vector hence the name and 322 are numericals which were significant to these scientists. Alternations and restructuring is done while deriving these plasmids from original natural plasmids. Genes for relaxed replication and genes for antibiotic resistance are inserted in them. pBR 322 is 4362 bp long DNA with genes for resistance to tetracycline and ampicillin. pBR 322 possesses on origin of replication and restriction sites for cleavage by variety of restriction enzymes.

Characteristics of pBR 322

1. It is much smaller in size than natural plasmid. Advantages of small size are : (i) easier to handle, (ii) less susceptible to physical damage, (iii) simpler restriction man, (iv) easy uptake by bacteria during transformation, (v) higher copy number therefore easy detection.
2. Bacterial origin of replication is present and this ensures that plasmid will be replicated in host.

3. Origin of replication is relaxed here, hence activity is not tightly linked to cell division. Resultantly plasmid replication will be initiated for more frequently than chromosomal replication. So more copy number of plasmid per cell is possible.
4. Two genes for different antibiotic resistance.
5. There are single restriction sites for number of enzymes, scattered on plasmid and can be used for cleavage and insertion. Presence of such sites within antibiotic resistance gene are important, as they cause respective loss of resistance.

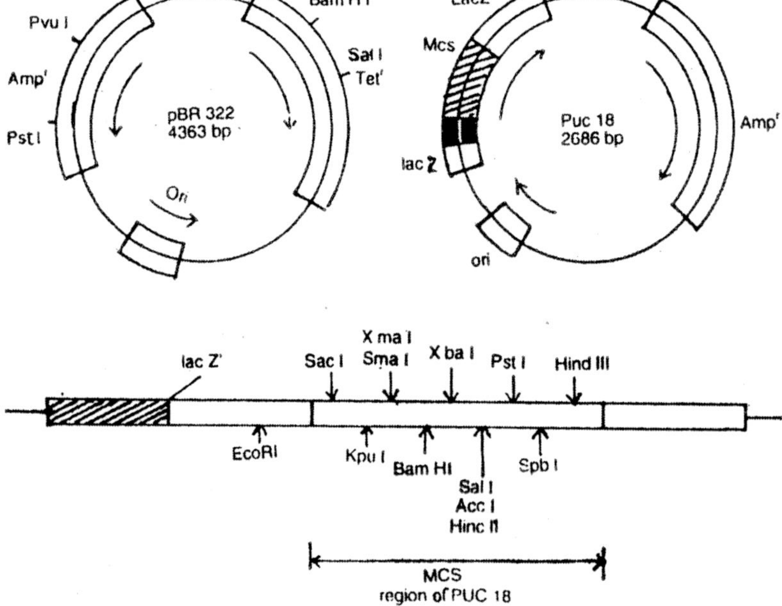

Fig. 7.7. Plasmids pBR 322 and pUC 18.

pAT 153 and *pXf 3* are two derivatives of pBR 322 having smaller size and higher copy number.

pBR 327

This is derived from pBR 322, by deletion of nucleotides between 1427 and 2516. These nucleotides are deleted to reduce size of vector and reduce interference. pBR 327 still has genes for resistance against tetracycline and ampicillin.

pUC Vectors

The name is derived from the place of their initial preparation (i.e., University of California). These vectors have 2700 base pairs and ampicillin resistance gene and lac Z gene. Insertion of foreign DNA in lac Z gene causes inactivation of lac Z gene. *E.coli* which is lactose –ve is used for transformation. When culture is grown in presence of IPTG (Isopropyl thiogalactoside) which induces the synthesis of β galactosidase) and X-gal substrate, *E.coli* (transformed with pUC vector with insert) gives white colonies while *E.coli* (transformed with pUC vector but without insert) gives blue colonies. (Substrate is chromogenic hence coloured product is formed if enzyme is active). Vectors of pUC family have a region that contains several unique restriction sites in a short stretch of DNA. This region is known as a polylinker or multiple cloning site (MCS).

Bacteriophages Vectors

Phage has a linear DNA molecule so a single break creates two fragments. Foreign DNA can be inserted between them and two fragments can be joined. Such phages when undergo lytic cycle in host will produce more chimeric DNA. Wild type λ phage could accommodate only 2.5 kb of foreign DNA. Phage vectors are restructured by removing nonessential genes and making vector DNA smaller so that larger insert can be accommodated in phage head during packing. Lambda phage (λ) such prepared has one Eco RI site and accommodates 20-25 kb of foreign DNA. They are used for preparing genomic library of eukaryotes. One other λ phage devised has two Bam HI sites that flank the I/E region.

Insertion vectors

Insertion phage vector has single restriction site which is used for insertion of foreign DNA. Smaller foreign DNA can be packed here.

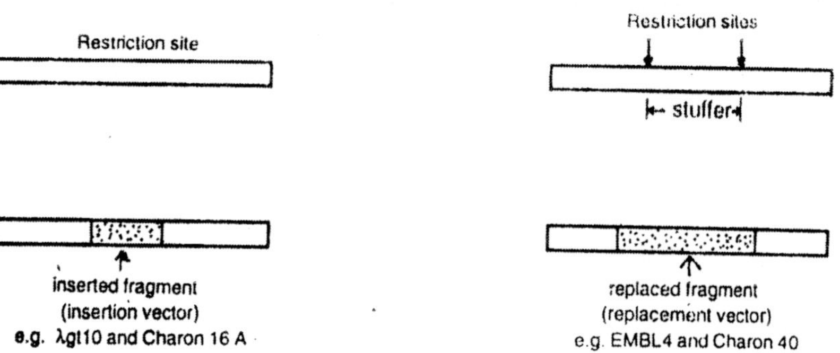

Fig. 7.8. Insertion and Replacement vectors.

Replacement vectors

Replacement vector has two restriction sites which flank a region known as stuffer fragment. Larger fragment of foreign DNA can be replaced between these two sites.

Cosmids

Cosmids are the novel cloning vectors which possess properties of both plasmid and λ phage. Cosmids first were developed in 1978 by Barbara Hohn and John Collins. Cosmids contain a cos site of λ phage (which is essential for packaging of nucleic acid into protein coat) plus essential features of plasmid and several unique restriction sites for insertion of DNA to be cloned. Cosmids can be perpetuated in bacteria in plasmid form, but can be purified by packaging *in vitro* into phages. When viral DNA is injected into the cell as a linear molecule it's both ends are cohesive and complementary to each other, 12 base in length.

Once inside the cell, these ends base pair and become permanently joined by ligation to form a region which is known as *cos-site* and give rise to circular DNA molecule. Replication of DNA by rolling circle mechanism results into a concatemer which is a long molecule made up of many copies of viral DNA linked end to end through cos sites. DNA is packaged by looping regions between cos sites into precursor of the viral head. When head is full, the cos sites should be at the mouth of the head, where they will be cleaved to generate a linear molecule with cohesive ends. Subsequently tail proteins are added to give infective particle. Thus only requirement for length of DNA to be packaged into viral heads is that it should contain cos sites spaced at correct distance. This distance ranges from 37 to 52 kb. Cos sites of λ possess 280 bp flanking sequence.

For cloning foreign DNA into cosmid vector, cosmid DNA is first linearised by cutting it with appropriate restriction enzyme. Then it is treated with the calf intestinal phosphatase to remove phosphate groups (5′) at its ends so as to prevent recircularisation of cosmid DNA. Foreign DNA which is to be cloned is also treated with the same restriction enzyme which was used for treating cosmid DNA. Subseque. tly cosmid DNA and foreign DNA fragments are mixed in presence of T4 DNA ligase. Many different products generate in the mixture.

A product may form where foreign DNA binds to cos site of cosmid in same orientation. This on *in vitro* incubation with phage head and tail proteins, cleavage of cos sites takes place and intervening

DNA is packed into phage particles. When susceptible bacteria are infected by such phage, hybrid DNA is injected into cell. Then it functions like plasmid and expresses antibiotic resistance gene. The infected cells can be selected and isolated on basis of antibiotic resistance gene expressed. Foreign DNA presence can be detected in antibiotic resistant bacterial colonies. Inserts of 40-50 kb length can be easily placed between cos sites and to get packageble forms of cosmids.

Advantage of using cosmid vector is that larger DNA can be cloned than what is possible with phage or plasmid. As larger inserts are possible genomic library can be created which is composed of fewer clones to be screened. Efficiency of cosmids is high enough to produce a complete genomic library of 10^6–10^7 clones from a mere 1µg of insert. Genomic libraries of Drosophila, mouse and several other organisations has been produced with cosmid vectors. Cosmid PLFR-5 has two cos sites, 6 restriction enzymes target sites, origin of replication and tetracycline resistance marker. It can insert 50 kb of DNA. Cosmid system in case of P_1 bacteriophage of *E.coli* can carry 85 kb of inserted DNA.

For *in vitro* packaging of recombinant DNA containing cosmid *packaging extract* is used. This is nothing but two strains of bacteria. These two strains when mixed with concatameric recombinant DNA under suitable conditions head, tail and DNA with proper cos site distance is available and phage particles are produced. These infective phages then can be used to infect *E.coli* cells.

Phasmid Vectors

Here combination of plasmid and λ phage is produced. Plasmid is inserted into phage λ genome by means of site specific recombination mechanism of the phage that is normally used by phage for insertion into bacterial chromosome during lysogen formation. This process is called 'lifting' the plasmid and combination is called phasmid.

Phasmid contain functional origins of replications of the plasmids and of λ and may be propagated as plasmid or phage in appropriate *E.coli* host strains. Plasmids can be released by several of lifting. Phage particles are easy to store, have infinitive shelf life and their screening as plaques by hybridization gives better results than screening of bacterial colonies. So plasmid with cloned gene can be lifted by phages and conveniently handled. Release of recombinant plasmid is easier. λ ZAP is highly developed phasmid containing λ, M_{13} and T_7 phages. λ ZAP is suitable for cloning cDNAs. λ ZAP has multiple unique cloning sites, can hold 10 kb of inserts, easy excision of cloned DNA possible.

Insertional inactivation of β galactosidase gives blue/white screening on X gal plates. Expression of hybrid polypeptides is analogous to that in λ gt 11.

Shuttle Vectors

Transfer of genes between unrelated species is one of the requirements of molecular biotechnology. Broad host range vectors exist in Gram negative bacteria and *Streptomyces* naturally. A shuttle vector however may be required having necessary replicon for maintenance in different combinations of unrelated hosts. Shuttle vectors have potential importance in the genetic manipulations of industrially important species. Shuttle vectors can exploit gene manipulative procedures of different hosts for example when *E.coli* amplification will be possible. Shuttle vectors exist for *E.coli* yeast cells combination. *E.coli–Agrobacterium* combination and *E.coli–B. subtilis. E.coli–Streptomyces lividans* and *E.coli–mammalian* cells combinations.

VEHICLES

A DNA molecule needs to display several features to be able to act as a vehicle for gene cloning. Most important, it must be able to replicate within the host cell, so that numerous copies of the recombinant DNA molecule can be produced and passed to the daughter cells. A cloning vehicle also needs to be relatively small, ideally less than 10 kilobases (kb) in size, as large molecules tend to break down during purification, and are also more difficult to manipulate. Two kinds of DNA molecule that satisfy these criteria can be found in bacterial cells: plasmids and bacteriophage chromosomes. Although plasmids are frequently employed as cloning vehicles, two of the most important types of vector in use today are derived from bacteriophages.

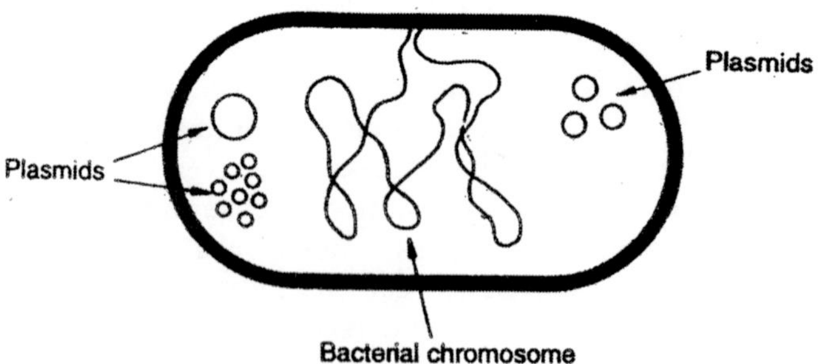

Fig. 7.9. Plasmids independent genetic elements found in bacterial cells.

Plasmids

Basic features of plasmids

Plasmids are circular molecules of DNA that lead an independent existence in the bacterial cell. Plasmids almost always carry one or more genes, and often these genes are responsible for a useful

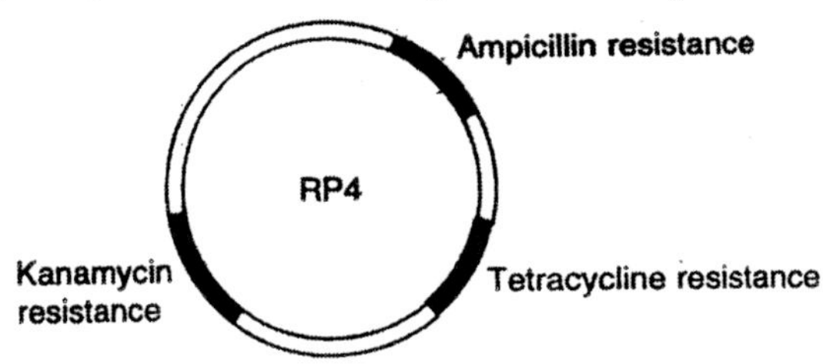

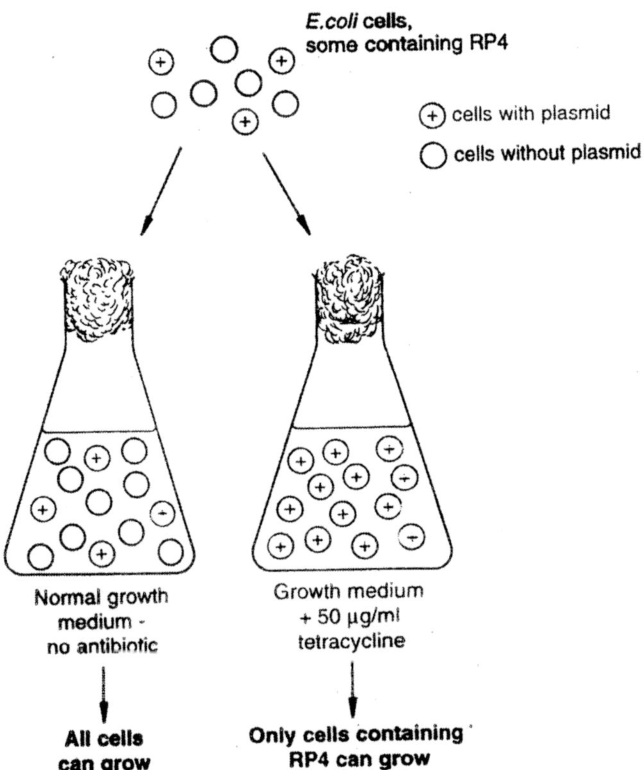

Fig. 7.10. The use of antibiotic resistance as a selectable marker for a plasmid.

characteristic displayed by the host bacterium. For example, the ability to survive in normally toxic concentrations of antibiotics such as chloramphenicol or ampicillin is often due to the presence in the bacterium of a plasmid carrying antibiotic resistance genes. In the laboratory antibiotic resistance is often used as a *selectable marker* to ensure that bacteria in a culture contain a particular plasmid.

All plasmids possess at least one DNA sequence that can act as an *origin of replication*, so they are able to multiply within the cell quite independently of the main bacterial chromosome. The smaller plasmids make use of the host cell's own DNA replicative enzymes in order to make copies of themselves, whereas some of the larger ones carry genes that code for special enzymes that are specific for plasmid replication.

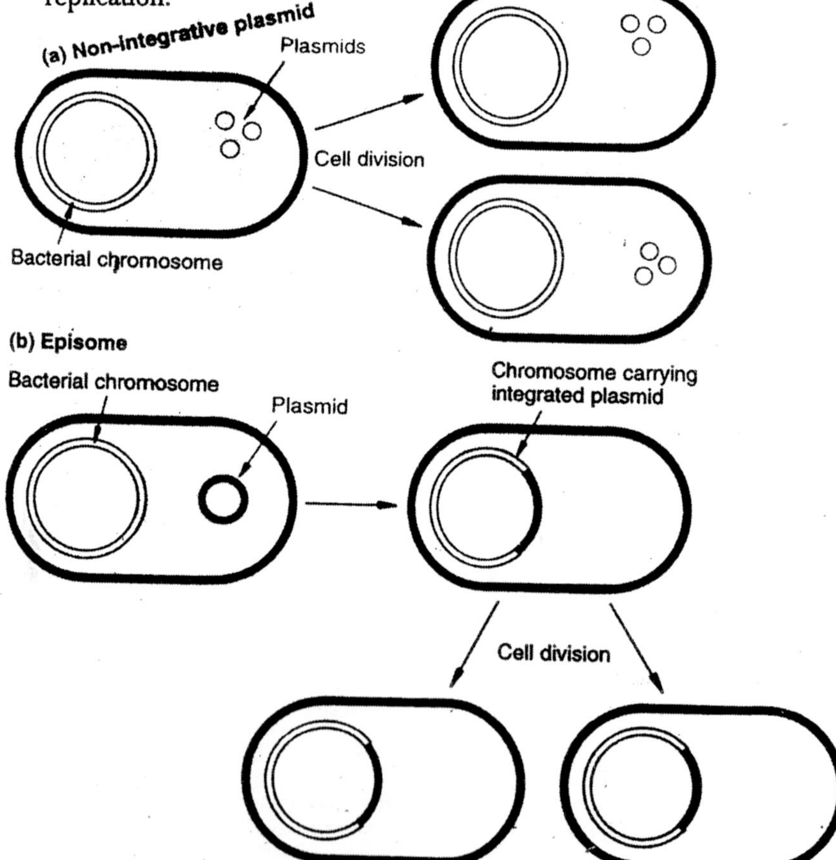

Fig. 7.11. Replication strategies for (a) a non-integrative plasmid, and (b) an episome.

A few types of plasmid are also able to replicate by inserting themselves into the bacterial chromosome. These integrative plasmids or *episomes* may be stably maintained in this form through numerous cell divisions, but will at same stage exist as independent elements. Integration is also an important feature of some bacteriophage chromosomes and will be described in more detail when these are considered.

Size and copy number

These two features of plasmid are particularly important as far as cloning is concerned. We have already mentioned the relevance of plasmid size and stated that less than 10 kb is desirable for a cloning vehicle. Plasmids range from about 1.0 kb for the smallest to over 250 kb for the largest plasmids, so only a few will be useful for cloning purposes. However, larger plasmids may be adapted for cloning under some circumstances. The *copy number* refers to the number of molecules of an individual plasmid that are normally found in a single bacterial cell.

The factors that control copy number are not well understood, but each plasmid has a characteristic value that may be as low as one (especially for the large molecules) or as many as 50 or more. Generally speaking, a useful cloning vehicle needs to be present in the cell in multiple copies so that large quantities of the recombinant DNA molecule can be obtained.

Table 7.5. Sizes of representative plasmids

Plasmid	*Size*		*Organism*
	Nucleotide length (kb)	*Molecular wt (MDa)*	
pUC8	2.1	1.8	*E.coli*
ColEI	6.4	4.2	*E.coli*
RP4	54	36	*Pseudomonas* + others
F	95	63	*E.coli*
TOL	117	78	*Pseudomonas putida*
pTiAch5	213	142	*Agrobacterium tumefaciens*

Conjugation and compatibility

Plasmids fall into two groups conjugative and non-conjugative. Conjugative plasmids are characterized by the ability to promote sexual *conjugation* between bacterial cell, a process that can result in a conjugative plasmid spreading from one cell to all the other cells in a bacterial culture. Conjugation and plasmid transfer are controlled by a set of transfer or *tra* genes, which are present on conjugative plasmids but absent from the non-conjugative type.

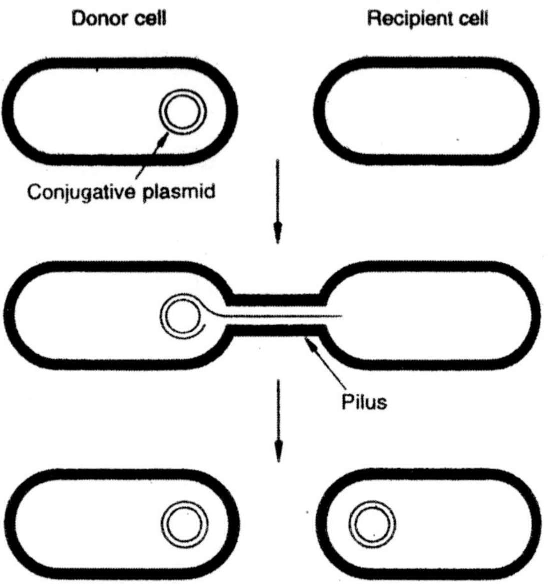

Fig. 7.12. Plasmid transfer by conjugation between bacterial cells.

A non-conjugative plasmid may, under some circumstance, be cotransferred along with a conjugative plasmid when both are present in the same cell. Several different kinds of plasmid may be found in a single cell, including more than one different conjugative plasmid at any one time. In fact, cells of *E.coli* have been known to contain up to seven different plasmids at once. To be able to coexist in the same cell, different plasmids must be *compatible*. If two plasmids are incompatible then one or the other will be quite rapidly lost from the cell. Different types of plasmid can therefore be assigned to different *incompatibility groups* on the basis of whether or not they can coexist, and plasmids from a single incompatibility groups are often related to each other in various ways. The basis of incompatibility is not well understood, but events during plasmid replication are thought to underlie the phenomenon.

Plasmid classification

There are five main types of plasmid which are as follows:

1. *Fertility* or *'F' plasmids* carry only *tra* genes and have no characteristic beyond the ability to promote conjugal transfer of plasmids, e.g., F plasmid of *E.coli.*
2. *Resistance* or *'R' plasmids* carry genes conferring on the host bacterium resistance to one or more antibacterial agents, such as chloramphenicol, ampicillin and mercury, R plasmids are very important in clinical microbiology as their spread through natural populations can have profound consequences in the treatment of bacterial infections; e.g., RP4, commonly found in *Pseudomonas*, but also occurring in many other bacteria.
3. *Col plasmids* code for colicins–proteins that kill other bacteria; e.g., ColE1 of *E.coli.*
4. *Degradative plasmids* allow the host bacterium to metabolize unusual molecules such as toluene and salicylic acid; e.g., TOL of *Pseudomonas putida.*
5. *Virulence plasmids* confer pathogenicity on the host bacterium; e.g., *Ti plasmids* of *Agrobacterium tumefaciens*, which induce crown gall disease on dicotyledonous plants.

Plasmids in organisms other than bacteria

Although plasmids are widespread in bacteria they are by no means so common in other organisms. The best characterized eukaryotic plasmid is the *2µm circle* that occurs in many strains of the yeast *Saccharomyces cerevisiae.* The discovery of the 2 µm plasmid was very fortuitous as it has allowed the construction of vectors for cloning genes with this very important industrial organism as the host. However, the search for plasmids in other eukaryotes (e.g., filamentous fungi, plants and animals) has proved disappointing and it is suspected that many higher organisms simply do not harbour plasmids within their cells.

Bacteriophages

Basic features of bacteriophages

Bacteriophages, or phages as they are commonly known, are viruses that specifically infect bacteria. Like all viruses, phages are very simple in structure, consisting merely of a DNA (or occasionally RNA) molecule carrying a number of genes, including several for replication of the phage, surrounded by a protective coat or *capsid* made up of protein molecules.

The general pattern of infection, which is the same for all types of phage, is a three-step process.

1. The phage particle attaches to the outside of the bacterium and injects its DNA chromosome into the cell.

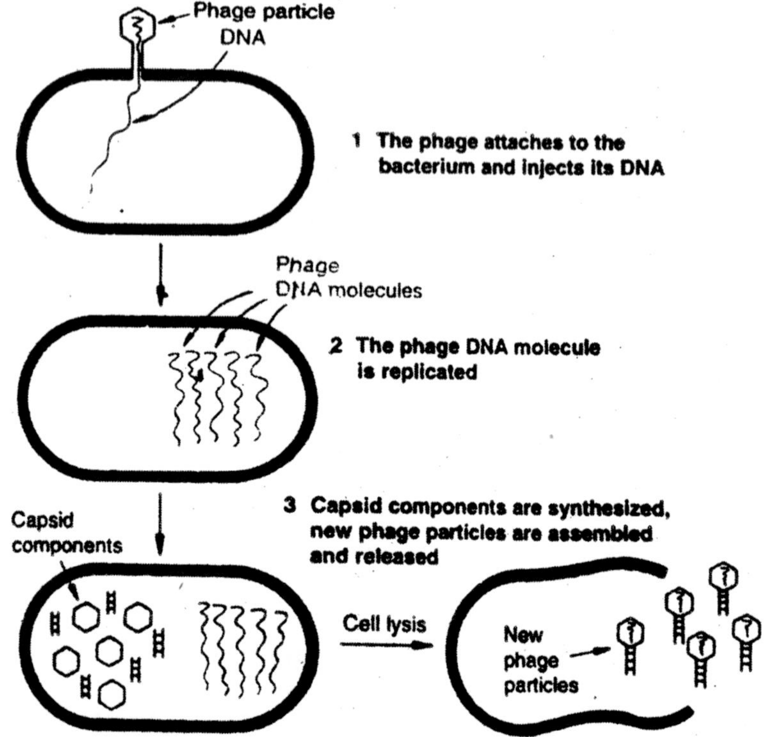

Fig. 7.13. The general pattern of infection of a bacterial cell by a bacteriophage.

2. The phage DNA molecule is replicated, usually by specific phage enzymes coded by genes on the phage chromosome.
3. Other phage genes direct synthesis of the protein components of the capsid, and new phage particles are assembled and released from the bacterium.

With some phage types the entire infection cycle is completed very quickly, possibly in less than 20 minutes. This type of rapid infection is called a *lytic cycle*, as release of the new phage particles is associated with lysis of the bacterial cell. The characteristic feature of a lytic infection cycle is that phage DNA replication is immediately followed by synthesis of capsid proteins, and the phage DNA molecule is never maintained in a stable condition in the host cell.

Lysogenic phages

In contrast to a lytic cycle, *lysogenic* infection is characterized by retention of the phage DNA molecule in the host bacterium, possibly

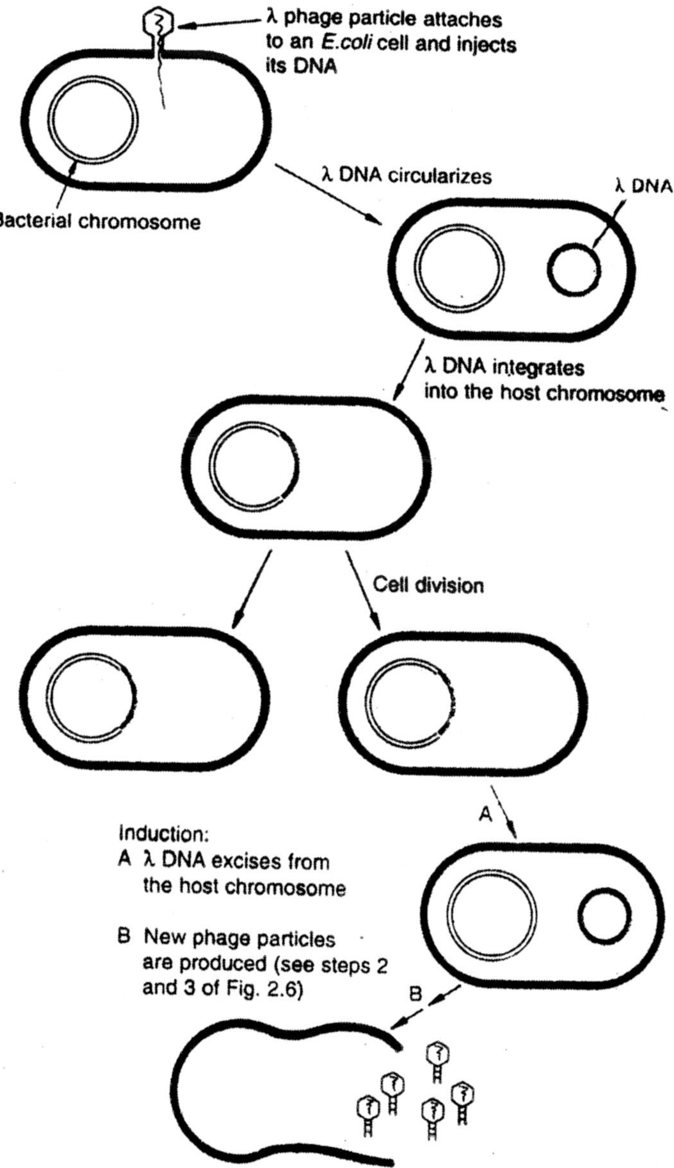

Fig. 7.14. The lysogenic infection cycle of bacteriophage λ.

for many thousands of cell divisions. With many lysogenic phages the phage DNA is inserted into the bacterial genome, in a manner similar to episomal insertion. The integrated form of the phage DNA (called the *prophage*) is quiescent, and a bacterium (referred to as a *lysogen*) which carries a prophage is usually physiologically indistinguishable from an uninfected cell.

The prophage is eventually released from the host genome and the phage reverts to the lytic mode and lyses the cell. The infection cycle of λ, a typical lysogenic phage is of this type. A limited number of lysogenic phages follow a rather different infection cycle. When M13, or a related phage, infects *E.coli*, new phage particles are continuously assembled and released from the cell. The M13 DNA is not integrated into the bacterial genome and does not become quiescent. With these phages, cell lysis never occurs, and the infected bacterium can continue to grow and divide, albeit at a slower rate than uninfected cells. Although there are many different varieties of bacteriophage,

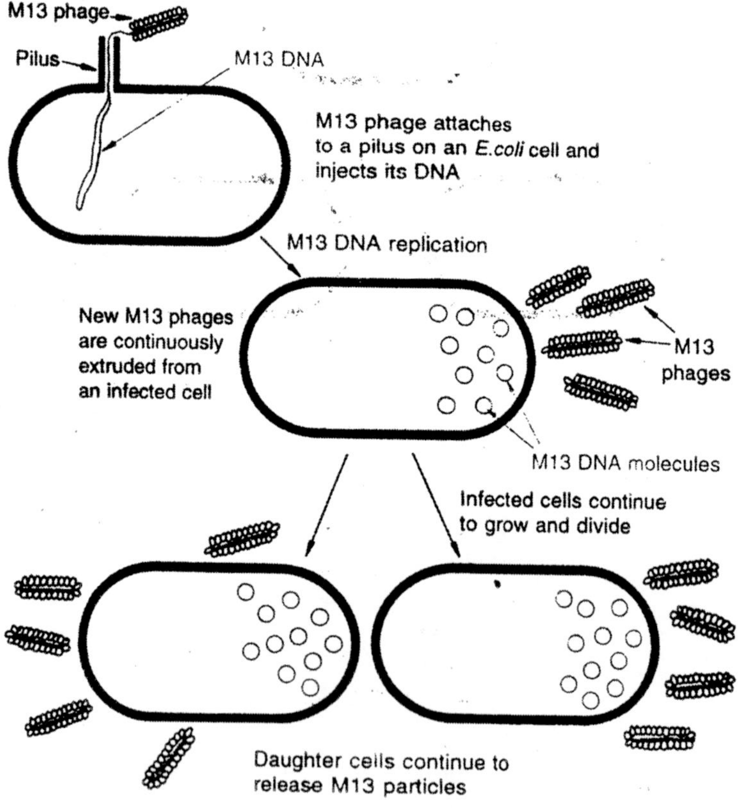

Fig. 7.15. The infection cycle of bacteriophage M13.

only λ and M13 have found any real role as cloning vectors. The properties of these two phages will now be considered in more detail.

(a) Gene organization in the λ DNA molecule

λ is a typical example of a head-and-tail phage. The DNA is contained in the polyhedral head structure and the tail serves to attach the phage to the bacterial surface and to inject the DNA into the cell. The λ DNA molecule is 49 kb in size and has been intensively studied by the techniques of gene mapping and *DNA sequencing.* As a result the positions and identities of most of the genes on the l DNA molecule are known.

A feature of the λ genetic map is that genes related in terms of function are clustered together on the genome. For example, all of the genes coding for components of the capsid are grouped together in the left-hand third of the molecule, and genes controlling integration of the prophage into the host genome are clustered in the middle of the molecule. Clustering of related genes is profoundly important for controlling expression of the λ genome, as it allows genes to be switched on and off as a group rather than individually. Clustering is also important in the construction of λ-based cloning vectors.

(b) The linear and circular forms of λ DNA

A second feature of λ that turns out to be of importance in the construction of cloning vectors is the conformation of the DNA molecule. The molecule is linear, with two free, ends, and represents ıe DNA present in the phage head structure. This linear molecule consists of two *complementary* strands of DNA, base-paired according to the *Watson-Crick rules* (that is, double-stranded DNA). However, at either end of the molecule is a short 12-nucleotide stretch, in which the DNA is single-stranded.

The two single strands are complementary, and so can base-pair with one another to form a circular, completely double-stranded molecule. Complementary single strands are often referred to as *'sticky' end* or 'cohesive' ends, because base-pairing between them can 'stick' together the two ends of a DNA molecule (or the ends of two different DNA molecules). The λ cohesive ends are called he *cos sites* and they play two distinct roles during the λ infection cycle.

First of all, they allow the linear DNA molecule that is injected into the cell to be circularized, which is a necessary prerequisite for insertion into the bacterial genome. The second role of the *cos* sites is rather different, and comes into play after the prophage has excised from the host genome. At this stage a large number of new λ DNA

(a) The linear form of the λ DNA molecule

Left cohesive end

Right cohesive end

CCCGCCGCTGGA

GGGCGGCGACCT

(b) The circular form of the λ DNA molecule

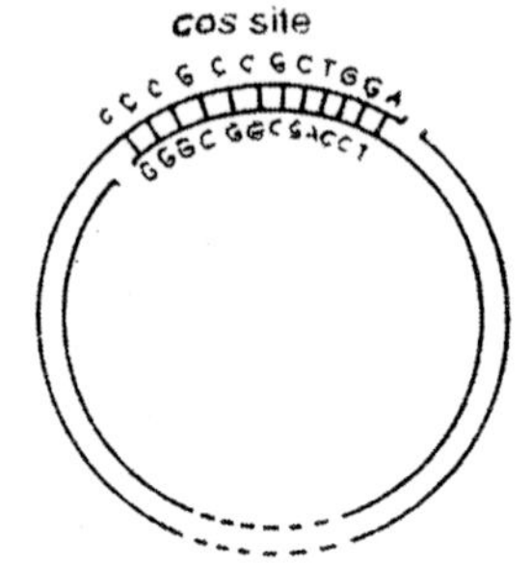

(c) Replication and packaging of λ DNA

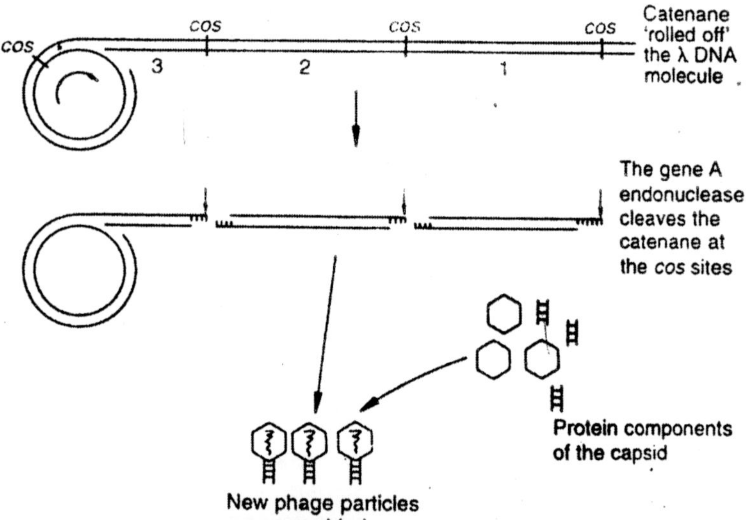

Fig. 7.16. The linear and circular forms of λ DNA.

molecules are produced by the rolling circle mechanism of replication in which a continuous DNA strand is 'rolled off' of the template molecule. The result is a catenane consisting of a series of linear λ genomes joined together as the *cos* sites. The role of the *cos* sites is now to act as recognition sequences for an *endonuclease* which cleaves the catenane at the *cos* sites producing individual λ genomes.

The endonuclease (which is the product of gene A on the λ DNA molecule) creates the single-stranded sticky ends, and also acts in conjunction with other proteins to package each λ genome into a phage head structure. The cleavage and packaging processes recognize just the *cos* sites and the DNA sequences to either side of them. Changing the structure of the internal regions of the λ genome, for example by inserting new genes, has no effect on these events so long as the overall length of the λ genome is not altered too greatly.

(c) M13–a filamentous phage

M13 is an example of a filamentous phage and is completely different in structure from λ. Furthermore, the M13 DNA molecule is much smaller than the λ genome, being only 6407 nucleotides in length. It is circular, and is unusual in that it consists entirely of single-stranded DNA. The smaller size of the M13 DNA molecule means that it has room for fewer genes than the λ genome. This is possible because the M13 capsid is constructed from multiple copies of just three proteins, whereas synthesis of the λ head-and-tail structure involves over 15 different proteins. In addition, M13 follows a simpler infection cycle than λ and does not need genes for insertion into the host genome. Injection of an M13 DNA molecule into an *E.coli* cell occurs via the *pilus*, the structure that connects two cells during sexual conjugation.

Once inside the cell the single-stranded molecule acts as the template for synthesis of a complementary strand, resulting in normal double-stranded DNA. This molecule is not inserted into the bacterial genome, but instead replicates until over 100 copies are present in the cell. When the bacterium divides, each daughter receives copies of the phage genome, which continues to replicate, thereby maintaining its overall numbers per cell. New phage particles are continuously assembled and released, about 1000 new phages being produced during each generation of an infected cell.

(d) The attraction of M13 as a cloning vehicle

Several features of M13 make this phage attractive as the basis for a cloning vehicle. The genome is less than 10 kb in size, well within the range that we stated was desirable for a potential vector. In addition, the double-stranded *replicative form* of the M13 genome behaves very much like a plasmid and can be treated as such for experimental purposes. It is easily prepared from a culture of infected *E.coli* cells and can be reintroduced by *transfection.*

Most importantly, genes cloned with an M13-based vector can be obtained in the form of single-stranded DNA. Single-stranded versions of cloned genes are useful for several techniques, notably DNA sequencing

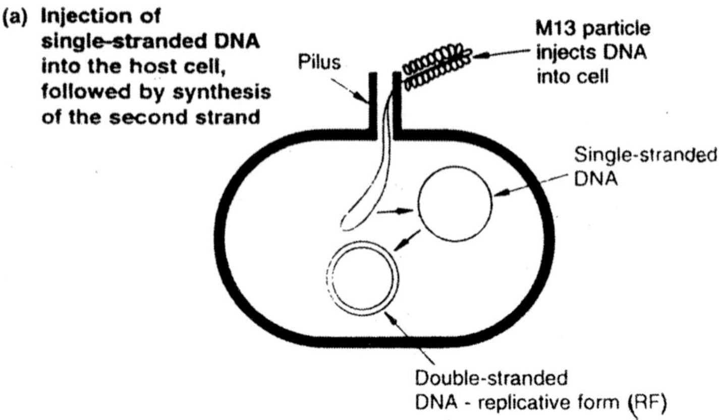

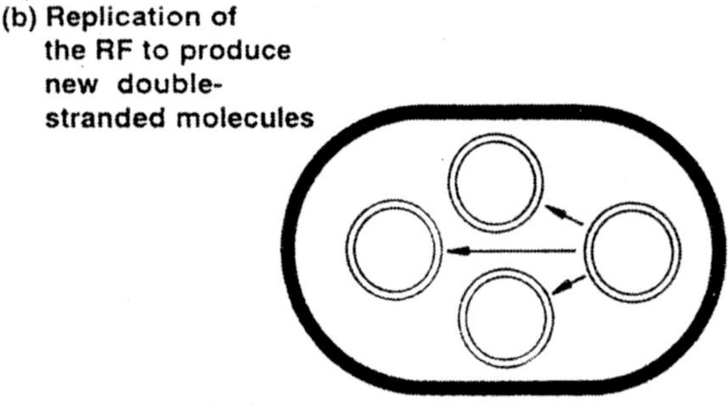

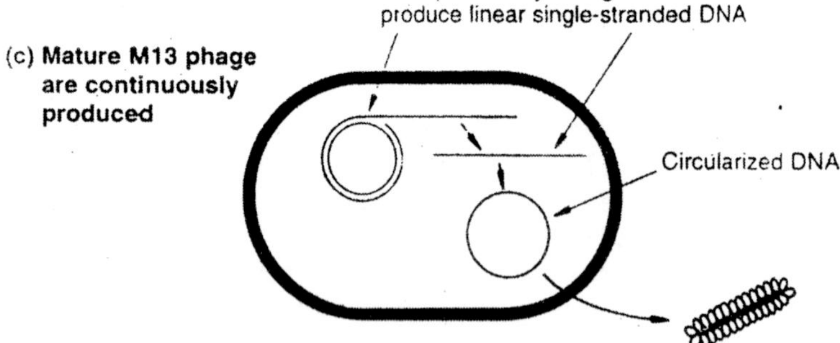

Fig. 7.17. The M13 infection cycle showing the different types of DNA replication.

and *in vitro* mutagenesis. Using an M13 vector is an easy and reliable way of obtaining single-stranded DNA for this type of work.

Viruses as cloning vehicles for other organisms

Most living organisms are infected by viruses and it is not surprising that great interest has been shown in the possibility that viruses might be used as cloning vehicles for higher organisms. This is especially important when it is remembered that plasmids are not commonly found in organisms other than bacteria and yeast. In fact viruses have considerable potential as cloning vehicles for animal cells. Mammalian viruses such as *simian virus 40 (SV40)* and *adenoviruses*, and the insect *baculoviruses*, are the ones that have received most attention so far, but others are also being studied.

The Role of Bacteria

Bacteria are one-celled organisms that reproduce very rapidly by binary fission (dividing into two cells) after a brief growth period. They live in practically every environment, but the species of greatest use to molecular biologists are those that can be easily cultivated in the laboratory by placing them on a suitable culture medium contained in a small dish. A single bacterial cell can give rise to tens or hundreds of thousands of identical cells (clones) in a single day. When a vector containing rDNA is inserted successfully into a bacterial cell, it is replicated, and hence the DNA fragment of interest is cloned.

Much of the earliest information about cell biology and molecular biology came from studies of bacteria. Today, they play a critical role in gene-cloning technologies. The most popular bacterium in all of this work has been *E.coli*, a normally harmless inhabitant of mammalian digestive tracts. More is known about the molecular biology of *E.coli* than about any other species on Earth. *E.coli* and some other bacteria are used to perform two functions in rDNA technology. First, they are used to clone genomic DNA or cDNA to obtain unlimited numbers of genes that can be used in further studies. Second, pure cultures, in which every bacterium contains a gene encoding a desired protein, can be created and put to work manufacturing huge quantities of that protein. Plasmids that contain specific sequences to direct protein synthesis are called *expression vectors.*

Gene Libraries

One of the major uses of rDNA technology is to create *gene libraries* for different species. In accordance with the procedure used to obtain the DNA, there are libraries of both genomic clones and cDNA clones. The number of "volumes" (DNA fragments or genes) required in the library to ensure that all gene sequences are represented is generally related to the size and complexity of the species. For *E.coli*, 1,500 fragments are necessary; for yeast, 4,600; fruit flies, 48,000; and humans, 800,000.

Researchers create a *gene library* for a particular species in the following sequence of events:

1. Genomic DNA is cut into thousands of fragments with appropriate restriction enzymes.
2. Each fragment, representing approximately one gene, is spliced into a phage-cloning vector.
3. The phages are replicated by the "host" bacteria, thus replicating (cloning) the inserted gene fragments.

Such a library is a repository of the entire genome where every gene of a species is represented. The space required for a typical gene library is a test tube or a set of small culture dishes that can be conveniently kept in a refrigerator pending further analysis. Molecular biologists have developed specialized techniques to "screen" gene libraries, that is, to determine where specific genes are located in the pieces of DNA. This involves isolating a gene and identifying the protein it encodes. Once a gene has been isolated and cloned, it is possible quickly to determine its entire base sequence. As you might expect, the greatest emphasis is currently on research related to the human gene library, and remarkable progress has been made in isolating human genes of medical importance. Nevertheless, the human gene library is so immense, and many genes are so difficult to locate, that a complete reading of our library probably lies one or more decades in the future.

Applications of Genetic Engineering

Recombinant DNA technology has revolutionized biology in the past decade. At present, its main uses are (1) facilitating the production of useful proteins, (2) creating bacteria capable of synthesizing economically important molecules, (3) supplying DNA and RNA sequences as a research tool, (4) altering the genotype of organisms such as plants, and (5) potentially correcting genetic defects in animals (gene therapy). Some examples of these applications follows:

Commercial Possibilities

Bacteria with novel phenotypes can be produced by genetic engineering sometimes by combining the features of several other bacteria. For example, several genes from different bacteria have been inserted into a single plasmid that has then been placed in a marine bacterium, yielding an organism capable of metabolizing petroleum; this organism has been used to clean up oil spills in the oceans.

Many biotechnology companies are at work designing bacteria that can synthesize industrially important chemicals. Bacteria have been designed that are able to compost waste more efficiently and to fix nitrogen (to improve the fertility of soil), and an enormous effort is currently being expended to create organisms that can convert biological waste to alcohol. A human insulin, synthesized in *E.coli*, is already commercially available. Altering the genotype of plants is an important application of recombinant DNA technology. Of great use is the bacterium *Agrobacterium tumefaciens* and its plasmid Ti, which produces crown gall tumors in dicotyledonous plants. These tumor result from disruption of the bacterium inside plant cells, release of bacterial DNA, and integration of a segment of the plasmid DNA into the plant chromosome.

It is possible by genetic engineering to introduce genes from one plant into this plasmid and then, by infecting a second plant with the bacterium, transfer the genes of the first plant to the second plant. (Actually genes are first cloned in an *E.coli* plasmid and then recloned in Ti). Attempts are being made to perform plant breeding in this way. An example is the attempted alteration of the surface structure of the roots of grains such as wheat, by introducing certain genes from legumes (peas, beans), in order to give grains the ability of the legumes to establish root nodules of nitrogen-fixing bacteria. If successful, this would eliminate the need for the addition of nitrogenous fertilizers to grain-growing soils.

The first engineered recombinant plant of commercial value was developed in 1985. An economically important herbicide (weed killer) is glyphosate, which inhibits a particular essential enzyme in many plants. However, most herbicides cannot be applied to fields growing crops because both the crop and the weeds would be killed.

The target gene of glyphosate is also present in the bacterium *Salmonella typhimurium*. A resistant form of the gene was obtained by mutagenesis and growth of *Salmonella* in the presence of glyphosate; the gene was cloned in *E.coli*, and then recloned in *Agrobacterium*. Infection of plants with purified Ti containing the glyphosate-resistance gene has yielded varieties of maize, cotton, and tobacco that are resistant to glyphosate. Thus, fields of these crops can be sprayed with glyphosate at any stage of growth of the crop. All weeds are killed, and the crop is totally unharmed. An interesting bacterium is a strain of *Pseudomonas fluorescens*, which lives in association with maize and soybean roots. A lethal gene from *Bacillus thuriengis*, a bacterium pathogenic to the back cutworm, has been engineered into this bacterium. The black cutworm causes extensive crop damage and is usually combatted with noxious insecticides. In preliminary studies inoculation of soil with the engineered *Ps. fluorescens* resulted in death of the cutworm.

Uses in Research

Recombinant DNA technology is extraordinarily useful in basic research. Several examples were already given in mutant bacteria in which particular genetic systems have been altered in an effort to make them more amenable to study (for example, the numerous mutants in the *lac* operon). Such mutants have usually been derived through standard genetic techniques. For simple mutants this procedure is straightforward, but for mutants required to have many genetic markers (which may be very closely linked) the frequency of production of mutants can be so low that their isolation becomes very tedious.

Recombinant DNA techniques can simplify mutant construction, since fragments containing desired genetic markers can be purified, altered, and combined in test tube, and then introduced into another cell. This saves time and labour, and often enables mutants to be constructed that cannot in practice be formed in any other way. An example is the formation of double mutants of animal viruses, which undergo crossing over at such a low frequency that mutations can rarely be recombined by genetic crosses.

The greatest impact of the new technology on basic research has been in the study of eukaryotes, in particular, of eukaryotic regulation. Experiments of the type, which study the regulation of operons in bacteria, have been made possible by the use of mutations in promoters, operators, and structural genes. However, this approach has not been feasible with eukaryotes, because eukaryotes are diploid and, hence, mutants are difficult to isolate.

Furthermore, except for the unicellular eukaryotes such as yeast, there is no simple and rapid way to do multiple genetic manipulations with eukaryotic cells. Cloning techniques have made it possible to study regulation of gene expression in eukaryotes by direct assays of mRNA molecules produced by particular genes. The approach is based on the success with which it has become possible to understand gene activity in bacteria by studying the synthesis of mRNA in a variety of conditions.

An excellent assay is DNA-RNA hybridization, in which either a primary transcript or mRNA is detected by hybridization to a DNA sample that has been enriched for the gene being studied. In microbial experiments specialized transducing particles would be the source of the DNA–for example, *E.coli* gal mRNA is assayed by hybridization of radioactivity labeled intracellular RNA with DNA of *l gal* transducing particles. However, transducing particles do not exist for eukaryotic systems, so recombinant DNA techniques have been used to clone a gene whose regulation is to be studied.

The usual procedure is to clone the gene first in a plasmid (or a phage) vector and then allow the cell containing the plasmid to multiply. Purified plasmid DNA provides the researcher with a large supply of the DNA of that gene. The vector containing that DNA sequence is called a DNA *probe*, because its DNA, in denatured form, can be added to a cell extract containing mRNA, to probe for a particular mRNA by renaturation. Since no natural genes of the vector are present in eukaryotic cells, the vector DNA can be regarded as a source of pure DNA copies of a particular eukaryotic gene because no other genes in the vector will participate in the renaturation.

A further useful technique is to apply denatured probe DNA to a nitrocellulose filter and then to use a filter-binding assay for the specific RNA species. DNA but not RNA binds to these filters. Radioactive mRNA can then be incubated with a filter containing the DNA, using conditions such that renaturation will result. Except where the mRNA has renatured to the denatured DNA, the RNA can be removed by washing the filter. Thus, presence of radioactivity in the filter after washing indicates that complementary mRNA has been bound to the DNA on the filter. This simple technique and variations such as Southern blotting have revolutionized the study of eukaryotic gene regulation.

Production of Eukaryotic Proteins

One of the most valuable applications of genetic engineering is the production of large quantities of particular proteins that are otherwise difficult to obtain. The method is simple in principle. The gene encoding the desired protein is cloned in a vector adjacent to a bacterial promoter, and tests are performed to ensure that the gene is oriented such that its coding strand is linked to the strand containing the promoter. A plasmid for which many copies are present in each cell, or occasionally an actively replicating phage (such as λ), is used as a vector; in both cases, cells can be prepared that contain several hundred copies of the gene, and this can result in synthesis of a gene product to reach a concentration of about 1 to 5 percent of the cellular protein. In practice, production of large quantities of a prokaryotic protein in a bacterium is straightforward. However, if the gene is from a eukaryote, special problems exist:

1. Eukaryotic promoters are not usually recognized by bacterial RNA polymerases.
2. The mRNA transcribed from eukaryotic genes may not be translatable on bacterial ribosomes.
3. Introns may be present, and bacteria are unable to excise eukaryotic introns.
4. The protein itself often must be processed (for example, insulin); and bacteria cannot recognize processing signals from eukaryotes.
5. Eukaryotic proteins are often recognized as foreign material by bacterial protein-digesting enzyme and are broken down.

Several approaches to solving these problems have been taken with some degree of success. One procedure uses a genetically engineered plasmid, called a *shuttle vector*, that consists of both *E.coli* and yeast DNA and replicates in both organisms.

Cloning is done in *E.coli*, which for a variety of technical reasons is easier than cloning in yeast, and then the plasmid is isolated and transferred to yeast for expression of the cloned gene. In yeast the five problems just stated are avoided, or at least substantially reduced. Another approach uses a plasmid containing the *E.coli lac* region, cleaved in the *lacZ* gene by a restriction enzyme making blunt ends. Either c-DNA or a synthetic DNA molecule whose sequence is known from the amino acid sequence of the protein product is inserted into the *lacZ* gene, so that the eukaryotic protein is synthesized as the terminal region of β-galactosidase, from which it can be cleaved.

The first example of this approach resulted in a synthetic gene capable of yielding a 14-residue polypeptide hormone–somatostatin– which is synthesized *in vivo* in the mammalian hypothalamus. The procedure, applicable to most short polypeptides, was the following.

By chemical techniques, a double-stranded DNA molecule was synthesized containing 51 base pair; the base sequence of the coding strand was

TAC-(42 base encoding somatostatin)-ACTATC

The base sequence of the corresponding mRNA molecule was

AUG-(42 bases encoding somatostatin)-UGAUAG

The AUG codon of the mRNA was not used for initiation, but specified methionine. The mRNA terminated with two stop codons UGA and UAG.

The vector was the plasmid pBR322 modified to contain the *lac* promoter-operator region and a portion of the *lacZ* gene encoding the amino-terminal segment of β-galactosidase. The vector was cleaved at a site in the *lacZ* segment by a restriction enzyme that leaves blunt ends; the synthetic DNA molecule was then blunt-end-ligated to the cleaved plasmid. When the *lac* operon was induced, a protein was made consisting of the amino-terminal segment of β-galactosidase ***coupled by methionine*** to somatostatin, and terminated at the repeated stop codons of the synthetic DNA. This protein was purified and treated with cyanogen bromide, a reagent that cleaves proteins only at the

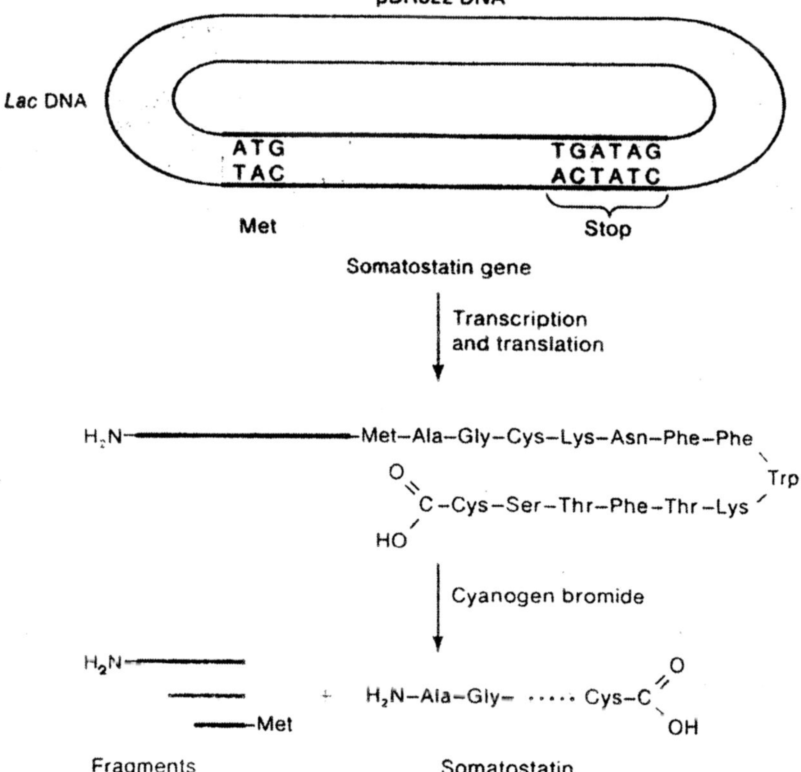

Fig. 7.18. Synthesis of somatostatin from a chemically synthesized gene joined to the plasmid pBR322 lac.

carboxyl side of methionine. In this way, the methionine linker remains attached to a β-galactosidase fragment and somatostatin is released. Use of methionine-coupling followed by cleavage with cyanogen bromide is a useful technique for separating any polypeptide from a bacterial protein to which it is fused, as long as the polypeptide itself does not contain methionine. Another more generally useful linker is $(Asp)_4$-Lys. In a sequence (Asp_4)-Lys-X, in which X is another amino acid, the enzyme enterokinase cleaves between lysine and X. Since such a sequence is not common, this linker is of more value than methionine.

Genetic Engineering with Animal Viruses

Retroviruses are RNA-containing animal viruses that have an unusual life cycle. These viruses, which contain the enzyme reverse transcriptase in their protein coats, use this enzyme to synthesize a double-stranded DNA copy of the viral RNA shortly after infection. As a normal part of the viral life cycle, this double-stranded DNA becomes inserted into the animal-cell chromosome, apparently at one of an enormous number of potential sites. Transcription occurs only after the DNA copy is inserted.

The infected host cell survives the infection, retaining the retroviral DNA in its chromosome. One of the best-understood retroviruses is *Rous sarcoma virus*, which causes tumors in chickens. Retrovirus DNA, either isolated from a cell or prepared synthetically, is useful vector with animal cells. Genetic engineering with retroviruses allows the possibility of altering the genotype of an animal cell. Since a wide variety of retroviruses are known, including two that can grow on human hosts, genetic defects may be correctable by these procedures in the future.

Many retrovirus species contain a gene that produces uncontrolled growth of a cell containing the retrovirus, thereby causing tumors in animals. If a retrovirus vector is to be used to change a genotype, the tumor-causing ability must be removed. This is possible by removal of the tumor-causing gene, which also provides the space needed for incorporation of foreign DNA. The recombinant DNA procedure employed with retroviruses consists of synthesis in the laboratory of double-stranded DNA from the viral RNA, through use of reverse transcriptase. The DNA is then cleaved with a restriction enzyme and, by the techniques already described, foreign DNA is inserted and selected.

Treatment of mutant cells in culture with the recombinant DNA, and application of a transformation procedure suitable for animal cells, yields cells in which recombinant retroviral DNA has been permanently inserted into an animal-cell chromosome. In this way, the genotype of the cell can be altered. Experiments have been done in which human cells deficient in the synthesis of purines have been obtained from patients with Lesch-Nyhan syndrome and grown in culture; these cells have been converted to normal cells by transformation with recombinant DNA. The exciting potential of this technique lies in the possibility of correcting genetic defects–for example, restoring the ability of a diabetic individual to make insulin or correcting immunological deficiencies. This technique has been termed *gene therapy.* However, it must be recognized that retroviruses are not well understood and are potentially dangerous. Gene therapy is not yet a practical technique, for major problems exist; for example, there is no reliable way to ensure that a gene is inserted in the appropriate target cell or target tissue. In addition, some means is needed to regulate the expression of the inserted genes. A major breakthrough in disease prevention has been the development of synthetic vaccines. Production of certain vaccines such as anti-hepatitis B has been difficult because of the extreme hazards of working with large quantities of the hepatitis B virus. The danger would be avoided if the viral antigen could be cloned and purified in *E.coli* or yeast, because the pure antigen could be given as a vaccine.

Several viral antigens have been cloned, but because of either thermal instability or poor antigenicity when pure, attempts to make vaccines in this way have generally been unsuccessful. However, the use of vaccinia virus (the anti-smallpox agent) as a carrier has been fruitful. The procedure makes use of the fact that viral antigens are on the surface of virus particles and that some of these antigens can be engineered into the coat of vaccinia. By 1985 vaccinia hybrids with surface antigens of hepatitis B, influenza virus, and vesicular stomatitis virus (which kills cattle, horses, and pigs) have been prepared and shown to be useful vaccines in animal tests. A surface antigens of *Plasmodium falciparum*, the parasite that causes malaria, has also been placed in the vaccinia coat; this may lead to an antimalaria vaccine.

Diagnosis of Hereditary Diseases

Restriction mapping has also been used in prenatal diagnosis–for example, in detecting sickle-cell anemia. This disease is caused by possession of an altered hemoglobin. The base change causing the mutation generates a cleavage site for a particular restriction enzyme. Cleavage of DNA from a sickle-cell patient produces a restriction map that differs from that of the DNA of a normal patient. The different restriction pattern can be detected by using Southern blotting and, as a probe, radioactive RNA prepared by transcribing cloned globin genes. The technique is so sensitive that the sickle-cell mutation can be detected in a small sample of cells obtained from the amniotic fluid surrounding a fetus, and detection is possible at a very early stage of development. Possibly in the future, gene therapy might be used to produce a normal child.

8

DNA Cloning

Cloning is the production of identical DNA, cell or organism. Identical twining is the natural cloning. For DNA cloning random fragments of source DNA can be generated by mechanical shearing. Mechanical shear can be caused either by high speed mixing in blender or sonication. Controlled mechanical shearing is possible in blender at 1500 rpm for 30 minutes. This gives fragments of 8kb mean size. Short single stranded regions–termini (overhangs) or blunt end fragments are produced. Intense sonication can reduce length of the fragment to about 300 nucleotide pairs. Shearing does not produce 5′ phosphate and 3′OH ends. Therefore the ends of the fragments must be repaired. This can be done either by action of S_1 nuclease which removes overhangs or filling in the ends with DNA polymerase. In both the cases the result is blunt-ended duplex DNA. Since fragments produced by mechanical shearing even after end repair do not have cohesive ends, ligation (joining) with the vector can be facilitated with the process of 'homopolymer tailing'. For example, a tail of dC residues can be added to 3′OH terminus of DNA of sheared fragment and a tail of dG residues to vector. Sonication is used as a method to get target DNA fragments in DNA sequencing. Fragments produced by mechanical shearing are not of reproducible nature.

Restriction Endonuclease

With the discovery of restriction endonuclease enzymes, cutting DNA at specific sites became possible. In 1972, for the first time restriction enzyme was used to cut the DNA specifically and fragments were ligated to form first recombinant molecule. Today restriction endonuclease have become most important tool for our recombinant

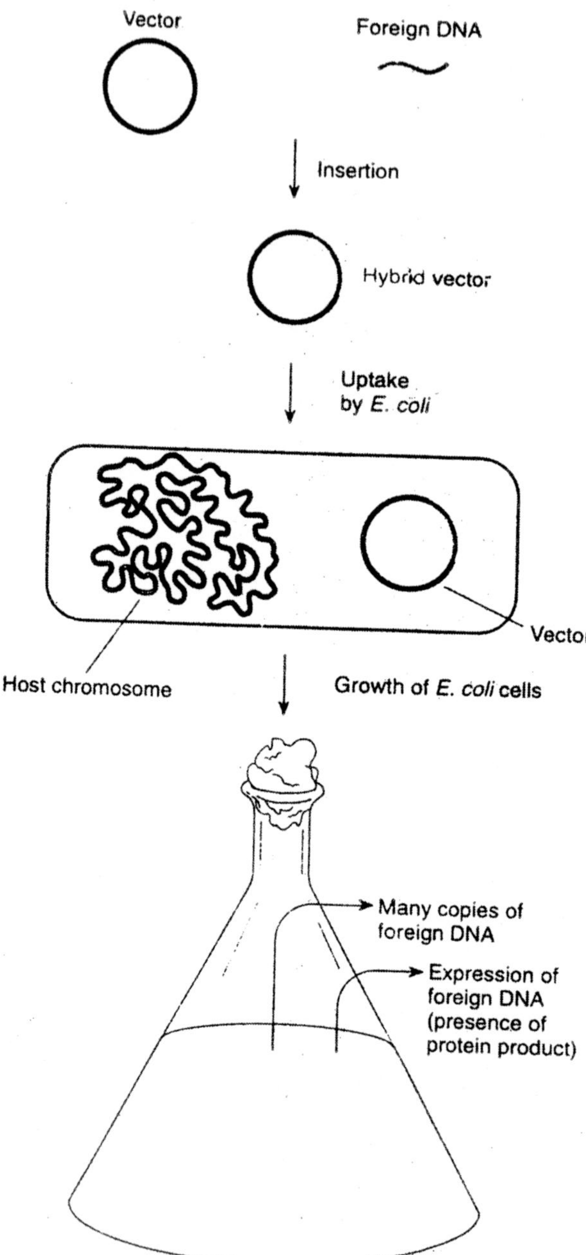

Fig. 8.1. Overview of recombinant DNA techniques.

DNA technology procedures. Large number of restriction enzymes which recognize and cut DNA within target sites of 4 or 5 or 6 or 7 nucleotides are known. Depending upon the number of target sites present, DNA may be cut into too small or too big size fragments. Thus target site for particular restriction enzyme may be too frequent or too sparse in distribution. Either of the extreme cases are not desirable.

It is important that cutting site falls on either end of the gene of interest and not in between. Restriction enzymes either produce sticky ends or blunt ends. Generally the same restriction enzyme is used for vector and the DNA of interest (to be cloned). The digestion carried out may be light, moderate or heavy producing from small number to large number of fragments. Digestion with restriction enzymes gives reproducibility in fragmentation.

Sticky ends produced are more efficient while joining two DNA fragments. *EcoRI* produces staggered cuts with sticky ends while *HaeIII* produces blunt ends. *BamHI, BglII* and *SauIII* each have different recognition sites and yet leave the same single stranded tails. They will have special significance in cutting and joining of DNA. The advantage of restriction enzyme which produces staggered cuts is that sticky ends are produced on both target and vector. After amplification in the recombinant DNA these target sites still exist and separation of DNA fragment which was cloned is possible.

Reverse Transcriptase

If eukaryotic gene is to be cloned and expressed in prokaryotic cell, then directly cutting the source DNA into suitable fragment alone will not be sufficient. The difference in the gene organisation of eukaryotes and prokaryotes is important here. Introns the segments of noncoding sequences are present in eukaryotic gene. These introns are transcribed into mRNA. Such precursor mRNA in eukaryotic cell undergoes post-transcriptional modification and removal of introns occurs to give rise to processed mRNA. Processed mRNA then give rise to protein product. In fact there are also post-transnational changes occurring in eukaryotic cell. Thus to get expression in form of protein product introns have to be removed.

Bacteria or yeasts do not have necessary splicing mechanism for removal of introns. Hence eukaryotic gene if directly cloned in bacteria or yeast will give rise to precursor mRNA but not the protein product at end. This difficulty can be overcome by cDNA route. The protein encoding parts of the gene are exons. Mature mRNA molecules from

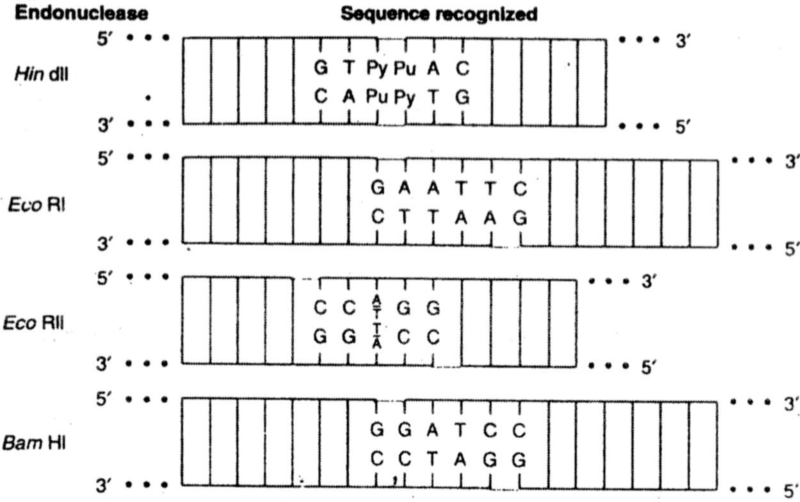

Fig. 8.2. Sequences cleaved by various type II restriction endonucleases.

animal cells do not contain sequences complementary to introns as they are removed by processing. These molecules then can be directly transcribed into DNA using an enzyme reverse transcriptase. cDNA thus produced for a particular protein can be joined to appropriate vector and cloned into a host.

Hybridization Method

Hybridization method depends on the principle that on mRNA forms a complex with complementary DNA segments from which it has been transcribed. This method is possible if protein encoding gene does not have introns. Here the total DNA from the donor organism is first isolated. This dS DNA is treated with heat or alkali and is converted to single stranded DNA by denaturation. Strands are then mixed with mRNA transcribed by the gene. mRNA pairs with cDNA portion to form DNA-RNA complex (Hybrid). This complex is then isolated and DNA is separated from RNA. ssDNA thus obtained can be converted to ds DNA by DNA polymerase I. This method is suitable for isolating genes which exist in multiple copies, e.g., ribosomal genes.

Chemical Synthesis

The method of chemical synthesis can be used to generate DNA fragment for cloning. This of coarse will be possible if amino acid sequence of protein of interest is known for which the sequence of nucleotide in mRNA and subsequently in DNA can be determined. Earlier this was difficult and tedious but with the introduction of triester method and automated synthesizers (gene machines) it has become possible. This method also enables us to make some specific alterations in the concerned gene to be produced. The technique is termed as reverse genetics. Kangas et al in 1982 synthesized a gene which produced proline-rich protein and it could be expressed in *E.coli.*

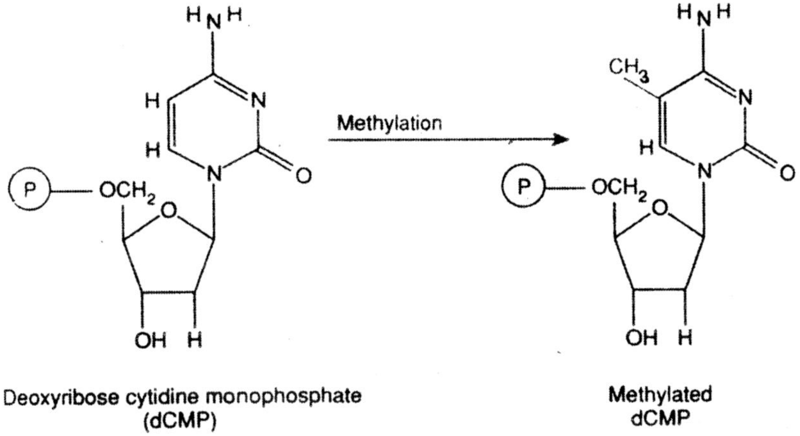

H N H
H
N
H
N
O
P
OCH2
O
OH H
Methylation
H N H
CH3
N
H
N
O
P
OCH2
O
OH H
Deoxyribose cytidine monophosphate
(dCMP)
Methylated
dCMP

Fig. 8.3. The addition of a methyl group to cytosine by a methylase enzyme converts it to 5-methylcytosine.

The Joining of DNA Molecules

In genetic engineering a particular DNA segment of interest is joined to a small, but essentially complete, DNA molecule that is able to replicate–a *vector* or *cloning vehicle.* By a transformation procedure, described below, this recombinant molecule is placed in a cell in which replication can occur. When a stable transformant has been isolated, the genes or DNA sequences on the donor segment are said to be *cloned.* In this section several types of vectors and few procedures used for joining molecules are described.

Vectors

A vector is a DNA molecule in which DNA fragment can be cloned; to be useful, it must have three properties:

1. A means of introducing vector DNA into a cell must be available.
2. The vector must have a replication origin–that is, be able to replicate.
3. Transformants must be selectable by a straightforward assay, preferably by growth of a host cell on a solid medium.

The types of vectors most commonly used at present are plasmids and *E.coli* phages λ and M13; several animal viruses are also gaining in use. These cloning vehicles can be detected in host cell by means of genetic features or particular markers made evident during colony or plaque formation. Plasmid and phage DNA can be introduced into cells by the *$CaCl_2$ transformation procedure.* With this technique cells are exposed to a series of $CaCl_2$ solutions at several temperatures, and by this regimen they gain the ability to take up free DNA.

Vector DNA, or any DNA that can replicate, is added to the cell at one stage of the protocol and, after a series of steps, the cells are plated on a solid medium. If the added DNA is a plasmid, colonies consisting of plasmid-containing bacteria will form, and transformants can sometimes be detected by the phenotype that the plasmid confers on the host cell. Selective procedures that prevent cells that lack the plasmid from growing are usually used. Alternately, if the vector is phage DNA, the "infected" cells are plated in the usual way to yield plaques. A variation of this procedure is used to transform animal cells with viral vectors.

Joining Fragments with Cohesive Ends

Circularization of fragments having complementary terminal bases was described; similarly, two DNA fragments can join together by the pairing of their complementary cohesive ends. Therefore, because a particular restriction enzyme produces fragments with the *same* cohesive ends, without regard to the source of the DNA, fragments from DNA molecules isolated from two *different* organisms (for example, a bacterium and a frog) can be joined. Furthermore, if DNA is treated with DNA ligase to seal the joint after base-pairing, the fragments will be joined permanently, and a molecule will be generated that may never have existed before.

The ability to join DNA fragment of interest to a vector is the basis of the recombinant DNA technology. Joining cohesive ends does not always produce a DNA sequence that has functional genes. For example, consider a linear DNA molecule that is cleaved into four fragments–A, B, C, and D–whose original sequence in the molecule was ABCD. Reassembly of the fragments can yield the original molecule, but since B and C have the same pair of cohesive ends, molecules with the fragment arrangements ACBD or BADC can also form with the same probability as ABCD. The problem of the scrambling of vector fragments is minimized by using a vector having only one cleavage site for a particular restriction enzyme. Plasmids of this type

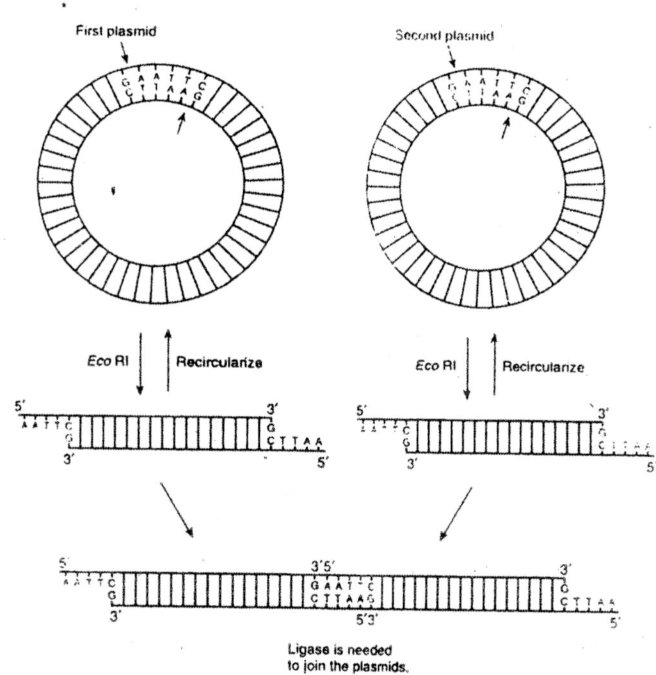

Fig. 8.4. Circular plasmid DNA with a palindrome recognized by Eco RI.

are available (most have been genetically engineered); usually they have sites for several different restriction enzymes, but only enzyme is used at a time.

Joining Fragments without Cohesive Ends

DNA molecule lacking cohesive ends can be joined. A direct method uses the DNA ligase made by *E.coli* phage T4. This enzyme differs from other DNA ligases in that it not only seals single-strand breaks in double-stranded DNA but can also join blunt-ended molecules. Another method uses the enzyme *terminal deoxynucleotidyl transferase*, an unusual DNA polymerase obtained from animal tissue. This enzyme adds nucleotides (by means of deoxynucleoside triphosphate precursors) to the 3′-OH group of an extended single-stranded segment of a DNA chain–without the need to copy a template strand.

In order to produce an extended single strand one need only treat a DNA molecule with a 5′-specific exonuclease to remove a few terminal nucleotides. If a mixture is prepared containing exonuclease-treated DNA, terminal deoxynucleotidyl transferase, and a single kind of nucleoside triphosphate–for example, deoxyadenosine triphosphate (dATP)–a DNA extension having adenine as the only base will form at both 3′-OH termini. Such extended segments are called ***poly(dA) tails***. If instead thymidine triphosphate (dTTP) were provided, the DNA molecule would have poly(dT) tails. Since a poly(dA) tail is complementary to a poly(dT) tail, any two DNA molecules can subsequently be joined if a poly(dA) tail is put on one molecule and a poly(dT) tail on the other. Gaps in the joined molecule can be filled by treatment with a DNA polymerase (*E.coli* polymerase I is commonly used), and the remaining single-stranded interruptions are again sealed by DNA ligase. The method, which is called ***homopolymer tail-joining,*** is a general method for joining any pair of DNA molecules.

INSERTION OF A PARTICULAR DNA MOLECULE INTO A VECTOR

For many genes, simple selection techniques are adequate for recovery of a vector containing that gene. For example, if the gene to be cloned were a bacterial ***leu*** gene, one would use a Leu^- host and select a Leu^+ colony on a medium lacking leucine. However, often the clone of interest is either so rare or so difficult to detect that it is preferable if, prior to joining, the fragment containing the DNA segment can be purified. One method has already been discussed: if the DNA of interest (for example, a viral gene) is known to be contained in a particular restriction fragment, that fragment can be isolated from a gel after electrophoresis and joined to an appropriate vector.

Eukaryotic cells contain about a million cleavage sites for a typical restriction enzyme, so direct isolation of a eukaryotic gene from a mixture of fragments separated by electrophoresis is not feasible.

The Use of c-DNA

In both prokaryotes and eukaryotes, selection of a vector containing a particular gene from a collection of vectors containing foreign DNA is not always straightforward. In theory, a gene is most easily detected by its expression–that is, the phenotype conferred by the gene, as mentioned above for the *E.coli leu* gene. However, all prokaryotic genes do not produce such easily detected phenotypes and for eukaryotes, a particular gene that has been inserted into a bacterial

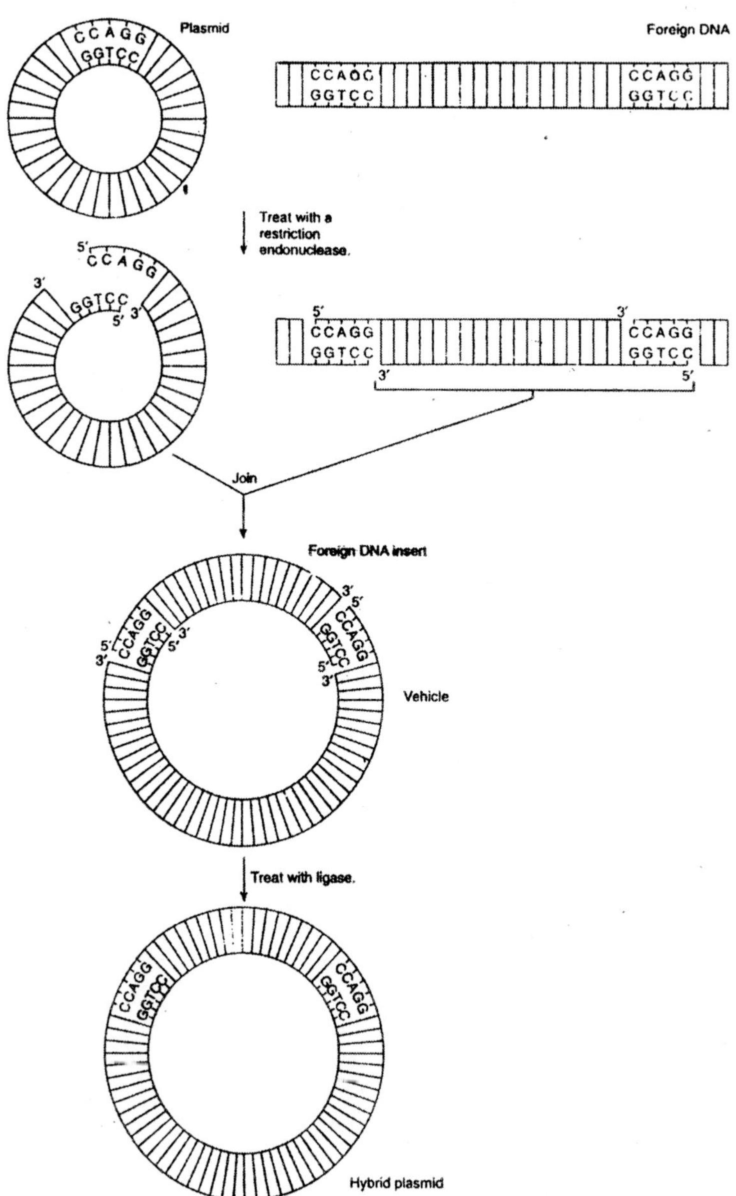

Fig. 8.5. Formation of a hybrid plasmid.

DNA vector may, for a variety of reasons, fail either to be transcribed or to be correctly translated into a functional protein.

Let us assume that the chicken gene for a ovalbumin is to be cloned into a bacterial plasmid. A recombinant plasmid could be formed by treating both chicken cellular DNA and the *E.coli* plasmid DNA with the same restriction enzyme, mixing the fragments, annealing, and finally transforming *E.coli* with the DNA mixture containing this recombinant plasmid. However, identification of a bacterial colony containing this plasmid could not be carried out by a test for the presence of ovalbumin, because a bacterium containing the intact ovalbumin gene will not synthesis ovalbumin.

Introns will be present in the gene and bacteria lack the enzymes needed to remove these non-coding sequences. Other tests might be performed to find the desired colony, such as hybridization of bacterial DNA with either the primary transcript or the mRNA of the ovalbumin gene isolated from chicken cells, but since only about one plasmid-containing colony in 10^5 would contain the ovalbumin gene, this method would be very tedious. A more convenient procedure is clearly desirable. Some specialized animal cells, such as those producing ovalbumin, make only one or a very small number of proteins.

In the cells the specific mRNA molecules, whose introns will have been removed by processing enzymes *if the mRNA is isolated from the cytoplasm*, constitute a large fraction of the total mRNA synthesized in the cell; consequently, mRNA samples can usually be obtained that consist predominantly of a single mRNA species–in this case, ovalbumin mRNA. If genes of this class–that is, those whose gene products are the major cellular proteins–are to be cloned, the purified mRNA of each type of cell can serve as a starting point for creating a collection of recombinant plasmids many of which should contain only the gene of interest. The technique depends on an unusual polymerase, *reverse transcriptase,* which can use a single-stranded RNA molecule as a template and synthesize a double-stranded DNA copy, called *complementary DNA* or *c-DNA*. If the template RNA molecule is an mRNA molecule (that is, if the introns have been removed from the primary transcript), the corresponding full-length c-DNA will contain an uninterrupted coding sequence. *This sequence will not be that of the original eukaryotic gene*; however, if the purpose of forming the recombinant DNA molecule is to synthesize a eukaryotic gene product in a bacterial cell, and if such processed RNA can be isolated, then c-DNA formed from processed mRNa is the material of choice to be inserted. Joining of c-DNA to a vector can be accomplished by homopolymer tail-joining or other available procedures for joining blunt-ended molecules.

Detection of Recombinant Molecules

When a vector is cleaved by a restriction enzyme and renatured with a mixture of all restriction fragments from a particular organism, many types of molecules result–some examples are a self-joined vector that has not acquired any fragments, a vector with one or more fragments, and a molecule consisting only of many joined fragments.

To facilitate the isolation of a vector containing a particular gene some means is need to ensure, first, that a vector established after established after $CaCl_2$ transformation does possess an inserted DNA fragment, and, second that it is the DNA segment of interest. In using the $CaCl_2$-transformation procedure to establish a plasmid in a bacterium, the initial goal is to isolate bacteria that contain the plasmid from a mixture of plasmid-free and plasmid-containing colonies. A common procedure is to use a plasmid possessing an antibiotic-resistance marker and to grow the transformed bacteria on a medium containing the antibiotic: only cells in which a plasmid has become established will form a colony.

A useful plasmid is pBR322. It is small, consisting of 4362 nucleotide pairs of known sequence, and has two different antibiotic-resistance markers–resistance to tetracycline (*tet-r*) and to ampicillin (*amp-r*). Thus, plasmid-containing transformants are easily detected by growth of a transformed culture on medium containing either one of these antibiotics. Also, pBR322 is very useful because it contains only one copy of each of seven different types of restriction-enzyme cleavage sites at which DNA can be inserted, so the position of inserted DNA is always known. In addition to a screening procedure for identifying plasmid-containing cells, a method is needed to identify plasmid in which DNA has been inserted.

Having two antibiotic-resistance markers, such as are available in pBR322, allows the use of a procedure for detecting insertion called *insertional inactivation*; this is carried out as follows. In pBR322 the *tet* gene contains sites for cutting by the restriction enzymes BamHI and SalI. Thus, insertion at either of these sites will yield a plasmid that is *amp-r tet-s,* because insertion interrupts and hence inactivates the *tet* gene. If wildtype (Amp-s Tet-s) cells are transformed with a DNA sample in which the cleaved pBR322 and restriction fragments have been joined, and the cells are plated on a medium containing ampicillin, all surviving colonies must be Amp-r and hence must possess the plasmid. Some of these colonies will be Tet-r and some Tet-s, and these can be identified by replica plating onto a medium containing tetracycline. Because unaltered pBR322 carries the *tet-r* allele, an Amp-r colony will also be Tet-r unless the *tet-r* allele has been inactivated by insertion of foreign DNA. Thus, an Amp-r Tet-s cell must contain not only pBR322 DNA but donor DNA as well. Other useful plasmids contain one antibiotic-resistance marker and the *E.coli lac* operon. Restriction sites are in the *lac* genes. The Lac^+ and Lac^- phenotypes can be distinguished by plating on a medium containing particular dyes, so insertion can be recognized by the appearance of colonies of a particular colour.

Screening for Particular Recombinants

The simplest procedure for detecting a cloned gene is complementation of a bacterial gene. Some eukaryotic genes can also be detected in this way. In the first such experiment reported, a cloned *his* (histidine synthesis) gene of yeast was detected by transforming a particular His^- *E.coli* mutant and selecting for growth on a medium lacking histidine. However, this method is not generally successful with animal or plant DNA cloned in bacteria, because either the genes are not expressed or there is no corresponding bacterial gene.

The following two procedures described in this section are more generally applicable and are commonly used for eukaryotic genes.

The *colony* or *in situ hybridization assay* allows detection of the presence of any gene for which radioactive mRNA is available. Colonies to be tested are replica-plated from a solid medium onto filter paper. A portion of each colony remains on the medium, which constitutes the reference plate. The paper is treated with NaOH, which simultaneously breaks open the cells and denatures the DNA. The paper is then saturated with ^{32}P-labeled mRNA, complementary to the gene being sought, and DNA-RNA renaturation occurs. After washing to remove unbound $[^{32}P]$mRNA, the positions of the bound radioactive phosphorus, usually detected by autoradiography, locate the desired colonies.

A similar assay is done with phage vectors; in this case, plaques are replica-plated. If the protein product of a gene of interest is synthesized, immunological techniques allow the protein-producing colony to be identified. In one test the colonies are transferred, as in colony hybridization, and the transferred copies are exposed to a

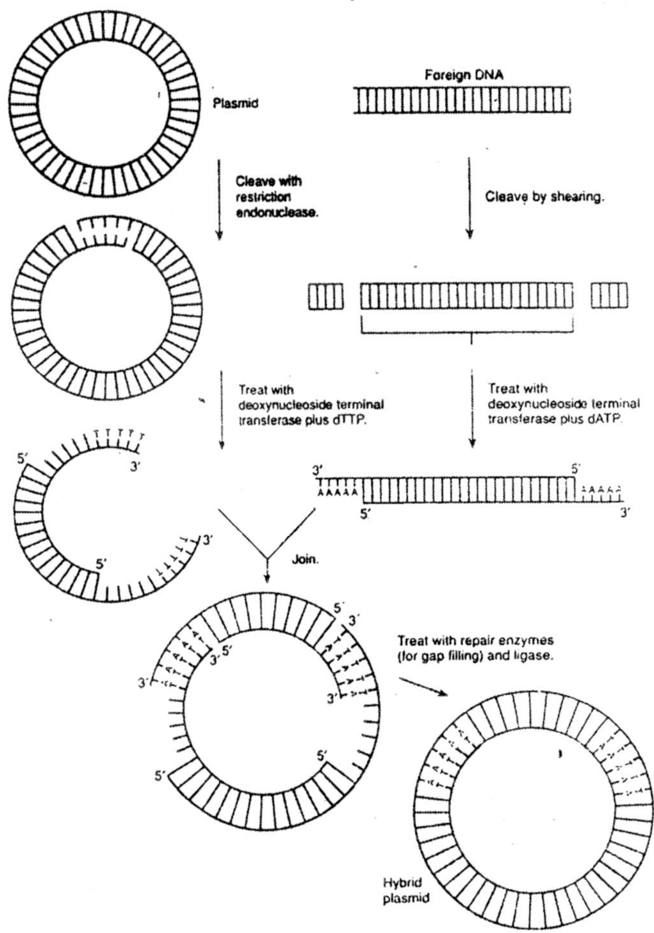

Fig. 8.6. Poly-dA/poly-dT technique for producing hybrid plasmids.

radioactive antibody directed against the particular protein. Colonies to which the radioactivity adheres are those containing the gene of interest. The radioactivity is detected by autoradiography.

DNA Cloning

The 1978 Nobel Prizes in physiology and medicine were awarded to W. Arber, H. Smith and D. Nathans for their pioneering work in the study of *restriction endonucleases.* These are enzymes that bacteria use to destroy foreign DNA, usually of viruses. The enzymes recognize certain nucleotide sequences found on the foreign DNA, usually from four to ten base pairs, then cleave the DNA at all sites containing that specific sequence. Three types of restriction endonucleases are known. Their grouping is based on the types of sequences recognized, the nature of the cut made in the DNA, and the enzyme structure. Type I and III restriction endonucleases are not useful for gene cloning because they cleave DNA at sites other than the recognition sites and thus cause random cleavage patterns. Type II endonucleases, however, recognize specific sites and cleave at just these sites. The sites recognized by type II endonucleases are inverted repeats; they have twofold symmetry. To see the symmetry, you must read outward from a central axis on opposite strands of the DNA. For example, the type II restriction endonuclease *Bam* HI recognizes

5′-GGA | TCC-3′
3′-CCT | AGG-5′

Reading out from the center (vertical line) is AGG on the top strand and AGG on the bottom strand. The sequence is, in a sense, a *palindrome*, a sequence read the same from either direction. The host cell protects its own DNA, not by being free of these sequences, but usually by methylating its own DNA in these regions. The same sequences that are attacked by the endonucleases in the unmethylated condition are protected when methylated.

After host DNA replication, new double helices are hemimethylated; that is, the old strand is methylated but the new one is not. In this configuration the new strand is quickly methylated. Restriction endonucleases are named after the bacterium from which they were isolated. *Bam* HI came from *Bacillus amyloliquefaciens,* strain H. *Eco* RI came from *E.coli*, strain RY13. *Hin* dII came from *Haemophilus influenzae*, strain Rd. *Bgl* I came from *Bacillus globigii.* From here on we will refer to type II restriction endonucleases simply as restriction enzymes.

Restriction enzymes cut the DNA in two different ways. For example, *Hin* dII cuts the recognition sequence down the middle, leaving "blunt" ends on the DNA example, by *Bam* HI leave "sticky" ends that can spontaneously reanneal by hydrogen bonding of the complementary base. The ability to reanneal these sticky ends opened up the field of gene cloning.

Hybrid Vectors

With current technology it is feasible to join together, *in vitro*, DNAs from widely different sources. One of the pieces of DNA involved in the annealing can be a plasmid, a piece of DNA that can replicate in a cell independently of the cellular chromosome. The *hybrid plasmid* can be transferred into a cell. A hybrid plasmid is also known as a *hybrid vehicle*, *hybrid vector*, or *chimeric plasmid*. The latter is named after the *chimera*, a mythological monster with a lion's head, a goat's body, and a serpent's tail.

A variety of procedure exist whereby this chimeric plasmid can be introduced into a host cell. For example, a bacterial cell can be made permeable to this, or any, plasmid by the addition of a dilute solution of calcium chloride. The foreign DNA will be replicated each time the plasmid DNA replicates. In the process of inserting a piece of foreign DNA, the restriction site is duplicated, with one copy at either end of the insert. This property makes it easy to remove the cloned insert at some future time, if needed, since it is inclosed by restriction sites.

Restriction cloning

A few conditions have to be met in order to succeed in cloning DNAs from different sources. The cleavage site of the plasmid must occur in a nonessential region of the plasmid or the plasmid will be rendered ineffective by the process. Also, the *passenger DNA* (foreign DNA) that is being fused into the vehicle plasmid should be a functional unit–that is, it should be at least a complete gene or intact region of interest. Other conditions also prevail. A plasmid vehicle must be cleaved at only one point by the endonuclease. If it is cleaved at more than one point, it will be fragmented during the experiment. Some phage vehicles must be cleaved at two points so that the foreign DNA can be substituted for a length of the phage DNA rather than simply inserted.

Common vehicles, derivatives of phage λ, have been named *Charon phage* after the mythical boatman of the River Styx. During normal phage infection, only DNA within a certain size range is packaged into λ heads. Thus for λ to be a useful vector, part of its DNA must be replaced by the foreign DNA. We note that λ can function quite well as a hybrid vehicle with a 15,000-base pair (15 kilobases or 15-kb) section replaced by foreign DNA because that section of phage DNA is used for integration into the *E.coli* chromosome, a nonessential phage function. Genetic engineers have created a λ DNA molecule with the nonessential region missing and containing and *Eco* RI cleavage site.

Hybrid DNA can thus be incorporated into phage heads because the diminished phage DNA without an insert is too small to be packaged properly. One disadvantage of cloning with normal *E.coli* plasmids is that they are unstable if the foreign DNA is very large, greater than 15 kb. That is, if a large chromosomal segment is cloned, the plasmid will tend to selectively lose parts of the clone as the plasmid replicates. Primarily for this reason, geneticists began using phage λ as a vector because these phages could successfully maintain foreign DNA as large as 24 kb. The phage chromosome is about 50 kb of DNA; within the phage head it is linear, within the cell it is circular. The DNA to fill the phage head is recognized because it has a small segment of single-stranded DNA at either end called a *cos* site (twelve bases).

Reannealing of *cos* site allows λ chromosomes to circularize when they enter a host cell. Cutting the DNA at the *cos* site opens the circle into a linear molecule. (*Cos* is derived from the term "cohesive ends.") Geneticists have taken advantage of these *cos* sites to clone even larger segments of foreign DNA because it turns out that even 24 kb is not adequate to study some eukaryotic genes or gene groups. Many eukaryotic genes are very large because of their introns and transcriptional control segments. DNA up to 50 kb can be cloned if *cos* segments are attached to either end with a plasmid origin of DNA replication. These *cos* site-containing plasmids are called *cosmids.* Not only do these cosmids allow the cloning of very pieces of DNA, they actually for large segments of foreign DNA because small cosmids are not incorporated into phage heads. Thus a range in size of foreign DNA, from 2.5 to 50 kb, can cloned plasmids, Charon phages or cosmids.

Selecting for hybrid vectors

In the methods, restriction enzymes separately cut both vector and foreign DNA. The two are then mixed in the presence of ligase. Many products are created, which can be generally divided into three categories: vectors with appropriate foreign DNA; vectors with inappropriate foreign DNA; and junk, including uncut vectors and vector fragments. In a later section we will discuss methods of finding a particular piece of foreign DNA in a vector. Charon phages are selected simply by their ability to infect *E.coli* cells. As we mentioned above, only λ DNA with a foreign insert will be packaged. Plasmids (including cosmids) that contain foreign DNA can be selected through screening for antibiotic resistance. A widely used cloning plasmid is pBR322. It contains genes for tetracycline and ampicillin resistance and various restriction sites. There is, for example, a *Bam* HI site in the tetracycline-resistance gene.

After cloning, there will be plasmids present with the foreign DNA and those without an insert. *E.coli* cells are then exposed to these plasmids and plated on a medium without antibiotics. Replica plating is then done onto plates with one or both antibiotics. Colonies resistant to both antibiotics have cells with plasmids having no inserts; those resistant only to ampicillin have a plasmid with an insert. Colonies resistant to neither antibiotic have cells with plasmids having no inserts; those resistant only to ampicillin have a plasmid with an insert. Colonies resistant to neither antibiotic have cells with no plasmids.

Poly-dA/Poly-dT and Blunt-end ligation methods of cloning

Restriction endonuclease treatment may not suffice for cloning: An endonuclease cut may fall in the wrong place, say in the middle of a desired gene, or the foreign DNA may have been isolated by other methods, such as physical shearing. In these cases, several other methods of cloning can be used. In the *poly-dA/poly-dT technique*, an addition of poly-dA (a sequence of deoxyadenylate nucleotides) is made to the 3′ ends of the passenger DNA, and poly-dT (deoxythymidylate nucleotides) is added to the 3′ ends of the vehicle, creating complementary 'sticky ends." To accomplish this, the foreign DNA is treated with the enzyme deoxynucleoside terminal transferase and dATP, whereas the cleaved plasmids are treated with deoxynucleoside terminal transferase and dTTP.

A more popular method of joining passenger and vehicle molecules that do not have sticky ends is called *blunt-end ligation*. Without adding poly-dA or poly-dT tails, blunt-ended DNA can be joined by the phage enzyme, T4 DNA ligase. Blunt ends can be generated when segments of DNA to be cloned are created by physically breaking the DNA or by certain restriction endonucleases, such as *Hin*dII, which form blunt ends. Since the ligase is nonspecific about which blunt ends it joins, many different, unwanted product result from its action. Cloning using restriction enzymes that produce sticky ends is the preferred method. A variation of blunt-end ligation uses linkers–short, artificially synthesized pieces of DNA containing a restriction endonuclease recognition site. When these linkers are attached to blunt pieces of

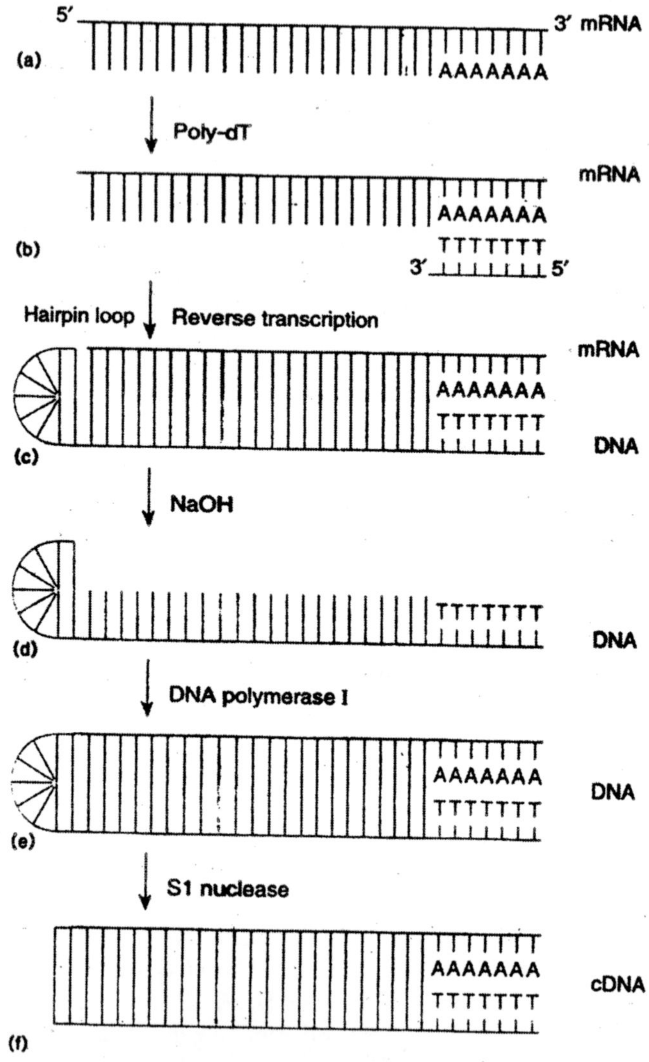

Fig. 8.7. Figure showing the reverse transcriptase.

DNA and then treated with the restriction endonuclease, sticky ends are created. The linkers are 12 bp (base pair) segments of DNA with an *Eco* RI site in the middle. They are attached to the DNA to be cloned with T4 DNA ligase. Subsequent treatment with *Eco* RI will result in DNA with *Eco* RI sticky ends.

DNA for cloning can generally be obtained in two ways: (1) a desired gene or DNA segment can be synthesized or isolated or (2) the genome of an organism can be broken into small pieces and the small pieces can be randomly cloned (originally referred to as a "shotgun" experiment). Then the desired DNA segment must then be "fished" for from among the various clones created. We look first at synthesizing or isolating a desirable gene before cloning it, and then we look at the process of locating a desired gene after it has been cloned.

Creating a Clonable Product

Creating DNA to Clone

In order to clone a particular gene (or DNA segment), and only that gene, a scientist must have a purified double-stranded piece of DNA containing that gene. There are numerous ways of obtaining that DNA, but most entail creating or isolating a single-stranded mRNA that is then enzymatically converted into double-stranded DNA. The problem is then reduced to obtaining the desired mRNA. The mRNA for a particular gene can be obtained several different ways, depending on the particular gene. If there are large quantities of the RNA available from a particular cell, it can be isolated directly. For example, mammalian erythrocytes have abundant quantities of a- and b-globin mRNAs. Also, ribosomal RNA and many transfer RNAs are relatively easy to isolate in quantities adequate for cloning.

Double-stranded DNA for cloning is made from the purified RNA with the aid of the enzyme reverse transcriptase, isolated from RNA tumor viruses. In the first step, a poly-T primer is added, which base-pairs with the poly-A tail of the mRNA. This short double-stranded region is now a primer for polymerase activity: a free 3′-OH exists. The primed RNA is then treated with the enzyme reverse transcriptase, which will polymerize DNA nucleotides using the RNA as a template. The result is a DNA-RNA hybrid molecule. Reverse transcriptase has the interesting property *in vitro* of creating a hairpin turn at the 5′ end of the template. It begins to double back on itself, displacing the RNA and using the newly synthesized DNA as a template. The hairpin loop is useful because it provides a 3′-OH structure, a primer, for continued DNA synthesis after the mRNA is removed with NaOH, a treatment that does not affect DNA. The resulting single stranded DNA is referred to as *complementary DNA* (cDNA). Since a primer structure exists, DNA polymerase I can use the cDNA as a template to create a double stranded DNA.

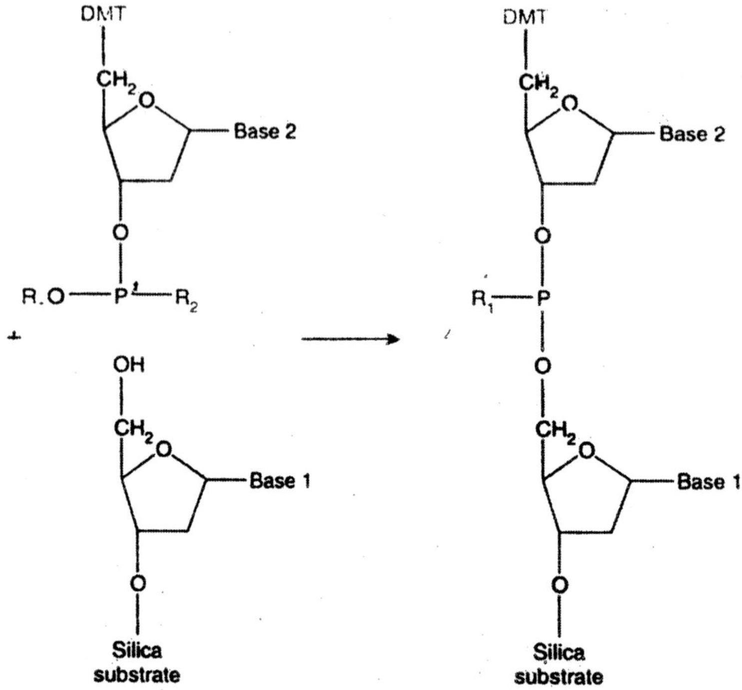

Fig. 8.8. The first cycle in automated, solid-phase DNA synthesis.

The hairpin loop can be removed by the enzyme SI nuclease, which degrades single-stranded DNA. Hence, starting with a piece of single-stranded mRNA, we have generated a piece of double-stranded DNA, which itself is often referred to as cDNA. This double-stranded DNA can then be cloned using the blunt-end methods described earlier. If the RNA is not available in large enough quantities, it is possible to directly synthesize DNA *in vitro* if the amino acid sequence of its expressed protein is known.

The nucleotide sequence can be obtained from the genetic code dictionary if the sequence of amino acids is known from the protein product of the gene. This method will probably not re-create the original DNA because there is redundancy in the genetic code. In other words, when a leucine is encountered in a protein, it could have been coded by any one of six different codons. Despite an element of guesswork, it is possible to synthesize a DNA that will code for a particular protein. Currently, DNA sequences of over one hundred bases can be synthesized by automated machines adding one base at a time in 10-minute cycles.

A nucleotide is anchored to a silica substrate by its 3′ end. In the first cycle, a second nucleotide is added. To avoid multiple (additional) attachments, the second nucleotide has its 5′ end blocked by a chemical group, DMT (dimethoxytrityl). Thus only one nucleotide can be added per cycle. Other groups on the nucleotides, such as the phosphates, are protected during the various chemical reactions with various R-groups (usually CH_3). The cycle is completed by washing out excess nucleotide and removing the DMT group. The dinucleotide is now ready for the next cycle.

After the desired oligonucleotide is synthesized, the R-groups are removed and the oligonucleotide is released from the substrate. You will notice that any method that uses the genetic code dictionary to produce DNA to clone will be missing the promoter of the gene and its transcriptional control sequences as well as other areas of the DNA not translated. Although techniques exist to attach a promoter to a synthetic gene, we will describe techniques to overcome this problem in which random pieces of the genome are cloned.

Direct Formation of Clonable DNA

The previous methods cannot be used when an RNA is not available; when desirable sections of the DNA, such as promoter sequences, are not transcribed or translated; or when the protein product itself has not been sequenced. In these cases another method is used to clone a gene. The DNA of an organism is broken into small pieces from which the desired gene, or DNA fragment, is isolated. The desired DNA can be isolated either before or after cloning.

The DNA is referred to as genomic DNA to differentiate it from cDNA. If the original DNA is isolated before cloning, then only that DNA need be cloned. Alternatively, a "shotgun" approach can be used in which the entire genome of an organism can be cloned (in small pieces, of course), creating a *genomic library*, a complete set of cloned fragments of the original genome of a species. In the case of a genomic library, a desired gene can be located after it is already cloned.

PROBING FOR A SPECIFIC GENE

When DNA segments are generated randomly be endonuclease digests or by physical shearing, a desired gene must be located. We look first at locating a specific gene in a DNA digest, before the DNA has been cloned.

Southern Blotting

To locate a specific gene in the midst of a DNA digest, a specific *probe* must be constructed. Probes are usually radioactively labeled

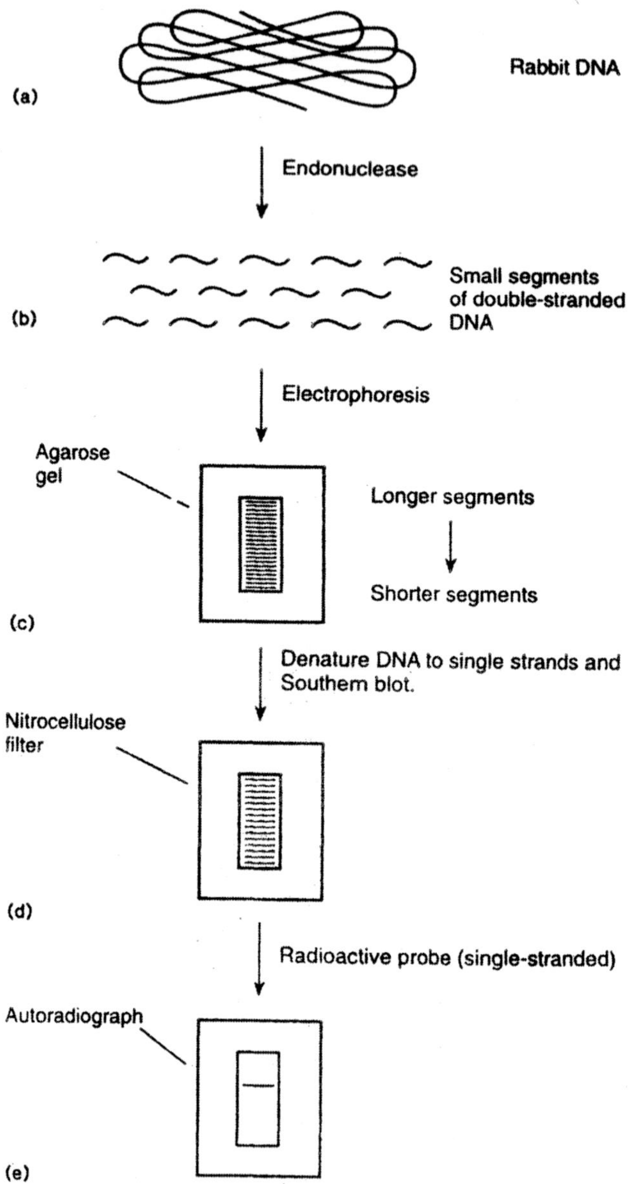

Fig. 8.9. Locating the rabbit β-globin gene within a DNA digest using the Southern blotting technique.

nucleic acid whose complementarity precisely locates a particular DNA sequence. Thus if we wished to locate the gene for β-globin, we could use radioactively labeled β-globin mRNA or radioactively labeled cDNA. RNA-DNA or DNA-DNA hybrids would form between the specific gene and he radioactive probe. Autoradiography would then locate the radioactive probe. Let us assume that we wanted to clone the rabbit β-globin gene. First, we would create a restriction digest or a DNA digest.

We would then subject this digest to electrophoresis on agarose to separate the various fragments according to their sizes. Agarose is a good medium for separating DNA fragments of a wide variety of sizes. In a digest of this kind, however, there are usually so many fragments that the result is simply a smear of oligonucleotides, from very small to very large.

To proceed further, we have to transfer the electrophoresed fragments to another medium for probing or the DNA fragments to another medium for probing or the DNA fragments would diffuse off the agarose gel. Nitrocellulose filters are excellent for hybridization because the DNA fragments bind to the nitrocellulose under high salt concentrations and will not diffuse off. The transfer procedure, first devised by E.M. Souther, is called *Southern blotting.* In this technique the double-stranded DNA on the agarose gel is first denatured to single-stranded DNA, usually with NaOH. Then the agarose gel is placed directly against a piece of nitrocellulose filter, which is then placed agarose-side down on a wet sponge. Dry filter paper is placed against

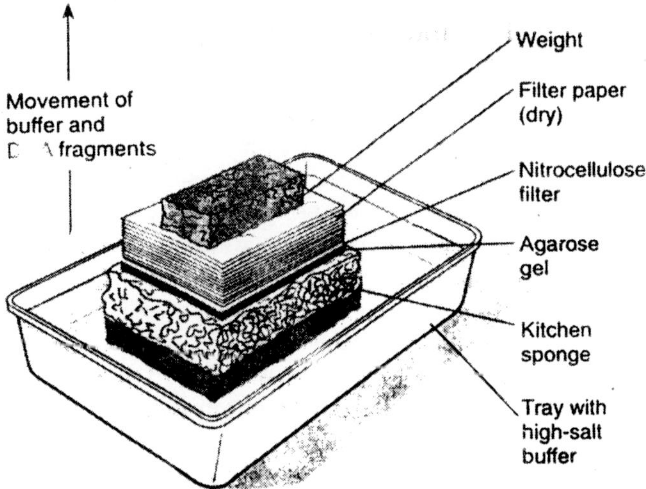

Fig. 8.10. Arrangement of gel and filters in the Southern blotting technique.

the nitrocellulose side. The high-salt buffer in the sponge is then drawn toward the dry filter paper, carrying the DNA segments from the agarose to the nitrocellulose.

The DNA digest fragments are then permanently bound to the nitrocellulose filter by heating DNA-DNA hybridization takes place on the filter (A similar technique can be performed on RNA in which it is called, tongue-in-cheek, *northern blotting*. Immunological techniques, not involving nucleotide complementarity, can be used to probe for protein in an analogous technique called *western blotting*). A radioactive probe can be obtained several different ways. In this example the easiest way to obtain a radioactive probe would be to isolate b-globin mRNA from rabbit reticulocytes and construct cDNA using the reverse-transcriptase method described previously.

The deoxyribonucleotides used during reverse transcription would contain radioactive phosphorus, ^{32}P. A single radioactive band locates a DNA segment with the β-globin gene. It should be noted that the probe, originating from mRNA, will lack the introns present in the gene. However, probing is successful as long as there is a reasonable partial match. To clone the β-globin gene, a second agarose gel would be run with a sample of the digest used. That gel, not subject to DNA-DNA hybridization, would have the β-globin segment in the same place. The band, whose location is known from the autoradiograph, could be cut out of the agarose gel to isolate the DNA, which could then be cloned.

Probing for a Cloned Gene

Dot Blotting

The methods previously described are also useful in locating genes already cloned within plasmids, for example after a genomic library has been constructed. To isolate yeast tRNA genes, for example, a library of the yeast genome was created using the *E.coli* plasmid pBR313. In this case the electrophoresis step is skipped since whole clones will be probed for a particular sequence of DNA rather than DNA segments within a digest. In one particular experiment, yeast cell were grown on a medium that contained ^{32}P in order to produce radioactive tRNA–the probes–that were then isolated Meanwhile 1,140 *E.coli* colonies, each resistant to ampicillin but sensitive to tetracycline–indicating the presence of a hybrid plasmid–were grown and transferred directly to twelve nitrocellulose filters, each holding 95 clones. In preparation for probing, the cells were lysed and their DNA denatured.

The plasmid DNA within the cells of each clone was then hybridized with the radioactive tRNA probes. Dark spots (radioactivity) indicate clones carrying yeast tRNA gene (those clone "light up" autoradiographically). This technique hybridization of cloned DNA without an electrophoretic-separation step, is referred to as *dot blotting.*

Western Blotting

An entirely different method used to locate particular cloned genes utilizes the actual expression of the cloned genes in the plasmid-containing cells. If a eukaryotic gene is cloned in an *E.coli* plasmid downstream from an active promoter, that gene may be expressed (transcribed and translated into protein). Plasmids that allow the expression of their foreign DNA are termed *expression vectors.*

A particular protein product can be located by a method completely analogous to either Southern blotting or dot blotting, called western blotting. In this technique, probing is done with antibodies specific for a particular protein rather than a radioactive oligonucleotide probe. These antibodies are located by a second antibody specific to the first, which has either a radioactive label or a colour marker. For example, assume we are looking for the expression of a particular protein in the clones. The clones would be transferred to a nitrocellulose membrane where they would be lysed (e.g., with chloroform vapour). Then an

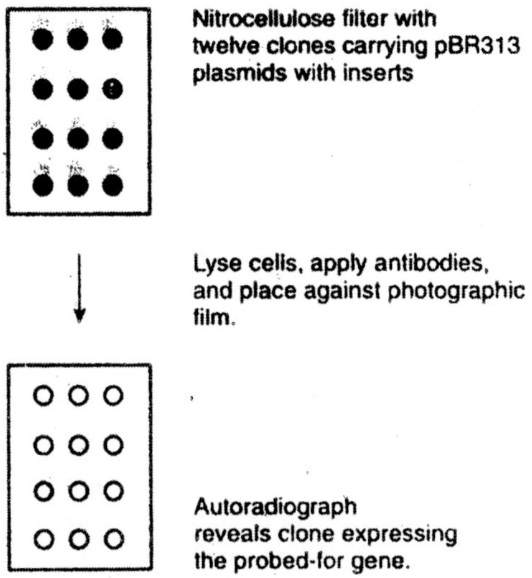

Fig. 8.11. Western blot technique to locate an expressed protein from among many cloned genes using radioactive antibodies.

antibody, specific for the particular protein, would be applied. A second antibody, specific for the first antibody and radioactively labeled, would be applied to the filters. Autoradiography would then locate the presence of the antibodies and thus indicate which of the clones contains the particular gene and is expressing it.

Chromosome Walking

Despite the limited size of any one inserted piece of foreign DNA, it is possible to learn about longer stretches of DNA by using a technique of overlapping clones called *chromosome walking.* For example, let's say that a particular gene (in region A) is located in clone 1, as discovered through probing.

The cloned insert can be removed, using the same restriction enzyme used to insert it in the vector initially, and broken into small pieces that are subcloned and used as probes themselves. The idea is to locate another clone with an inserted region that overlaps the first one. The second clone is now treated the same way–subcloned with segments used to probe for yet another overlap farther down the chromosome. In this way, relatively long segments of a chromosome can be available for study in overlapping clones. One obvious use of chromosome walking is to discover what genes lie next to each other on eukaryotic chromosomes, assuming that a genomic library exists. To date, regions as long as 250 kb have been walked. The technique is very tedious and is halted at certain areas not amenable to walking such as repeated sequences that are found in the DNA of eukaryotes. Once an overlapping probe contain a commonly repeated sequence, it hybridizes to many clones that do not contain adjacent segments. This "cross-referencing" lessens the value of the technique. Currently, newer techniques (termed *chromosome jumping*), designed to bypass regions not amenable to walking are being developed.

These techniques depend on the ability to locate the two ends of a segment without having to walk through the middle. Ends of a segment can be located if the regions has been inverted or if a large region is cloned and the middle part later removed, leaving just the ends. A probe of the ends allows the investigator to locate clone with first one end and then the other, effectively jumping over the intervening region.

Heteroduplex Analysis

Another useful tool in the molecular geneticist's bag of techniques is *heteroduplex analysis* or *heteroduplex mapping.* This procedure allow the direct electron microscopic visualization of cloned products by observing the heteroduplex DNA formed when the clone is hybridized with another piece of nucleic acid, usually its probe or the plasmid without any clone. Heteroduplex mapping was used to study strains of the bacterium *Clostridium tetani*, which causes tetanus by producing a protein neurotoxin.

Based on the known amino acid sequence of the amino end of the toxin, an eighteen-nucleotide DNA probe was synthesized. This probe hybridized with DNA in a plasmid from a toxic strain but did not hybridize with DNA in plasmids from nontoxic stains; the plasmid containing the toxin gene was designated pCL1. In the course of the work, a second plasmid, designated pCL2, was derived from pCL1. The pCL2 hybridized with the probe but did not produce the tetanus toxin. Researchers hypothesized that pCL2 arose as a deletion of part of the tetanus toxin gene in the plasmid. This conjecture was verified using heteroduplex analysis. First, both pCL1 and pCL2 were treated with *Bam* HI endonuclease, which cleave each plasmid only once. Before hybridization studies are done, circular plasmids are usually made linear so they can unwind fully and hybridize. Clearly, pCL2 is a plasmid with a region of about 22 kb missing. The location of the toxin gene is thus localized on the plasmid.

Eukaryotic Vectors

There are three types of eukaryotic vectors as given below:

Yeast Vectors

Yeasts, small eukaryotes that can be manipulated in the lab like prokaryotes, have been extensively studied. Baker's yeast, *Saccharomyces cerevisiae*, has a naturally occurring plasmid about 2 μm in length. In addition, bacterial plasmids have been introduced into yeast. Unfortunately, these plasmids eventually tend to be lost from the cells. This plasmids eventually tend to be lost form the cells. This tendency has been overcome, however, by constructing bacterial plasmids that contain a piece of yeast centromere (CEN) and the origin of yeast DNA replication. The plasmid are then carried from one generation to the next by the yeast.

The plasmids can have telomeric sequences inserted and can then be linearized by cutting with an endonuclease at the telomeric sequences, or the plasmids can be linearized first and then have telomeric sequences added to their ends. These plasmids are then called *yeast artificial chromosomes* (YACs). The particular advantage of the YACs is that they are capable of having very large piece of DNA inserted. Remember that a cosmid can hold about 50 kb; a YAC can hold as much as 800 kb. The ability to clone this much DNA is valuable when working with the large genes found in eukaryotes. Recombinant DNA studies in yeast have increased our knowledge of gene regulation in eukaryotes of the way in which the centromere works, and of the way in which the tips of eukaryotic linear chromosomes are replicated. In addition. YACs have allowed us to analyze and sequence very large segments of eukaryotic DNA.

Animal Vectors

The vehicle most commonly used in higher animals is the DNA tumor virus SV 40 (SV, or simian vacuolating virus, was first isolated in monkeys; however, it can transform normal mouse, rabbit, and hamster cells. Unlike the use of the word transformation in bacteria, transformation in eukaryotes refers to the changing of a normal cell into a rapidly growing, cancerous one). SV 40 is an icosahedral particle with a small (5,224-bp) chromosome, which is a circular, double-stranded DNA molecule. Like λ vectors, SV 40 virions can have part of their DNA replaced by foreign DNA.

The viruses can then be used in recombinant DNA studies in one of two ways. They can replicate and complete their life cycle with the help of nonrecombinant viruses, or they can replicate in the host without making active virus particles by existing as circular plasmids in the cytoplasm or by being integrated into the host's chromosomes. SV40 has become a valuable tool in mammalian genome studies. For example, the rabbit β-globin gene was cloned in SV40, and enhancer sequences were discovered in SV40. Much new understanding of transformation (oncogenesis) has come from using DNA tumor viruses as vehicles.

Plant Vectors

The best studied system of introducing foreign genes into plants is the naturally occurring crown gall tumor system. The soil bacterium *Agrobacterium tumefaciens* causes tumors, known as crown galls, in many dicotyledonous plants. In essence, the crown gall is made of transformed plant cells. These cells have been transformed by a plasmid within the bacterium called the *tumor-inducing*, or Ti, plasmid. Transformation occurs when a piece of the plasmid called T-DNA (for transferred DNA) is integrated into the host (plant cell) chromosome.

Crown gall cells produce odd amino acid derivatives termed opines, which are used by the *A. tumefaciens* cells. By manipulating this system, geneticists have begun to understand the transformation process in plants as well as to develop manipulatable system for introducing foreign genes into plants. The study of genetics in plants has been helped a great deal by the availability of model organisms similar to *E.coli*, yeast, and fruit flies.

Recently, much attention has been turned to the meadow weed, *Arabidopsis thaliana*. This small plant is deal for studying the genetics of plants because the genome is small, approximately 100 million base pairs located in only five chromosomes (2n = 10). This is about the five times the genome of yeast or twenty times the genome of *E.coli*. Thus, in terms of size of genome, it is quite manageable and joins the ranks of organisms whose genomes are stated to be completely sequenced in the near future. The plants are also easy to grow in very large numbers and each plant produces as many as 10,000 seeds. Hence this organism compares very favourably with fruit flies and yeast as an organism to study questions of gene control in a eukaryote, in this case a plant.

Expression of Foreign DNA in Eukaryotic Cells

Foreign DNA can be introduced into eukaryotic cells in methods similar to bacterial transformation, but is called *transfection* because, as we just described, the term transformation in eukaryotes has come to mean cancerous growth. Eukaryotic organisms that take up this foreign DNA are referred to as *transgenic*. Most of the techniques described here transcend taxonomic lines. Animal cells, or plant cells with their walls removed (protoplasts), can take up foreign chromosomes or DNA directly from the environment with a very low efficiency.

The efficiency can be greatly improved by directly injecting the DNA. For example, transgenic mice are now routinely prepared by injecting either oocytes or one or two-celled embryos. They are obtained from female mice after appropriate hormone treatment. After injection of about two picoliters (2×10^{-12} liters) of cloned DNA, the cells are reimplanted into receptive female hosts.

Using modifications of nuclear transplant techniques developed in the 1950s, DNA is injected directly into the mouse oocyte or embryo. In about 15% of these injections, the foreign DNA is incorporated into the embryo. Transgenic animals are useful both to study the expression and control of foreign eukaryotic genes and also in their own right as recipients of successful therapy and manipulation. For example, a mutant mouse lacking the gonadotropin-releasing hormone, and thus sterile, was restored to normalcy by being transfected with an intact gonadotropin-releasing hormone gene that was incorporated into its genome and its germ line. In 1988, a transgenic mouse that is prone to cancer was the first genetically engineered animal to be patented. This mouse provides an excellent model for studying cancer.

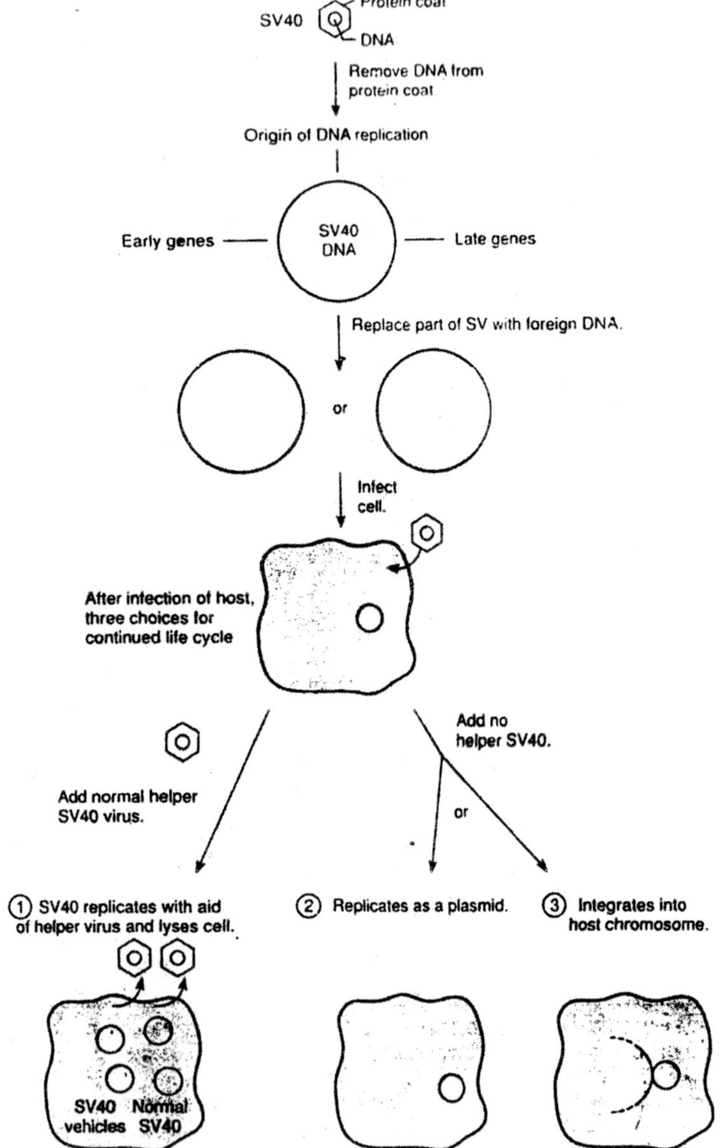

Fig. 8.12. SV40 virus can be used as a gene-cloning vehicle. Although part of the virus is replaced by inserted DNA during cloning, it can still replicate with the aid of normal helper viruses (nonrecombinant SV40).

A controversy has arisen as to whether engineered higher organisms should be patentable. Currently, work is under way to transfect agriculturally important animals (cattle, sheep, pigs) with growth hormone genes, producing "super" organisms. Mice have already been successfully transfected with a rat growth hormone gene, and transgenic sheep have been produced that express the gene for a human clotting factor. Transfection can also be mediated by retroviruses (RNA viruses containing the gene for reverse transcriptase). For example, human white blood cells lacking the enzyme adenosine deaminase were repaired by infection with a retroviral vector.

A retrovirus responsible for a form of leukemia in rodents, the Moloney murine leukemia virus, was engineered such that all genes of the virus were removed and replaced with an antibiotic marker (neomycin resistance) and the human adenosine deaminase gene. The virus binds to the cell surface and is taken into the cell, its RNA is converted to DNA by reverse transcription, and the DNA is incorporated into one of the cell's chromosomes.

It is not possible for this highly modified virus to attack and damage the cells unless a helper virus is added. Unlike the SV40 viruses, the modified Moloney viruses cannot initiate a successful infection without the helpers because vital genes have been removed. Three other recent techniques for the delivery of recombinant DNA to eukaryotic cells are worth mentioning: electroporation, liposome-mediated transfer, and "biolistic" transfer.

In *electroporation*, exogenous DNA is taken up by cells subjected to a brief exposure of high-voltage electricity. Presumably, this electric field creates transient micropores in the cell membrane, allowing exogenous DNA to enter. Liposome-mediated transfection is a technique in which foreign DNA is encapsulated in artificial membrane-bound vesicles, called liposomes. The liposomes are then used to deliver their DNA to target cells.

In one experiment, 50% of mice injected with these DNA-containing liposomes were successfully transfected–they expressed the proteins encoded by the transfecting DNA. Last is the development of techniques to deliver foreign DNA into mitochondria and chloroplasts, which have proven difficult targets for genetic engineering because, among other reasons, they have double-membrane walls that have not proven amenable to delivery of recombinant DNA.

Recently, transfection has been successful in both mitochondria and chloroplasts using a *biolistic* (biological ballistic) process in which recombinant DNA was delivered coated on tungsten microprojectiles that were literally shot into these organelles. When a plant is infected with *A. tumefaciens* containing the Ti plasmid, a crown gall tumor is induced when the Ti plasmid transfects the host plant, transferring the T-DNA region. Those cells transfected with the T-DNA are induced to grow as well as produce opines that the bacteria feed on. Much recent research has been concentrated on engineering Ti plasmids to contain other genes that are also transferred to the host plants during infection, creating transgenic plants. One series of experiments has been especially fascinating. Tobacco plants have been transfected by Ti plasmids containing the luciferase gene from fireflies. The product of this gene catalyzes the ATP-dependent oxidation of luciferin, which emits light. When a transfected plant is watered with luciferin, it glows like a firefly. The value of these experiments is not the production of glowing plants, but rather the use of the glow to "report" the action of specific genes. In further experiments, the promoters and enhancers of certain genes were attached to the luciferase gene. Luciferase thus would only be produced when these promoters were activated, thus the glowing areas of the plant show where the transfected gene is active.

The luciferase gene is a "reporter" system used to monitor the activity of the gene of interest, and thus this system has much promise. *Agrobacterium tumefaciens* does not successfully attack all plants, being notably ineffective in attacking cereals. However, transfection has been done successfully in cereals (including transfecting the luciferase system) by using electroporation. Hence the same experiments can be done in almost any plant, using Ti-plasmid transfection in some and electroporation in others. Soybeans have also been transfected by the use of particle acceleration (biologistics) using DNA-coated gold particles.

Restriction Mapping

The number of cuts that a restriction enzyme makes in a segment of double-stranded DNA depends on the size of that DNA, its sequence, and also the number of base pairs in the recognition sequence of the particular enzyme. That is, one with only three base pairs in the recognition sequence will cut more times than one with six base pairs in the sequence since the probability of a sequence occurring by chance is a function of the length of that sequence. A sequence of three bases occurs more often by chance ($1/4^3 = 1/64$ bp) than a sequence of six bases ($1/4^6 = 1/4,096$ bp). *Hin* dII, for example, cuts the circular DNA of the tumor virus SV40 into eleven pieces; some restriction enzymes can cut *E.coli* DNA into hundreds of pieces.

The product of the action of a restriction enzyme is called a *restriction digest.* Using electrophoresis, we can separate the fragments of a restriction digest by size. With techniques described below, we can locate the restriction sites on the original gene or piece of DNA. That is, we can construct a map of the restriction recognition sites that will give us the physical distance between sites, in base pairs. This *restriction map* is extremely valuable for several reasons. Since it is a representation of the physical structure of the DNA, various sites can be localized to their physical positions on the DNA. For example, when tritiated thymidine, a radioactive nucleotide, was added for a very short period of time during the beginning of DNA replication in SV40 viruses, the radioactivity always appeared in only one restriction fragment. This demonstrated that SV40 replication started from a single, unique point. In addition, a restriction map often allows researchers to make a correlation between the genetic map and the physical map of a chromosome.

Certain large physical changes in the DNA, such as deletions, insertions, or nucleotide changes at restriction sites, can be localized on the genetic map. These changes can be seen as changes in size, or the absence, of certain restriction fragments when compared with wild-type DNA that has not suffered these alterations. This information not only allows us to see changes in the DNA but also has been found useful in looking at the evolutionary distances between species. The differences in fragment sizes are called *restriction fragment length polymorphisms* (*RFLPs*)–usually pronounced "ruflups" or "riflips"–and have proven valuable in pinpointing the exact location of genes and determining identity or relatedness of individuals. A restriction digest is also useful for isolating short segments of DNA that can be sequenced easily.

Constructing a Restriction Map

The restriction enzyme makes three cuts in the DNA, generating fragments that are 200, 50, 400 and 100 base pairs (bp) long. The banding pattern on the gel is the result of the electrophoresing of the digest. The sizes of the segments are determined by comparison with standards of known size. The order of these segments on the chromosome is not obvious from the gel.

Several methods can be used to determine the exact order of the restriction segments on the original piece of DNA. Before restriction enzyme digestion, the 5′ ends of the DNA can be radioactively labeled with ^{32}P using the enzyme polynucleotide kinase. Since the enzyme is acting on double-stranded DNA, both ends will be labeled. Upon electrophoresis after digestion of the DNA, the 200-bp and 100-bp bands will be radioactively labeled, indicating that these segments are the termini of that piece of DNA. The order of the other segments can be determined by slowing down the digestion process to produce a *partial digest.* If the reaction is cooled or allowed to proceed for only a short time, not all the restriction sites will be cut. Some pieces of DNA will not be cut at all, some will be cut once, some twice, and some cut at all three restriction sites. From this gel we can reconstruct the order of segments. This gel contains the four original segments plus six new segments each containing at least one uncut restriction site. From the total digest gel, we know that the 200- and 100-bp segments are on the outside because they were radioactively labeled and the 50- and 400-bp segments are on the inside.

In the partial digest, we find a 250-bp segment but not a 150-bp segment, which tells us that the 50-bp segment lies just inside and next to the 200-bp terminus. There is a 500-bp segment but not a 600-bp segment, which tells us that the 400-bp segment lies adjacent to the 100-bp terminus. An unlabeled 450-bp segment confirms that the 400- and 50-bp segments are adjacent and internal in the DNA. We thus unequivocally reconstruct the original DNA, giving a map of sites of restriction enzyme recognition regions separated by known lengths of DNA.

Double Digests

In practice, restriction mapping is usually done with several different restriction enzymes. Using the same methodology just outlined, we can show that the order of the B segments is 350, 250 and 150 bp arising from two cuts by endonuclease B. What we do not know is how to overlay the two maps. Do the B segments run left to right or right to left with respect to the A segments? We can determine the unequivocal order by digesting a sample of the original DNA with both enzymes simultaneously, thus producing a *double digest.*

The two orders *a* and *b* are used to make different predictions about the double digest. From the first order (a), we predict a 200-bp end segment, radioactively labeled. From the second order (b), we predict that the labeled 200-bp segment will be cut back to 150 bp: There should not be labeled 200-bp segment. The double digest shows a labeled 200-bp segment, verifying order (a).

Restriction mapping thus provides us with a physical map of piece of DNA, showing restriction endonuclease sites separated by known lengths of DNA. This technique gives us short DNA segments of known position that can be sequenced as well as a physical map of the DNA that can be compared with the genetic map and upon which mutations and other particular markers can be located.

Restriction Fragment Length Polymorphisms

DNA Fingerprinting

Restriction fragment length polymorphisms (RFLPs), obtainable from restriction digests, are proving to be very valuable genetic markers in two areas of study: human gene mapping and forensics. In a restriction digest of the whole human genome, there might be thousands of fragments from a single restriction enzyme. Probes have been developed from virtually any small segment of DNA for Southern blotting of these digests. Most probes recognize a single band. Genetic variation usually comes in the form of a second allele that, due to a mutation, lacks a restriction site and is therefore part of a larger piece of DNA. RFLPs with only two alleles do not have the genetic variation needed for the studies described in the following paragraphs. However, some probes have uncovered *hypervariable loci* with many alleles (any one person has, of course, only two of the many possible alleles).

The variations within a population is generated because these hypervariable loci contain many tandem repeats of short (10- to 60-bp) segments. Due presumably to unequal crossing over, much variation is generated by one of these loci, termed *variable-number-of-tandem-repeats* (VNTR) loci. As a result of this variation, probing for one of these VNTR loci in a population will reveal many alleles. In cases in which a number of different loci are recognized by the same probe, each individual will have many bands on a Southern blot, with most people having unique patterns. In one system, developed by A. Jeffreys, a single probe locates fifty or more variable bands per person.

The Southern blots create a genetic "fingerprint" of extreme value of forensics. DNA extracted from blood or semen samples left by the offender can be compared with DNA patterns of suspects. If Jeffreys' probes are used to compare the patterns, the likelihood that the two patterns would randomly match is infinitesimally small. This technique thus has a greater power to positively identify individuals than using their fingerprints. Recently, DNA fingerprinting has been accepted in some cases in courts of law and used to positively identify rapists and murderers.

Although the theory of using this technique is not in doubt, courts have been somewhat reluctant to accept it because of its practical application. Notably, one of the companies doing the tests has been criticized for some of its methods in declaring matches between suspects and samples when the Southern blots were flawed. On an electrophoretic gel, or the Southern blot form that gel, differences in quantity of samples in different lanes or variation in the gels can cause differences in migration or distortion of bands, and thus corrective analyses frequently are done. These latter techniques have caused DNA fingerprinting evidence to be dismissed in some cases. With refinements of techniques and greater sophistication by courts, these techniques should become used routinely in legal situations.

Linkage analysis

RFLPs are now proving useful in locating genes in the human genome through linkage analysis RFLPs are especially valuable when a genetic disease is known but its protein product and gene location are unknown. By locating the gene techniques such as chromosome walking will allow the eventual localization and isolation of the gene. With the gene in hand, its sequence and protein product an be determined, a first step in medical treatment. To locate a gene this way, pedigrees are needed of families with a genetic disease, as well as RFLPs of known location. By linkage analysis of the disease gene and the RFLP variants, the disease gene can be located. Currently the genes for cystic fibrosis, Alzheimer's disease, Marfan syndrome, neurofibromatosis (NF), fragile-X syndrome, and myotonic muscular dystrophy (among other genes) are being localized with these methods. The blot reveals two alleles, labeled "1" (of a 1.3-kb fragment) and "2" (of a 2.3-kb fragment). In the pedigree in every case in which genotypes have been determined, a person with myotonic muscular dystrophy also has the "1" allele. Thus, in this pedigree, the "1" allele is very tightly linked to the LDR152 probe and therefore the gene for myotonic muscular dystrophy mut be very close to this probe on chromosome 19. Localizing genes this way is of great medical value and also a forerunner of the DNA sequencing of the whole genome.

Polymerase Chain Reaction

In many instances, a DNA sample is available, but it is in such small quantity or is so old as to be considered useless for further study. That situation was changed in 1983 when Kary Mullis, a biochemist working for the Cetus Corporation, devised the technique

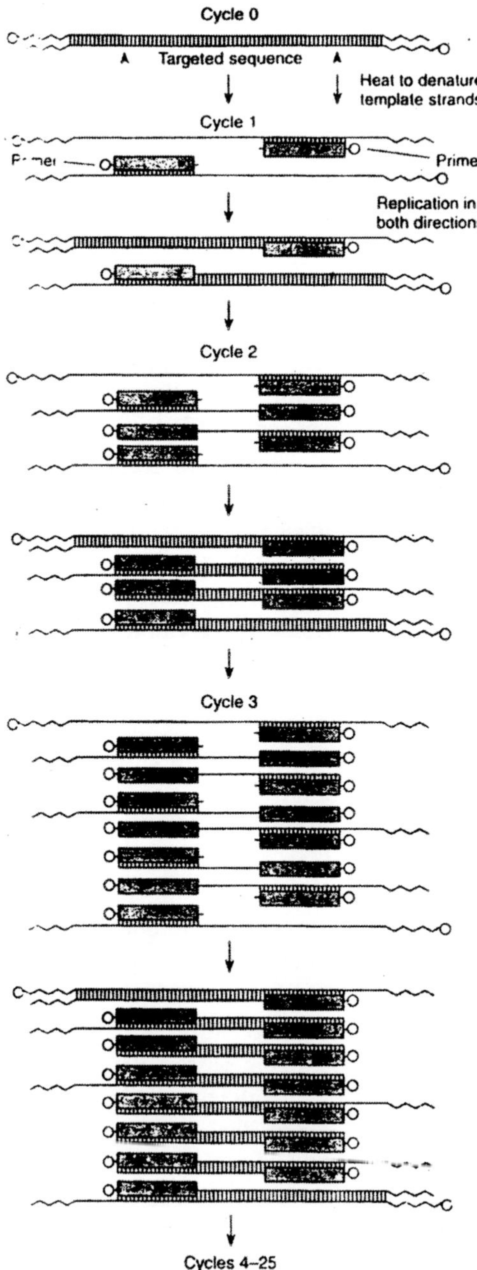

Fig. 8.13. Polymerase chain reaction.

we now refer to as the *polymerase chain reaction* (PCR). It can be used to amplify whatever DNA is present, however small in quantity or poor in quality.

The only requirement is that the sequence of nucleotides on either side of the sequence of interest is known. That information is needed to construct primers on either side of the sequence of interest. Once that is done, the sequence between the primers can be amplified. In the technique, the primers and the ingredients for DNA replication are added to the sample. Then, the mixture is heated. This separates the DNA strands leading to their replication, starting from either side of the sequence of interest. After a few minutes of DNA replication, the process is stopped by heating; the mixture is denatured to separate the strands and a new cycle of replication is initiated.

The various stages in the cycle are controlled by changes in temperature since the temperatures for denaturation, primer annealing, and DNA replication are different. In about twenty cycles, a million copies of the DNA are made; in thirty cycles, a billion copies are made. The technique is aided by using DNA polymerase from a hot-springs bacterium, *Thermus aquaticus*, that can withstand the denaturing temperatures. Thus after each cycle of replication new components do not have to be added to the reaction mixture. Rather, the cycling can be continued without interruption in PCR machines that are simply programmable water baths that can accurately and rapidly change the water temperature that surrounds the reaction mixture. Depending on the availability of primers, any DNA can be amplified in this technique. For example, mitochondrial DNA from the remains of a 40,000-year-old woolly mammoth was amplified using PCR. The mammoth's mitochondrial DNA was very similar in sequence to that of modern elephants. Currently a project is underway to analyze Abraham Lincoln's genetic makeup.

There still exist several dried blood and tissue samples from clothing taken at the time of Lincoln's assassination, as well as skeletal remains. Those samples have enough DNA to determine, for example, whether Lincoln had Marfan's syndrome, a connective-tissue disease whose symptoms include some physical attributes that Lincoln had. Although not necessarily lethal, the disease can lead to death from heart and aortic problems. In a more mundane fashion, PCR is now a routine tool in the laboratories of molecular geneticists who use it to rapidly amplify DNA regions of interest.

We now turn our attention to a major result of recombinant DNA technology, DNA sequencing. Recombinant DNA technology, with its ability to isolate and amplify small, well-defined regions of chromosomes, has allowed the development of these techniques.

DNA SEQUENCING

Paul Berg of Stanford, Walter Gilbert of Harvard, and Fred Sanger of the Medical Research Council in Cambridge, England, shared the 1980 Nobel Prize in chemistry. Berg won for creating the first recombinant DNA molecules when he spliced the SV-40 genome into phage λ. Gilbert and Sanger were awarded the prize for independently developing methods of sequencing DNA. Gilbert, along with Allan Maxam, developed a method of DNA sequencing called the chemical method. It involves chemically breaking down the DNA at specific bases. Sanger, who won a Nobel Prize in 1959 for sequencing the insulin protein, later took part in developing methods for sequencing RNA. His method of sequencing DNA synthesis and was called the plus and minus method. Recent advances developed by Sanger, Coulson, and S. Nicklen, using specific chain-terminating nucleotides, have led to a modification of the plus-and-minus method known as the dideoxy method.

The Dideoxy Method

In the *dideoxy method*, newly synthesized DNA is sequenced. DNA synthesis occurs at a primer configuration, one in which double-stranded DNA ends with a 3′-OH group on one strand. The other strand continues as single-stranded DNA. In the dideoxy method, a primer configuration of the DNA to be sequenced is created and replication proceeds. A trick, using chain-terminating nucleotides, is used to stop DNA synthesis at known positions.

Chain termination is achieved by using nucleotides whose sugars lack OH groups at both the 2′ and 3′ carbons (hence the term 'dideoxy'). Without a 3′-OH group a dideoxynucleotide cannot be used for further DNA polymerization. Because of chain-terminating nucleotides, synthesis can be stopped at a known base. The sample is elongated separately in four different reaction mixtures, each having all four normal nucleotides but also having a proportion of one of the chain-terminating dideoxy nucleotides. For example, if the pool of thymine precursors contains a portion of the dideoxythymidine triphosphate molecules, then synthesis of the growing strand will be terminated some of the time when adenine appears on the template, thus creating

Deoxythymidine triphosphate (dTTP)

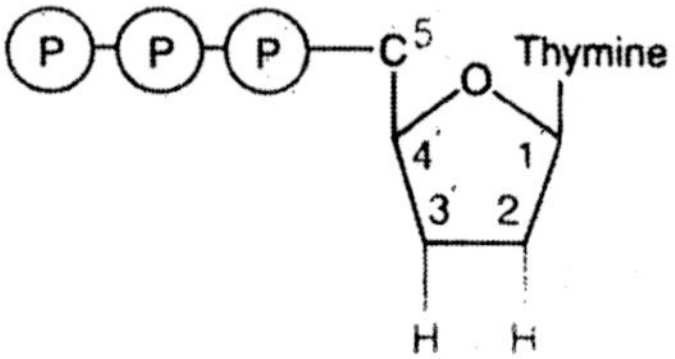

Dideoxythymidine triphosphate (ddTTP)

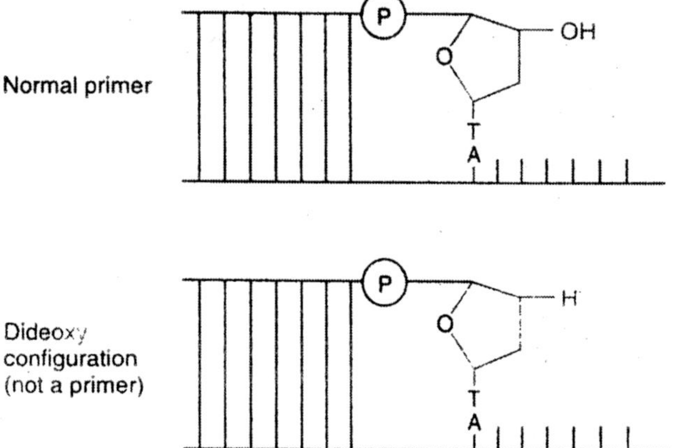

Fig. 8.14. Dideoxy nucleotides cause chain termination during DNA replication.

fragments that end in thymine. Similar reactions are carried out for each of the other nucleotides, producing fragments that terminate when the respective complementary nucleotide is present. The resulting fragments from each reaction are electrophoresed, generating a pattern from which the sequence of the newly synthesized DNA can be read directly off the gel. Let's go through an example.

In order to sequence a small segment of DNA, it is necessary to get one strand of this double-stranded segment into the configuration.

The DNA to be sequenced must be the template for new DNA synthesis. Having created the necessary primer configuration, we take four subsamples of it, each of which includes all four nucleotide triphosphates plus DNA polymerase I. At least one of the nucleoside triphosphates is radioactively labeled, usually with ^{32}P. This label will allow us to identify newly synthesized DNA by autoradiography. To each of the four subsamples one of the dideoxynucleotides is added–one subsample gets ddTTP, one gets ddATP, one gets ddCTP, and one gets ddGTP. These dideoxynucleotides are added in addition to the regular deoxynucleotides so that there will be some probability that chain termination will occur at every appropriate position. If the dideoxynucleotide were added in place of the deoxynucleotide, then the chain would be terminated the first time–and only the first time–that the complement of that base appeared in the template strand.

By mixing the dideoxy- and deoxynucleotides, that termination will occur in every appropriate position. For example, if in reaction subsample 1, 50% of the thymine-containing nucleotides were dTTP and 50% were ddTTP, then there would be a 0.5 probability of chain termination every time a thymine was required during DNA synthesis. If the ratio were shifted to more dTTP, then the average synthesized segment would be larger. If the ratio were shifted toward ddTTP, then the average synthesized segment would be shorter. Note that the template has two adenines. Therefore, in the ddTTP reaction mixture, adenine's complement (thymine) is needed twice.

There are two possible points for ddTTP to be incorporated, two possible chain terminations, and therefore two fragments possible ending in dideoxythymidine, of two and seven bases, respectively. Similarly, there are three possible fragments ending in adenine, of one, three, and eight bases; three ending in cytosine, of four, five, and nine bases; and one ending in guanine of six bases. After DNA synthesis is completed, the old primer is removed by method mentioned below, leaving only newly synthesized DNA fragments.

Newly replicated segments of varying length from each reaction mixture are placed in separate slots and then electrophoresed on polyacrylamide gels to determine the lengths of the segments present. Since only newly synthesized DNA segments are radioactive, radioautography lets us keep track of newly synthesized DNA. As you can see from the autoradiograph of the gel, each subsample will produce segments that begin at the primer configuration (beginning of synthesis) and end with the chain-terminating dideoxy base. By starting at the bottom and reading up back and forth across the gel, the exact sequence of the DNA segment can be determined directly. Because they have the appearance of stepladders in each lane, the gels are usually referred to as *stepladder gels* or *ladder gels.*

This technique was first used to sequence the genome of the DNA phage φX174. That phage was used because it lent itself to the sequencing method. It has single-stranded DNA within the phage coat, yet its DNA become double-stranded once it enters the bacterium. Creating a primer configuration was thus relatively easy.

The double-stranded circle from within the host could be treated with a restriction endonuclease to produce double-stranded fragments. These fragments could then be denatured. From this mixture, a particular fragment could be isolated by electrophoresis. The isolated strand would reanneal to the single-stranded DNA taken from phage heads, forming a primer for new growth. The same restriction endonuclease would free the new growth after it had taken place. Thus the dideoxy method previously described was relatively easy to apply to the 5,387-base chromosome of φX174.

Creating a General-Purpose Primer

In order for the dideoxy method to be generally applicable, a routine method to create a primer was needed. A general primer was created for routine sequencing work by recombinant DNA engineering of an *E.coli* vector, the single-stranded DNA phage M13. This phage is similar to φX174 in that both are packaged as single-stranded DNA and both are replicated to double helices within the host. Therefore, the double-stranded form within the host, called the replicating form (RF), can be engineered by stranded methods and the single-stranded form can be used for sequencing. The system works as follows.

By very clever engineering, J. Messing and his colleagues created cloning sites for a variety of restriction enzymes in a bacterial gene (*lacZ*) that had been inserted into M13. The gene is for the β-galactosidase enzyme that normally breaks down lactose. It also breaks down an artificial substrate of the enzyme. X-gel, which is normally colourless. When cleaved by β-galactosidase, X-gel becomes blue.

In the presence of the functional *lacZ* gene, M13 plaques are blue. If the gene is disrupted by a cloned insert, X-gel is not broken down, and hence plaques are colourless. (M13 doesn't form true plaques because it doesn't lyse the *E.coli* cells. It does form turbid sites due to reduced growth of the bacteria.) After cloning, phages released from the host contain single-stranded DNA. An oligonucleotide primer has been synthesized that is complementary to a region of the phage DNA upstream from the cloning sites.

Single-stranded phage DNA containing a cloned insert in isolated and hybridized with the synthetic oligonucleotide. This operation creates the primer configuration for dideoxy sequencing of the cloned DNA. Virtually any clonable segment of DNA can be sequenced using this very general method. Stepladder gels, however, are effective only up to about 400 bp. To sequence regions larger than that requires sequencing overlapping segments and reconstituting the sequence by the overlap pattern, similar to the methods we described for amino acid sequencing. Overlapping segments of DNA to sequence are obtained either by using two or more restriction enzymes or by sequencing segments created by shearing the DNA into small, blunt-ending pieces and then using linkers or blunt-end ligation to clone them. The most recent innovation in DNA sequencing involves using fluorescent dyes. Four dyes are used, each fluorescing at a different wavelength (505, 512, 519, and 526 nm).

The methodology is essentially the same as described previously except that each of the four dideoxy nucleotides has a different one of the four dyes attached. After the newly synthesized fragments are isolated, the products from all four reactions are run together in the same lane of a polyacrylamide gel. The gel is then scanned with an argon laser that excites the dye molecules. An instrument records the colour of peaks, reading the sequence directly and automatically. This method greatly simplifies sequencing since it is automated. It also alleviates the necessity for radioactive tags. Even when atuomated DNA sequencers are not used, radioactive tags can now be replaced with chemiluminescent tags of the dideoxy nucleotides, again alleviating the need for radioactive isotopes.

In addition, it is possible to combine PCR and dideoxy sequencing to get simultaneous amplification and sequencing of a particular DNA sequence. Innovative techniques make it possible to sequence DNA more quickly, cheaply, and efficiently. Either the dideoxy method, or the other method of sequencing, the chemical method, which we have not discussed, allows us to read the sequence of hundreds of nucleotides on a single gel. Whole viral genomes, prokaryotic and eukaryotic genes, and numerous regions of interest in prokaryotes, eukaryotes, and viruses have been sequenced. Sequence information on about one million base pairs per year is accumulating.

Practical Benefits from Gene Cloning

Throughout the chapter we have mentioned applications of genetic engineering. Here we summarize some of the accomplishments and future directions in the medical, agricultural, and industrial arenas.

Medicine

In biomedicine, genetic engineering has had remarkable successes. On the one hand, basic knowledge about how genes work (and do not work) has been advanced tremendously. On the other hand, recombinant DNA methodology has made available large quantities of substances previously in short supply. The latter include insulin, interferon (an antiviral agent), growth hormone, growth factors, blood-clotting factors, and vaccines for diseases such as hepatitis B, herpes, and rabies. Advances in AIDS and cancer research will be discussed in later. Genetic engineering is marking it possible to manufacture antibodies to order and to diagnose and treat disease. The sequencing of the human genome will further aid medicine by identifying the genes for various diseases, a first step in discovering cures.

DNA Libraries

DNA libraries are collecting of cloned DNAs. There are two types of libraries–(1) Genomic library and (2) cDNA library.

Genomic Library

Genomic library represents an entire genome of individual animal, plant, bacteria, virus under study. Gene libraries are constructed by isolating the complete genomic DNA from cell. Isolated genomic DNA is then cut into fragments of average desired length. This may be done either randomly by shearing or with suitable restriction enzymes. The fragments thus obtained are then cloned into suitable vectors.

A set of cloned fragments of this short is called genomic library. These chimeric cloning vectors are then perpetuated indefinitely and retrieved when wanted Screening, identification and characterization of fragments (functional genes) will be possible from this set of clones (i.e. Genomic library) with suitable probes. If enough clones are produced there will be a very high chance that any particular gene will be present in at least one of the clones. The perfect genomic DNA library would contain DNA sequence representatives of an entire genome in a stable form, as manageable number of overlapping clones.

Library should be easy to construct and easy to screen. Fragments of about 10 kb in length are needed for prokaryotic libraries while fragments of 40 kb in length are required for mammalian libraries so as to keep the number of clones in a library manageable. For 99% probability of getting each one fragment per clone, 1500 cloned fragments are needed with *E.coli* 4,600 with yeast and 800,000 for mammals. For most organisms prior information about genome is required for library construction.

Table 8.1. Genomic library size for various organisms

Organism	*Genome size*	*Number of clones with*	
		20 kb inserts	45 kb inserts
1. *Escherichia coli*	4×10^3	6×10^2	2.7×10^2
2. *Saccharomyces cerevisiae*	1.4×10^4	2.1×10^3	9.3×10^2
3. *Arabidopsis thaliana* (Simple plant)	7.0×10^4	1.1×10^4	4.7×10^3
4. *Drosophila melanogaster* (fruitfly)	1.7×10^5	2.5×10^4	1.1×10^4
5. *Homo sapiens* (Human)	3.0×10^6	4.5×10^5	2.0×10^5

$$N = \ln(1-P)/\ln(1-a/b)$$

Where, N = Number of clones required

P = desired probability

a = average size of DNA fragment to be cloned

b = size of the genome

Construction of genomic library

Thus, construction of genomic library involves following steps–

1. Isolation of chromosomal DNA of interest.
2. Cutting fragments of suitable size.
3. Cloning the fragment in suitable vector.
4. Screening, identification, characterization of clones.
5. Maintenance of set of clones (i.e. genomic library).

Following are the steps for isolation of high molecular weight eukaryotic DNA from cells grown in tissue culture:

(a) Wash the cells in monolayer with Tris buffer saline of pH 7.4.

(b) Centrifuge and collect the pellet.

(c) TE + EDTA (0.5 M) + Proteinase K (100 μg/m) + Sarcosyl (0.5%) is added to the pellet.
(d) Lysed cells are kept at 50°C for 3 hours.
(e) Phenol extraction.
(f) Collected aqueous extract. Dialyze it against 50 mM Tris-Cl.
(g) Treat the sample with RNase (free of DNase)
(h) Extraction with phenol-chloroform.
(i) Dialyze with TE.

Fragments of genomic DNA may be generated by mechanical shear or by use of restriction enzymes.

Fragmentation by shearing–important features are:

(a) Technical difficulties.
(b) No exclusion caused.
(c) Chromosome walking is possible.
(d) Random shearing may result in infinite fragments which can be used as hybridization probes for chromosome walking.
(e) Fragments are relatively large, therefore less clones are to be screened.
(f) Knowledge of distribution of sites for restriction enzymes around sequence of interest is not required.

Fragmentation by use of restriction enzymes–important features are:

(a) It becomes easier to insert restriction fragment into the vector.
(b) Drawback is, sequence of interest may have restriction site which will cause more fragmentation of that sequence.
(c) If sequence is present in fragment larger than what vector can accommodate then it may not be cloned at all.
(d) Since normally fragments produced will be ~4 kb in size, large number of recombinant vectors will have to be screened.
(e) Knowledge of distribution of sites for restriction enzymes around sequence of interest in required.

Cloning in suitable vector

Earlier strategy used for cloning was to obtain highly enriched sequence of interest before actual cloning. For this total genome is cut with restriction enzyme. Fragments obtained are then electrophoresed and subjected to column chromatography. Fractions of interest are identified by hybridization with probes and then cloned. This results in 100-300 fold enrichment and reduces further screening work on clones. Subsequently when efficient *in vitro* packaging systems and *in situ* hybridization was developed, initial enrichment is not considered necessary. Here fragment obtained are directly cloned and then identification (screening) done by hybridization with probes.

	Lambda vectors		*Cosmid vectors*
1.	Insert size is 20-25 kb.	1.	Insert size is 45-50 kb.
2.	More clones required for complete library.	2.	Less clones required for complete library.
3.	λ libraries are easier to make.	3.	Cosmid libraries are difficult to make.
4.	λ libraries are simpler for screening and more efficient too. Plaque hybridization technique is used	4.	Cosmid libraries are more difficult for screening and less efficient. Colony hybridization technique is used.
5.	Yield or recombinants per mg of starting DNA is higher with λ vectors.	5.	Yield of recombinants per µg of starting DNA is lower with cosmid vector.
6.	With λ vectors more steps are required to walk along the genome.	6.	With cosmid vectors less steps are required to walk along the genome.
7.	Long eukaryotic gene may not be included in single DNA fragment that is cloned.	7.	Long eukaryotic genes can be included in a single DNA fragment that is cloned.
8.		8.	If genes of unknown size are to be isolated from small number of libraries, with some chromosome walking then cosmid cloning is a choice.
9.	λ phage stock consisting library of recombinant clones in total contains most of the sequence of DNA. It is concentrated by CsCl density gradient. Maintenance of phage stocks is tedious.	9.	Cosmid libraries are maintained in form of transformed bacteria. If each bacterium in population carries the recombinant cosmid, 3,50,000 transformants are to be generated.

Lambda and cosmid vectors are more commonly used for construction of genomic library. Materials like vector arms, packaging extract etc. and efficient protocols are now commercially available in 'ready for use' state. Lambda and cosmid vectors have their own advantages and disadvantages.

Screening of clones

There are three methods which can be used for screening of clones from genomic library. There are: (a) Hybridization with probe followed by detection of probe label; (b) Immunological screening of protein product; (c) Screening of protein activity.

(a) Using colony hybridization technique with radioactive or non-radioactively labelled probes, colonies of transformed cells can be screened for particular DNA.

(b) Procedure of immunological screening is parallel to colony hybridization. Here first primary antibody is added on plate of colonies and then enzyme-labelled anti-antibody is added. Enzymes bound are then allowed to cause conversion of colourless substrate to coloured products. This colours colony indicates presence of particular DNA giving that protein products.

(c) If the target gene produces an enzyme that is not normally made by host cell, a plate assay can be carried out to detect the clone (of transformed cells) processing that particular enzyme activity. If the gene that is being searched is coding a product essential for growth of mutated host cell then clones can be screened by transferring them on minimal media.

The cDNA Library

cDNA library represents the complete mRNA complement of single type of cell. mRNA from tissues which are actively synthesizing proteins is used. cDNA libraries are useful for study of tissue specific gene expression.

A certain number of cDNA clones is required. So that a library becomes a faithful representation of total sequence complexity present in the initial mRNA pool. Number of independent clones for different proteins and mRNA will be different. It is 1.7×10^5 clones from standard fibroblast mRNA while it is 1/5 to 1/10 of this for low level proteins. Thus, cDNA library is collection of cDNAs, prepared by starting from mRNA of particular cell and these cDNAs are cloned and maintained either in phage or in plasmid (in *E.coli*).

Many genetic engineering applications require expression of eukaryotic gene in bacteria. If eukaryotic gene product is to be obtained in bacteria, DNA isolated directly from genome is not suitable but DNA obtained by using mRNA as the source has to be used. Original copy of DNA has intron and it is used for cloning in bacteria, primary mRNA will be produced. Since bacteria do not have machinery to cut away introns, functional mRNA coding for protein will not be obtained. If product synthesis is expected in bacteria then cDNA representing functional protein-coding mRNA should be used.

In eukaryotic cells post transcriptional modification occurs to form functional protein-coding mRNA from primary mRNA, but that is not the case if cloned in bacteria. Hence, single stranded cDNA is first obtained using mRNA as starting material. Then double stranded cDNA can be prepared, cloned mapped and can be used as probe for identifying corresponding RNA or genome DNA sequences.

Construction of cDNA library

1. Synthesis of cDNA eukaryotic functional mRNA.
2. Cloning cDNA population in suitable vectors.
3. Screening of cDNA library.
4. Maintenance of clones (i.e., cDNA library).

Synthesis of cDNA from eukaryotic functional mRNA. mRNA from tissues which are actively synthesizing proteins is used. Roots and leaves of plants, ovaries or reticulocytes in mammals may be used.

(i) Extracted cellular eukaryotic RNA is passed through column packed with cellulose beads to which short chains of thymidine residues (about 15) are bound.

(ii) The tRNA and rRNA pass through column whereas mRNA with poly A tails bind to beads with oligo dT.

(iii) mRNA is then eluted from columns by treatment with buffer that breaks. A:T hydrogen bonds.

(iv) mRNA molecule is then converted to double stranded DNA using following procedure:

(a) Purified mRNA plus short unbound sequences of oligo dT are taken along with the enzyme reverse transcriptase and four deoxyribonucleotide (dATP, dTTP, dCTP, dGTP) in the reaction mixture.

(b) Oligo dT provide priming as they combine with poly A tails of mRNA.

(c) Reverse transcriptase causes formation of complementary DNA strand with hairpin loop at end. mRNA is used as the template and deoxyribonucleotide are incorporated.

(d) Second DNA strand is synthesized by addition of klenow fragment of *E.coli* DNA polymerase which uses first DNA strand as template.

(e) At the end of second strand synthesis, treatment with T_4 DNA polymerase ensure that linkers can be blunt-end ligated to the cDNA. These linkers make cDNA ends compatible with ends present on cloning vector.

(f) After the reaction is complete, the sample is treated with the enzyme RNAse which degrades mRNA molecules. S nuclease is used which opens hairpin loops and degrades ss DNA extensions.

Cloning of cDNA population in suitable vectors. cDNA population obtained can then be cloned by blunt end ligation of other joining mechanisms into a plasmid cloning vector to form cDNA library. Alternately cDNA may be cloned into phage. Packaging of cDNA bearing phage or transformation of *E.coli* with plasmids bearing different cDNA is the first step towards actual cDNA library.

Screening of cDNA library. Procedures here will be similar to that used for genomic library. cDNA library can be screened by either DNA hybridization or immunological assays to identify the clone that carries a specific plasmid cDNA construct.

9

IMMUNOGENETICS

For centuries it has been recognized that individuals who have recovered from an infections disease, such as smallpox or influenza, are less likely to succumb to the same disease a second time. As a result of prior exposure, these individuals have acquired the ability to resist the infectious agent; they are said to be *immune* to the agent. The process of becoming immune is called the *immune response*. The recognition that immunity can be acquired has led to the development of vaccines, harmless in themselves, that produce immunity to many diseases. Immunity results from the ability of the body to recognize viruses, bacteria, or other foreign substances that may invade the body and to attack and destroy them.

Most large molecules, and almost all viruses and cells, have the ability to elicit an immune response. Such immunity-evoking substances are called *antigens*. When blood transfusions first began to be administered around 1900, it was quickly recognized that the blood of the donor and the recipient had to match in some way in order for the transfusion to be successful. In many cases transfusions could be carried out successfully with no complications, and when this occurred, the donor and the recipient were said to be *compatible*. In other cases, upon transfusion, recipients went into severe shock and many died; in these cases the donor and the recipient were obviously *incompatible*. However, blood from the same donor might be compatible with some recipients and at the same time incompatible with other recipients.

Additional studies revealed that the basis of incompatibility was the presence of certain genetically determined antigens present on the surface of the red blood cells of the donor, and laboratory tests for compatibility were soon developed. These studies led to the first rule of immunogenetics:

> A recipient individual will produce an immune response against any antigen that the individual himself does not possess.

That is, if the antigens present on the red blood cells of the donor are also present on the red blood cells of the recipient, then the donor and the recipient will be compatible for transfusion. Shortly after blood transfusion were first attempted in humans, skin transplants began to be carried out in highly inbred, and therefore virtually homozygous, strains of mice. Usually, transplants between individuals of the same inbred strain were accepted. The small transplanted piece of skin would grow on the recipient mouse just as well as if the transplanted skin had been taken from another place on the body of the recipient. However, with individuals of different inbred strains the skin around the transplant become inflamed and the transplanted skin would be not heal; this phenomenon was called *rejection.* Such rejection reactions were also shown to result from the presence of genetically determined antigens present on the transplanted skin–but these were antigens of a different type than those implicated in blood transfusion.

When two different inbred strains were crossed to produce an F_1, and the F_1 was used as donor or recipient of skin transplants, the following results were obtained: (1) Transplants using either parental strain as the donor and the F_1 as the recipient were almost always accepted. (2) Transplants using the F_1 as the donor and either parental strain as the recipient were almost always rejected. Such results led to a second rule of immunogenetics.

> Antigens are genetically determined by dominant or codominant alleles, so the cells of an individual will possess all antigens corresponding to the alleles the individual has inherited from its parents.

The two rules of immunogenetics account for the observations about transplantation. Because the F_1 must inherit one allele from each homozygous parent, the F_1 will possess the antigens of *both* parental strains. Consequently, transplants from an inbred parent to the F_1 will be accepted, because the parental strain will possess no antigens not also possessed by the F_1. However, transplants from the F_1 to either parental strain will be rejected, because the F_1 will possess antigens inherited from the other parental strain, which are not present in the parental recipient of the transplant.

Transfusion incompatibility and skin-graft rejection exemplify a division of functions of the immune system arising from different characteristics of two classes of white blood cells–B cells and T cells *B cells* secrete proteins called *antibodies* capable of combining with antigens. Stimulation of a B cell by a suitable antigen causes secretion of an antibody, which circulates in the blood and lymph and combines with the antigen, clumping the antigen molecules together and marking them for destruction by other classes of white blood cells.

Unlike B cells, which attack foreign antigens indirectly by producing circulating antibodies. *T cells* attack foreign antigens directly, releasing substances that cause a local inflammation, and attract other types of white blood cells to aid in the destruction of the invasive antigens. Individuals with T-cell deficiency will accept skin transplants and are prone to viral infections. T-cell deficiency occurs in several rare genetic disorders.

In the remainder of this chapter three items will be considered: (1) The genetic determination of the red-blood-cell antigens that are important in transfusions, (2) the genetic determination of antigens implicated in tissue compatibility, and (3) the genetic basis of B-cell and T-cell responses.

Components of the Immune System

Three different types of white blood cells play central roles in the immune response in vertebrates. These cells are (1) *B lymphocytes* (called B cells because they are produced in bone marrow), (2) *T lymphocytes* (called T cell because they are produced in the thymus gland), and (3) *macrophages*. Antibodies are synthesized by *B lymphocytes* and are either secreted or remain membrane-bond on the surface of the B cell depending on the conditions. During the *bumoral immune response*, these antibodies bind to free antigens in the circulatory system and agglutinate them. The resulting *antibody-antigen complexes* are then ingested and degraded by *macrophages*. *T lymphocytes* mediate the cellular immune response. The T cell synthesize antigen receptors that recognize antigens on cell surfaces and trigger the lysis of the antigen-containing cells by the activated T cells. Different T lymphocytes perform this function in slightly different ways. However, in general, the attack of the T cell on the antigen-carrying cell requires both the *specific T. cell receptor* and one or more *histocompatibility antigen receptors*.

Antibodies

Antibodies are immunoglobulin proteins that are present in serum and tissue fluids of all mammals. Antibodies are produced when host lymphocyte cells come into contact with a foreign substance called an *antigen.*

A specific antibody is produced to neutralize the antigen that induced its creation. In 1890, von Behring and Kitasato first reported that antibodies from immunized animals could neutralize diphtheria and tetanus toxins. Each toxin was neutralized only by its own specific antibody or antitoxin. It was first believed that all human body cells were capable of producing antibodies, but in 1957, Burnet and Talmage independently presented the *clonal selection theory of acquired immunity,* which said that all antibody molecules synthesized by given lymphocyte are identical. In other words, it would take a separate lymphocyte to generate each type of antibody. Once a noncommitted lymphocyte was stimulated to produce a given antibody, called the *primary immune response,* all cells produced from that dividing lymphocyte would also produce that same specific antibody–a clonal population of cells for the production of a single specific antibody.

Over time, only a few of these clonal cells persist in an individual's system, although these do retain a "memory" of the first encounter with the antigen. If the antigen is again encountered, a higher level of the antibody would be produced in a shorter period of time. This is because the previously programmed cells can immediately produce that antibody. This accelerated response is called the *secondary immune response.* How, though, can any random lymphocyte commit itself to produce a single specific antibody against an antigen that it has never met? How can any random lymphocyte react to any one of an almost infinite number of antigens and produce only that antibody?

Lymphocytes

The human immune system contains about one trillion (1×10^{12}) lymphocytes. Lymphocytes travel among most other cells of the body. They circulate in blood and lymph tissue and occur in large numbers in human thymus, bone marrow, lymph nodes and appendix. By 1968, the lymphocytes had been divided into two classes: the *B lymphocytes* (*B cells*), derived from and mature in bone marrow, and the *T lymphocytes* (*T cells*), that mature in the thymus. The B lymphocytes produce the different antibodies, but require the help of certain T lymphocytes. Learning that only B cells produce antibodies was very important, but because different antibodies are secreted into and mixed with all others in the serum, these cells could not be readily separated for further study. A system was needed in which the antibodies from a single, or monoclonal, lymphocyte could be isolated for study.

Monoclonal Antibodies

A *monoclonal antibody* is a single specific antibody produced form a clone of B lymphocytes. The clone of cells is derived from an immunized lymphocyte fused to a cancer cell to produce a somatic cell hybrid called a *hybridoma.* The immunized lymphocyte (a cell previously subjected to a given antigen and producing a given antibody to that antigen) offers the hybridoma cell its ability to make a specific antibody.

The cancer cell offers the hybridoma perpetual growth. The fabrication of the mouse hybridoma cell offered research scientists a rapidly proliferating cell line that produce a pure or single type antibody in "unlimited" quantity. This exciting breakthrough in immunology was accomplished in 1975 by Kohler and Milstein, who shared the 1984 Nobel Prize for their work. The original hybridomas were constructed using immunized mouse lymphocytes fused to mouse myeloma cells. Today, the same procedure is used to make pure human antibody. It involves the use of human lymphocytes from a cancer patient and a human myeloma cell.

Synthesis of monoclonal antibodies

In the production of either human or mouse monoclonal antibodies, the researcher begins with hypoxanthine guanine phosphoribosyl transferase mutant myeloma cells ($HPRT^-$). To obtain immunized mouse lymphocytes for fusing with the mouse myeloma cells, the researcher injects a mouse with a given antigen to which the mouse produces a specific antibody. The mouse spleen is removed, and the lymphocytes are extracted, mixed with mouse myeloma cells, and briefly incubated in polyethylene glycol to promote somatic cell fusion. The cell mixture is then placed in HAT growth medium.

Unfused myeloma cells and myeloma-myeloma hybrids, which lack the enzyme HPRT, cannot survive in the HAT medium. Unfused spleen lymphocyte cells and spleen-spleen lymphocyte hybrids die naturally after a few replications. The remaining spleen cell-myeloma hybrids are assayed for antibody production against the antigen. Positive hybridomas are cloned. The hybridomas can then be frozen and stored. To amplify antibody production, mice may be injected with hybridomas. Such injections produce ascites tumors.

The tumors generate large amounts of monoclonal antibody. For humans, the lymphocytes are taken from a lymph node of a person with cancer and mixed with myeloma cells. Hopefully, the lymph node cells will produce an antibody against the patient's cancer; if not,

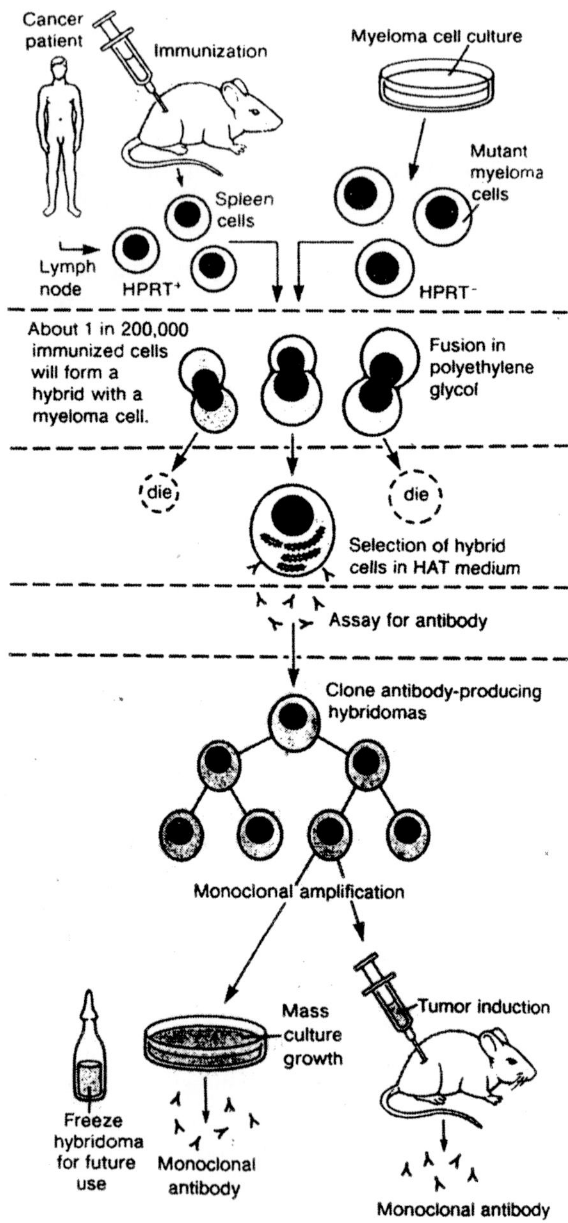

Fig. 9.1. Production of monoclonal antibodies.

the procedure will not work. The human mixed-cell culture is treated as for mice. Amplification of the monoclonal antibody for humans is done in tissue culture or in immunosuppressed animals (animals having a drug-suppressed immune system). In mid-1986, the Food and Drug Administration approved the first monoclonal antibody for use in preventing rejection in human kidney transplants. Production of a single type antibody has allowed scientists to research its structure and diversity, thereby answering the question of how a B cell produces a single specific antibody to one of millions of possible antigenic challenges.

Hypothesis: Genetic Basis of Antibody Diversity

Past attempts to explain the genetic basis of antibody diversity can be roughly grouped into three different hypotheses.

1. The "*germ line*" *hypothesis* stated that there is a *separate germ line gene for each antibody.*
2. The "*somatic mutation*" hypothesis stated that there is only one or a few germ line genes specifying each major class of antibodies and that the *diversity is generated by a high frequency of somatic mutation*–mutation occurring in the antibody-producing somatic cells or in cell lineages leading to antibody-producing cell.
3 The "*minigene*" *hypothesis* stated that the diversity is generated by the *shuffling of many small segment of a few genes into a multitude of possible combinations.* The shuffling would occur by recombination processes in somatic cells.

We now known that the minigene hypothesis explains a great deal of the observed diversity. However, we also know that somatic mutation contributes additional diversity. Finally, we know that one segment of each antibody chain is specified by a "gene" or "gene segment" that is present in the genome in only a few copies. Thus, all three hypothesis were correct in certain respects.

Structure of Antibodies

Antibodies belong to the class of proteins called *immunoglobulins.* Each antibody is a tetramer composed of four polypeptides, *two identical light chains and two identical heavy chains,* joined by disulfide bonds. The light chains are about 220 amino acids long, and the heavy chains are about 440-450 amino acids long. Every chain, heavy and light, has an animo-terminal *variable region,* within which the amino acid sequence varies among antibodies specific for different antigens, and a carboxyl-terminal *constant region,* within which the amino acid sequence is the same for all antibodies of a given immunoglobulin (Ig) class, regardless of antigen-binding specificity. The variable regions of all antibody chains are about 110 amino acids long. Regions of proteins that carry out particular functions are called *domains.*

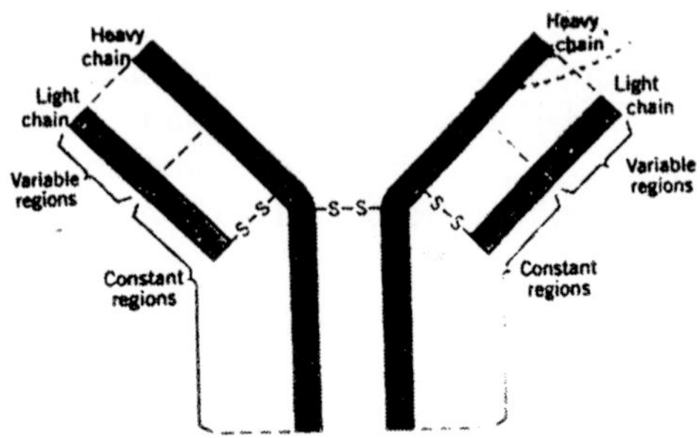

Fig. 9.2. Schematic diagram of antibody structure.

Each antibody has *two antigen-binding sites* or *domains*, each of which is formed by the variable regions of one light chain and on heavy chain. In addition, the constant regions of the two heavy chains interact to form a third domain, called the *effector function domain*, which is responsible for the proper interaction of the antibody with other components of the immune system. There are five classes of antibodies. IgM, IgD, IgG, IgE, and IgA. The class of which an antibody belongs, and thus the function that it carries out, is determined by the structure of its heavy chain constant region (i.e., the structure of its effector function domain). For example, IgD antibodies usually remain bound to the surface of the cells in which they are synthesized, whereas IgG antibodies are usually secreted and circulate through the body in the bloodstream.

The light chains of antibodies are of two types, *kappa* and *lambda*, with type being determined by the structure of the light chain constant region. As we shall see, antibodies may have the same antigen-binding specificity, as determined by the variable regions of the four chains, but different immunological functions, as determined by the constant regions of the two heavy chains.

If these polypeptides were synthesized from colinear nucleotide-pair sequences of genes, one gene per polypeptide chain, the genome would have to contain a vast array of genes with highly variable sequences at one end and essentially identical sequences at the other end. However, this is not the case. Recombinant DNA techniques have made it possible to isolate and sequence many of the segments of chromosomal DNA of mice and humans coding for antibody chains. The results of these studies have provided an elegant explanation for the generation of proteins with great diversity in certain regions and constancy in other regions.

ANTIBODY DIVERSITY: GENOME REARRANGEMENT DURING B-LYMPHOCYTE DIFFERENTIATION

Very simply, the *genetic information coding for antibody chains is stored in bits and pieces, and these bits and pieces are put together in the appropriate sequences by genome rearrangements occurring during the development of the antibody-producing cells (called B lymphocytes) of the body.* Each B lymphocyte produces only a single type of antibody, that is, all the antibodies produced by a given B lymphocyte have the same antigen-binding specificity.

Each antibody chain is synthesized using information stored in several different "genes" of "gene segments." Note that the classical concept of one gene-one polypeptide is not adequate, at least in its simplest form, to explain gene-antibody relationships.

Kappa Light Chains

Synthesis of the kappa light chain is controlled by three different gene segments: (1) a V_k gene segment, coding for the N-terminal 95 amino acids of the variable regions; (2) a J_k gene segment (J for joining segment), coding for the last (constant region-proximal) 13 amino acids of the variable region; and (3) a C_k gene segment, coding for the C-terminal constant region. A fourth gene segment, the L_k segment, codes for an N-terminal *hydrophobic leader sequence* 17-20 amino acids long, which is essential for the transport of the antibody chain through the cell membrane.

The leader sequence is cleaved off the chain as it passes through the membrane, and thus is not part of the final antibody. In mice and humans, all the kappa chain gene segments are located on the same chromosome (chromosome 2 in humans). The same is true for the lambda gene segments (chromosome 22 in humans) and the heavy chain gene segment (chromosome 14 in humans).

There are a large number; probably about 300 of V_k gene segments, each with a nearby L_k gene segment. On the other hand, there is only one C_k gene segment. Five J_k gene segment are located between the V_k gene segments and the C_k gene segment. In germ line cells, the five J_k segments are separated from the V_k segments by a long noncoding sequence and from the C_k segment by an approximately 2000-nucleotide-pair-long noncoding sequence.

During the development or a B lymphocyt the particular kappa light chain gene that will be expressed in that cell is assembled from one L_k–v_k segment, one J_k segment, and the single C_k segment by a process of somatic recombination. This process joins any one of the approximately 300 L_k–V_k segments with any one of the five J_k segments, with the deletion of all intervening DNA. It yields a fused V_kJ_k gene segment coding for the entire variable region of the kappa chain. The noncoding sequence between the J_k gene-segment cluster and the C_k gene segment, and the C_k-proximal J_k segments, if any, remain between the fused V_kJ_k segment and the C_k segment in the differentiated B lymphocytes. This entire DNA sequence (L_k–V_kJ_k–noncoding-C_k) is transcribed and the noncoding sequences are removed during RNA processing just like the noncoding sequence or introns of any other eukaryotic gene.

Lambda Light Chains

Lambda light chain genes are also assembled from separate segments during B lymphocyte development. The major difference is that each J_λ gene segment comes with its own C_λ gene segment, that is, the genome rearrangements required for lambda chain synthesis join L_λ–V_λ segment to J_λ–C_λ segments. Mice have only four J_λ–C_λ gene segments, whereas humans have six. This correlates with the fact that only 5 percent of the antibodies of mice are of the lambda type, whereas 40 percent of the antibodies of humans have lambda light chains.

Heavy Chains

The genetic information coding for antibody heavy chains is organized into L_H–V_H, J_H, and C_H gene segments analogous to those for kappa light chains; but there is one additional gene segment, called D for diversity, that codes for 2-13 amino acids of the variable region. The variable region of the heavy chain is thus encoded in three separate gene segments that must be joined during B lymphocyte development. In addition, there are from one to four C_H gene segments for each Ig class. In the mouse, there are a total of eight C_H gene segments, all functional, arranged on the chromosome in the sequence $C_{H\mu}$, C_{Hd}, C_{Hg3}, C_{Hg1}, C_{Hg2b}, C_{Hg2a}, C_{He}, C_{Ha}. $C_{H\mu}$, C_{Hd}, C_{He}, and C_{Ha} code for the heavy chain constant regions of IgM, IgD, IgE, and IgA, respectively. Four gene segments, C_{Hg3}, C_{Hg1}, C_{Hg2b}, and C_{Hg2a}, code for IgG heavy chain constant regions.

In humans, there are 9 to 10 function C_H gene segments: $C_{H\mu}$, C_{Hd}, C_{Hg1}, C_{Hg2}, C_{Hg3}, C_{Hg4}, C_{He1}, probably C_{He2}, C_{Ha1}, and C_{Ha2}. The human C_H gene cluster also contains two nonfunctional "gene," commonly called ***pseudogenes***, with very similar structures. Pseudogenes are partial duplicates of structural genes that have incorporated sufficient changes that they are not biologically active and usually are not transcribed. Pseudogenes, are turning out to be quite common in eukaryotes.

In mouse germ line cells, there are about 300 L_H–V_H gene segments, something like 10-50 D gene segments, 4 J_H gene segments, and 8 C_H gene segments, arranged on the chromosome in the preceding order. During the development of a B lymphocyte from a stem cell (a mitotically active somatic cell from which other types of cells "stem" or arise by cell division and differentiation), somatic recombination joins one L_H–V_H gene segment with one D gene segment and one J_H gene segment, deleting the two intervening sequences of DNA to form one continuous DNA sequence (V_HDJ_H) that codes for the entire heavy chain variable region.

Class Switching

At the time that antibody synthesis begins in the developing B lymphocyte, all the C_H gene segments are still present, separated from the newly formed L_H–V_HDJ_H gene segment by a short noncoding sequence. At this stage, all antibodies synthesized have IgM heavy chains ($C_{H\mu}$ gene-products). If an antigen is recognized and bound to an antibody on the surface of a developing B lymphocyte, however, that cell is stimulated to differentiate into a mature B lymphocyte. During this differentiation some B lymphocytes will switch from producing antibodies of class IgM to producing antibodies of another class. This phenomenon, called *class switching*, often involves further genome rearrangements during which the C_H gene segments closest to the previously joined L_H–V_HDJ_H gene segments are deleted. The class of antibodies produced after class switching is determined by which gene is brought into the closest proximity with the L_H–V_HDJ_H gene segment.

Antibody Diversity: Alternate Pathways of Transcript Splicing

Another type of class switching during B lymphocyte differentiation occurs at the level of RNA processing ("splicing"). Certain mature B lymphocytes produce both IgM and IgD antibodies. It should be emphasized, however, that these antibodies differ only in their effector function domains; they have identical antigen-binding domains, specified by the same V_kJ_k (or $V_\lambda J_\lambda$) and V_HDJ_H fused gene segments. In these cells, a primary transcript that extends through both the $C_{H\mu}$ and $C_{H\delta}$ gene segments is synthesized. During processing, the V_HDJ_H transcript sequence may be spliced to either the $C_{H\mu}$ sequence or the $C_{H\delta}$ sequence, such that both types of heavy chain are synthesized in the small cell.

A further complexity observed in antibody synthesis is the *sequential production of membrane-bound and secreted forms of a given antibody.* The first antibodies to appear in developing B lymphocytes are membrane-bound IgM molecules. Subsequently, these cells switch to the production of a secreted form of IgM. These two forms of IgM differ only in the C-terminal portions of their heavy chains. The heavy chain of the membrane-bound form is 21 amino acids longer than that of the secreted form. The *membrane-bound heavy chain* has a *41-amino acid-long hydrophobic sequence* at the C terminus that is probably responsible for anchoring it to the cell surface. This hydrophobic sequence is replaced by a *20-amino-acid-long hydrophilic sequence* in the *secreted form.*

The coding sequence (exons) of the C_H gene segments are interrupted by noncoding sequences (introns) just like those of many other eukaryotic gene. The C_H gene segments contain four to six exons and three to five introns. In membrane-bound antibodies, the heavy chain constant regions are produced by splicing all exons together. The last two exons code for the hydrophobic tails of the membrane-bound heavy chains. During synthesis of the membrane-bound form, the fifth C_H exon is spliced to a site 20 codons from the end of the fourth exon, thus changing the amino acid sequence of this portion of the heavy chain constant region. In secreted antibodies, the heavy chain constant regions are therefore the product of the first four exons.

The use of alternate pathways of transcription and RNA processing to synthesize membrane-bound and secreted forms has been firmly established for the IgM class of antibodies. Recent evidence suggests that similar alternate pathways of transcription and splicing are responsible for the production of the membrane-bound and secreted forms of the other classes of immunoglobulins as well.

Signal Sequences Govern Genome Rearrangements

How are the genome rearrangements that occur during B lymphocyte development regulated? What controls the somatic recombination events such that a V gene segment is joined to a J segment and not to another V segment or directly to a C segment? Several long segments of chromosomal DNA carrying clusters of V gene segment, D gene segments, and J gene segments of both mice and humans have now been sequenced, and the resulting nucleotide-pair sequences suggest the presence of specific V–J, V–D, and D–J joining signals. The same *signal sequences* are found adjacent to all V gene segments. Similarly, all J gene segments have identical signal sequences located adjacent to their coding sequences; however, their signal sequence is different from that adjacent to V gene segments. Likewise, D and C gene segments have their own adjacent signal sequences.

The signal sequences controlling V–J, V–D, and D–J joining contain 7-base-pair (heptamer) and 9-base-pair (nonamer)-long sequence separated by spacers of different, but specific lengths. For V_k–J_k joining, the spacer in the V_k signal sequence is 12 nucleotide-pairs long, whereas that in the J_k signal sequence is 22 nucleotide-pairs long. The heptamer and nonamer sequences located "after" the V_k gene segments are complementary (with the exception of one base-pair) to those "preceding" the J_k gene segments. These signal sequences have the potential to form "stem and loop" structures, thus bringing the V_k and J_k gene segments into juxtaposition for joining. Apparently, joining will occur only when one signal sequence contains a 12-base-pair spacer and the other contains a 22-base-pair spacer. This requirement would supposedly be enforced by the specific protein(s) mediating the joining process. Very similar signal sequences appear to control V_H–D and D–J_H joining, whereas somewhat different signal sequences mediate class switching.

Antibody Diversity: Variable Joining Sites and Somatic Mutations

A comparison of the diversity of amino acid sequences present in antibody molecules with that predicted from the sequences of gene segments that encode these antibodies revealed that there is more variation in amino acid sequences at the V–J junctions than is predicted by the nucleotide sequences. Subsequent studies showed that *much of this additional diversity could be explained by variation in the exact site of recombination during the* V–J *joining events.* An example of the use of alternate sites of joining of V_k and J_k gene segments in the mouse. During the joining of gene segments V_{k41} and J_5, recombination has been shown to occur between four adjacent nucleotide positions at the junction sites.

These recombination events produce four different nucleotide sequences that encode three distinct amino acids at position 96 in the mouse kappa light chain. Since amino acid 96 occurs in a region of the antibody chain that is involved in antigen binding, alternate V–J joining events of this type undoubtedly contribute significantly to the great diversity of antibody specificity that is observed in vertebrates. Similar alternate joining events have been documented for V_λ–J_λ and V_H–DJ_H joining reactions. Thus the use of alternate sites of recombination during the joining events that are involved in the assembly of mature antibody genes ***provides an additional mechanism for generating antibody diversity.***

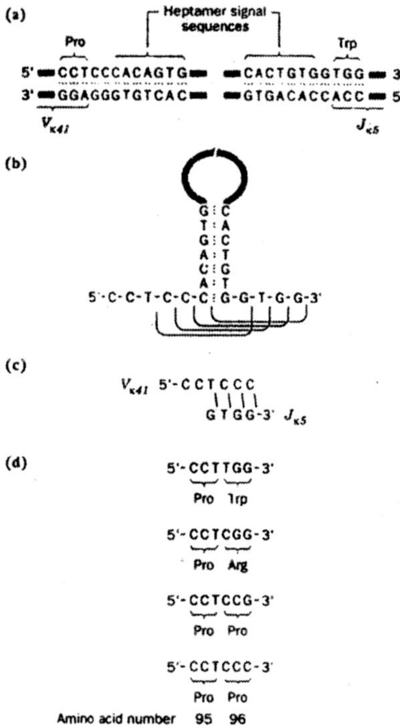

Fig. 9.3. Diversity at the V_kJ_k junction produced by variation in the exact position of the joining reaction.

Despite the vast array of antibody diversity produced by (1) the joining of large families of V, D, and J gene segments and (2) the use of alternate positions of recombination during the joining reactions, considerable data demonstrate that still another mechanism must be involved in the generation of antibody diversity. This has been established by comparing (1) the nucleotide-pair sequences of expressed gene with the sequences of germ line gene segments and (2) the actual amino acid sequences of antibody chains with the amino acid sequence predicted from the nucleotide sequences of the genes. For example, when the actual amino acid sequences of different mouse λ_1 chains were compared with the predicted amino acid sequences (based on the nucleotide-pair sequences of λ_1 gene segments) of λ_1 light chains, differences were found in the variable regions away from the junction sites. Similar observations have been made in studies of heavy chain variable regions. In essentially all cases, the changes have resulted from single nucleotide-pair substitutions. Such substitutions may represent 1-2 percent of the nucleotide-pairs of the segments encoding the variable regions of antibodies. These nucleotide-pair substitutions are presumed to occur by some mechanism of ***somatic mutation*** that is restricted to the DNA sequences encoding the variable regions of antibody chains. Because these changes in the variable segments of antibody genes occur at such a high frequency, the process by which they occur is often called *somatic hypermutation*. The mechanism by which somatic hypermutation occurs is unknown.

Somatic hypermutation of regions of antibody genes that encode antigen-binding sites may be of great value to the organism. Without this mechanism for generating antibody diversity, the range of available antibody specificity would be fixed in terms of the sequences present in the genome at birth and the combinations that could be produced by the various levels of gene segment joining reactions. Viruses and other pathogens are constantly evolving and producing new variants with new antigenic determinants. To provide an adequate defense against the changing antigenic composition of these viruses and other components of the environment, the immune system must also be capable of rapidly responding to these changes. What better way to provide this safeguard than to endow antibody genes with their own mechanism for rapid adaptation to new antigens that might evolve in the future–somatic hypermutation?

How Many Combinations?

One can readily see that a large amount of diversity can be generated by the joining of antibody gene segments as just described. For example, consider the number of different kappa light chains possible in humans: 300 V_k gene segments × 5J_k segment = 1500 fused V_kJ_k gene segments. The heavy chain variable region provides even greater diversity because of the multiple D gene segments. If there are 300 V_H gene segments, 25 D gene segments, and 6J_H gene segments in human germ line cells, 45,000 different heavy chain variable regions could be assembled. Using these estimates, 67,500,000 different antigen-binding sites could be produced using just kappa light chains. Lambda light chains produce another level of diversity. Clearly, these antibody gene-segment fusions provide for a vast amount of antibody diversity. We now know, however, that further diversity is generated in two additional ways: (1) somatic mutation and (2) variability in the sites at which V–J, V–D, and D–J joining events occur. In total, the possible range of antibody diversity seems almost limitless.

Regulation of Transcription: A Tissue-Specific Enhancer

It has been known for several years that germ line antibody genes are not transcribed or are transcribed at very low levels. Yet, in antibody-producing B lymphocytes, 10 to 20 percent of the mRNA molecules are antibody gene transcripts. What, then, is responsible for the activation of transcription of antibody genes that undergo rearrangements and become active? In the case of the heavy chain genes, the answer appears to be that *the rearrangement process bring the promoters located upstream of* L_H–V_H *gene segments into the range of influence of a strong enhancer* element located in the intron between the J_H gene segments and the $C_{H\mu}$ gene segment. Each L_H–V_H gene segment contains an upstream promoter. However, prior to the genomic rearrangement events that lead to heavy chain synthesis, this enhancer is over 100,000 nucleotide-pairs away from the closest L_H–V_H promoter. Presumably, this enhancer cannot activate transcription from promoters that are located that far away. However, rearrangement events occurring during B cell differentiation move the promoter of the closest L_H–V_H gene segment to within less than 2000 nucleotide-pairs from the enhancer. The enhancer can now activate transcription from the promoter located upstream from the L_H–V_H gene segment. The *enhancer involved in the activation of heavy chain synthesis is tissue specific*, it activates transcription only in lymphocytes and has no effect in cells derived from other tissues. Presumably, the activation process requires the presence of a transcriptional-activating factor that is synthesized in lymphocytes, but not in other types of cells.

A similar enhancer element has been found in the intron between the light chain J_k gene segment cluster and C_k coding sequence. Thus, it seems likely that the movement of antibody gene promoters into the range of influence of tissue-specific enhancers may be a general mechanism of activation of antibody genes during the differentiation of B lymphocytes.

Clonal Selection Theory

Various theories have been proposed regarding the formation of antibodies, although only the clonal selection theory seems to be acceptable in most situations. According to this theory, lymphocyte cells are able to make situations. According to this theory, lymphocyte cells are able to make a specific antibody against a specific antigen. The cell may not be making the antibody continuously but may get transformed into an antibody synthesizing cell when the antigen contacts the proper lymphocyte cell, B cells, while not engaged in active synthesis of antibodies, carry specific antibodies on their surfaces. Due to these surface bound antibodies, a specific antigen may get bound with specific B cells and may thus trigger the proliferation and maturation of that particular B cell. The cell responds with repeated cell divisions until a clone of antibody synthesizing cells called ***plasma cells*** is formed. These cell then form large number of same antibodies, at a rate of 10,000 molecules per second per cell.

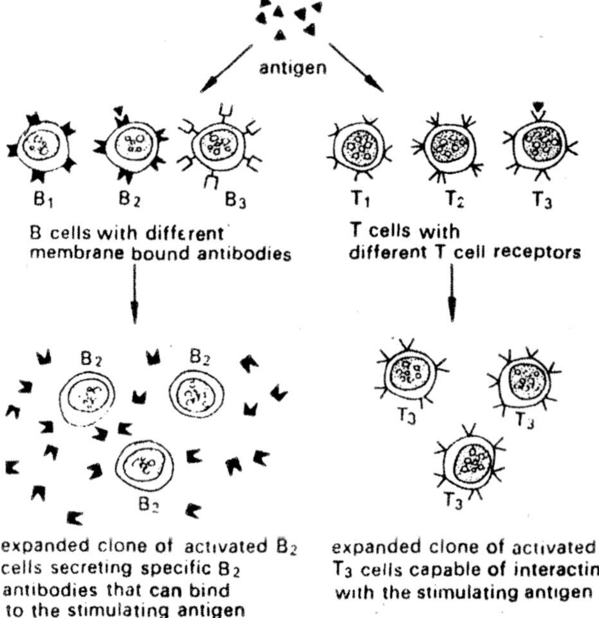

Fig. 9.4. The theory of clonal selection as it applies to B cells and T cells.

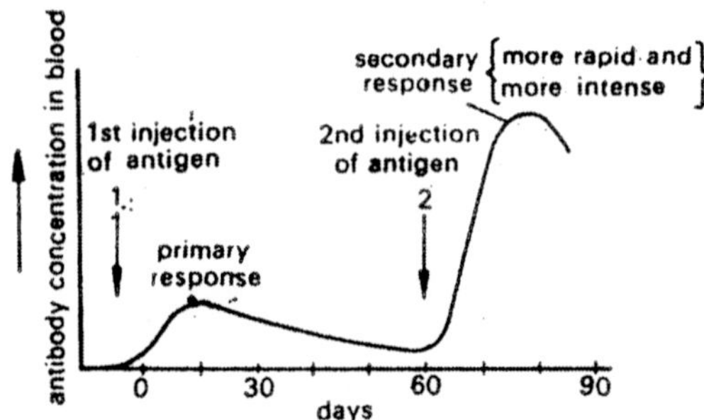

Fig. 9.5. A graph showing the primary and secondary responses of antibody production in a rabbit.

The antibodies are produced by B-lymphocytes, and there is a specific antibody for each specific antigen. Antibodies are also produced by other cells in the body, such as *1-lymphocytes, macrophages, granulocytes* and *must cells*, which, like B-lymphocytes also differentiate from *stem cells in bone marrow*. A concerted effort of all these cells is required to produce the immune response. When an animal first encounters an antigen, it produces a low *primary response*, since only few B cells or T cells can respond to antigen. However, the next encounter with the same antigen leads to a far more rapid and intense response called *secondary response*, since many cells are now available for response. Due to first exposure to antigen, not only mature antibody producing B cells are produced, but *immature B* and *T memory cells* are also produced, which can respond on a second exposure to the antigen.

About 10,000 different antibodies are known to exist in a human body at a given time (there are 10^{12} lymphocytes with 10^{6}–10^{8} specificities). Obviously thousands of *plasma cells* throughout the circulatory system are in operation for the production of antibodies at any given time. About 90-95% of the protein produced in plasma cells gives rise to antibodies, which are sometimes stored in large quantities in these cells, and can be visible as crystals within the cell. With the introduction of an antigen inside the body, the antibody producing plasma cells increase in number, their RNA content increases and entire protein synthesizing machinery is activated for use in antibody production.

Allelic Exclusion

Consider one final point about the genetic control of antibody synthesis. Each B lymphocyte makes only *one* type of antibody. Why? Mammalian cells are diploid; they carry two sets of genetic information

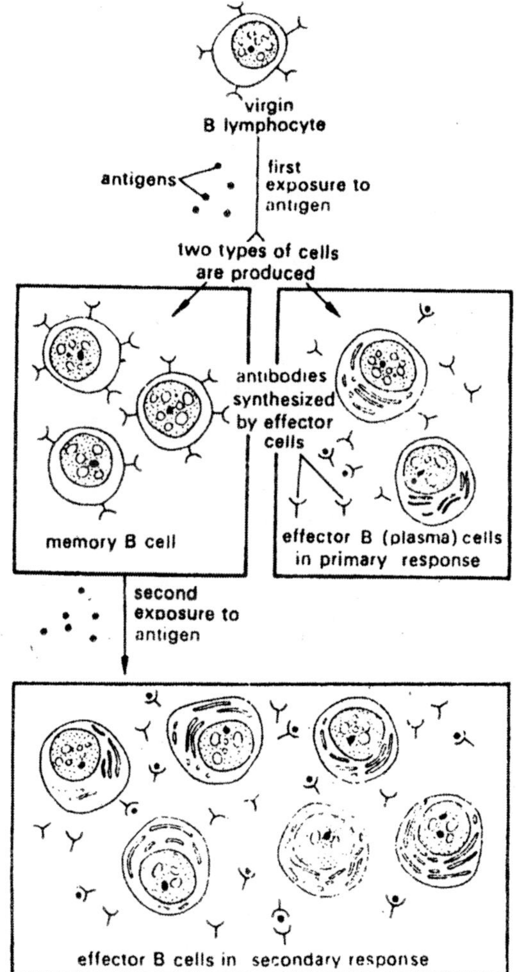

Fig. 9.6. Cellular basis for greater speed and intensity of a secondary antibody response.

coding for each of the antibody chains. But only *one productive genome rearrangement of light chain coding sequences and one productive genome rearrangement of heavy chain coding sequences occur in each B lymphocyte!* This phenomenon is called *allelic exclusion* because one of the "*alleles*" *is excluded from being expressed.* How? Why? At present, we still don't know. Clearly, there must be some type of a feedback mechanism that arrests the recombination process(es) involved in these antibody gene rearrangements once a productive rearrangement has occurred and the cell has started to synthesize a functional antibody. The simplest mechanism would involve inhibition of this process by the mature antibody itself. However, further work will be required to establish the mechanism by which allelic exclusion occurs.

T Cell Receptor Variability

T lymphocytes mediate the cellular immune response. The T cells recognize antigens on the surface of cells and kill the cells carrying these antigens. Like the antibodies produced by B lymphocytes. T cells can recognize and destroy cells carrying an amazing variety of antigens. Thus, the T cell response also exhibits a phenomenal degree of specificity. How is this specificity produced? The answer is that *T cells produce membrane-bound receptors that are very similar to the antibodies* produced by B lymphocytes. Moreover, *the diversity of T cell receptor specificity is produced by genome rearrangements analogous to those involved in antibody production.* But how do T lymphocytes avoid interacting with free antigens to avoid duplicating the function of B cells in the immune response? As it turns out, T cells must simultaneously recognize both the offending antigen on the cell surface and another protein that occurs only attached to cell surfaces. This second cell surface protein that the T cell must recognize is the product of one of many genes in the *major histocompatibility complex (MHC).* The MHC locus encodes a complex group of proteins that are present on all the cells in the body of a human (or a mouse). Thus, T cells are able to recognize and destroy any cell that is producing a given antigen (e.g., the coat protein of a virus) in any tissue of the body. The interaction of the T cell receptor with the two types of cell-surface antigens (the offending antigen and the histocompatibility antigen).

The T cell receptors are composed of two polypeptide chains α and β, each encoded by L–V, D, J, and C gene segments just like antibody chains. The α- and β-polypeptides, like antibody chains, contain variable regions that form the antigen-binding sites and constant regions that anchor the receptors on the cell surface. The variable regions of the T cell receptors are encoded by multiple L–V, D, and J gene segments; the constant regions are encoded by a small number of C gene segments. The T cell receptor genes are assembled by genomic rearrangements that occur during the differentiation of T lymphocytes from stem cells just as in the case of antibody genes in developing B lymphocytes. The α and β receptor proteins are encoded by gene segments that are lined up in clusters on chromosomes similar to those encoding antibody chains. In humans, the α and β gene segment clusters are located on chromosomes 14 and 7, respectively. (Another gene segment cluster encodes a third type of T cell receptor polypeptide designated γ that is present on a specific type of T cell. There are several distinct types of T cells that carry out different functions during the immune response).

The structures of the T cell receptor gene clusters are very similar in humans and in mice. Heptamer and nonamer signal sequences that are very similar to those that control antibody gene rearrangements are also present at essentially the same locations in the T cell receptor gene clusters. Their presence in both type of gene clusters suggests that the same mechanism of gene segment joining is employed during rearrangements of both antibody genes and T cell receptor genes. There probably is somewhat less total variability in T cell receptors than in antibodies. T cell receptor variable regions are encoded by only about 30 V gene segments, whereas there are about 300 V gene segments for both kappa light chains and heavy antibody chains. However, there are more J gene segments in the T cell receptor gene clusters. For example, there are 12 functional J gene segments for β receptor polypeptides. Moreover, we still do not know whether or not the variable segments of T cell receptor genes undergo somatic hypermutation. In any case, it is clear that *T cell receptors exhibit a great amount of diversity*, and that *this diversity is generated by genome rearrangements during T lymphocyte differentiation* in a manner analogous to those involved in the production of antibody diversity in B lymphocytes.

Major Histocompatibility Complex

The immune response in mammals is a very complex process involving a large number of different macro-molecules and different cell types. Our discussion to this point has been restricted to the genetic control of the synthesis of antibody chain and T cell receptor proteins. Many of the other components of the immune response, such as the *transplantation antigens* that are largely responsible for the rejection of foreign tissues in transplant operations, are controlled by a multigene complex called the *major histocompatibility complex (MHC)*. In humans, the MHC proteins are encoded by the HLA (for Human Leukocyte Antigen complex) locus on chromosome 6, in the mouse, the MHC locus is designated H-2 (Histocompatibility locus 2) and is on chromosome 17. In both mice and humans, the MHC locus is very large ($>2 \times 10^6$ nucleotide-pairs) and contains a large number of genes. Moreover, there is a very large number of distinct alleles for many of these genes such that the probability of any two individuals being identical for all the MHC genes is extremely small. The MHC genes are said to be *highly polymorphic* because of the large number of alleles of individual genes that are usually segregating in a given population.

The MHC genes encode three different classes of proteins that are involved in different aspects of the immune response. The structure of the human MHC (HLA) locus and the relative locations of genes that encode the different classes of histocompatibility antigens are shown in fig. The class I genes encode the transplantation antigens mentioned. The class I proteins exist as glycoprotein anchored as integral membrane proteins with the antigenic determinants exposed on the outside of cells. They are present on all cells of an organism and permit T lymphocytes to distinguish "self" from "foreign". The MHC class I proteins are the antigens that usually are responsible for the rejection of foreign tissues in tissue and organ transplants. These antigens play a key role in the recognition and destruction of cells carrying foreign antigens by cytotoxic T lymphocytes. A single T cell receptor is believed to recognize both the foreign antigen and the class I histocompatibility antigen during the cytotoxic T cell immune response.

The MHC class II genes encode polypeptides that are located primarily on the surfaces of B lymphocytes and macrophages. MHC class II proteins provide a special type of T lymphocyte called the "T helper cell" with the capacity for self-recognition and facilitate communication between the different types of cells involved in the immune response. Finally, the MHC class III genes encode complement proteins that interact with antibody-antigen complexes and induce cell lysis. The MHC class I and class II antigens are anchored in the cell membrane and have structures very similar to the structure of T cell receptors. However, the diversity of MHC antigens is much less than that of antibodies and T cell receptors, and so, far as is known, no genomic rearrangement is involved in the genetic control of MHC antigen diversity. Instead, the observed diversity results from the presence of a large number of highly polymorphic MHC genes.

Overview of B Cell Antibody Production

The precursors of antibody-producing cells are the *stem cells* of bone marrow. Stem cells do not express an immunoglobulin. During the differentiation of a stem cell into a pre-B cell, stem cells first express the mu (μ) heavy chain, then, as a B cell, they express one of the two light chains. Together the first antibody IgM is made in all B

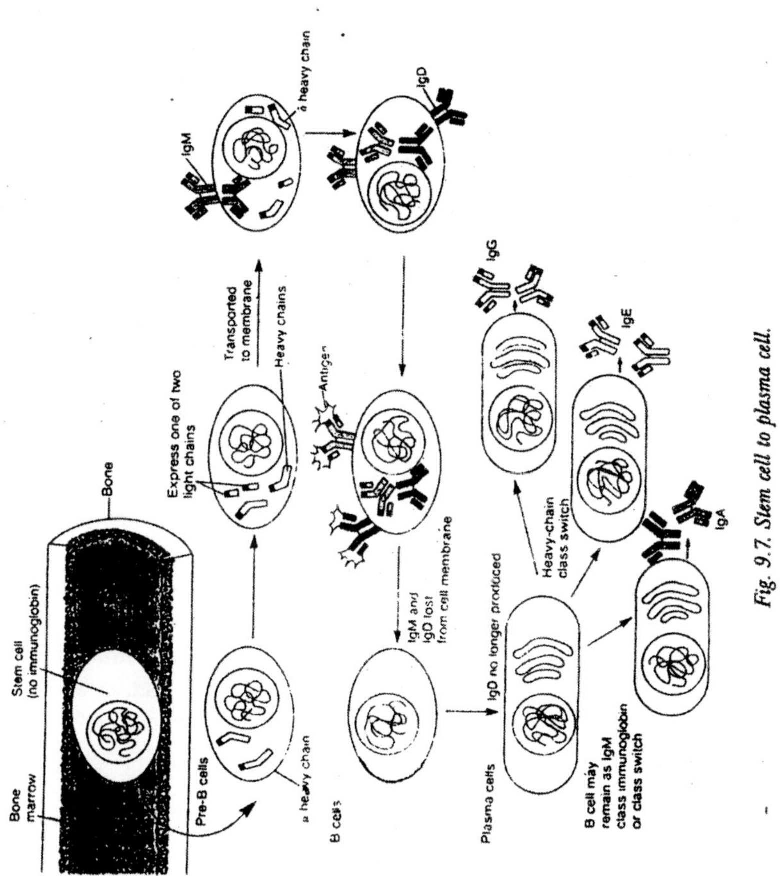

Fig. 9.7. Stem cell to plasma cell.

cells. This IgM-antibody molecule is transported into the membrane and projects as an external receptor molecule form the membrane. Nest, molecules of IgD are produced to form external receptor sites on each B cell. Once a B cell contains both immunoglobulins, it is considered competent. When competent B cells encounter an antigen, the B cell is triggered to divide into a monoclonal-type antibody-producing plasma cell. Some of the progeny of the selected clones remains as circulating B lymphocytes. They serve as the immune system's memory, providing a faster response to any subsequent exposure to the same antigens. The memory cells are responsible for the immunity that develops following many infections or as a result of vaccination.

Other members of the selected B-cell clones undergo *terminal differentiation*: they grow larger, stop reproducing, and dedicate all their resources to the production of antibodies. In this state, they are called *plasma cells*, and although they live for only a few days, they secrete large quantities of immunoglobulins. During the differentiation into the plasma cell, both IgM and IgD are lost from the cell membrane. The cell now begins to produce and secrete only IgM. The clone of cells derived from the initial antigen-stimulated B cell may continue to produce IgM or may switch over (*class switching*) to produce one of the other heavy-chain classes. That is, the same variable region of the antibody is joined to a different constant-region domain, which, by definition, presents a new immunoglobulin class. Heavy-chain class switching provides the final antibody product with a means of performing its essential activities after it has attached itself to an antigen someplace within the body. Thus, a given competent B cell may produce different Ig classes with the same antibody-attachment site for specific function in separate areas of the body.

With respect to the genetic control of antibody production, it can be said that a human responds to antigens by generating an entirely new antibody gene from sequences of minigenes that are spatially separated along the chromosome. What is inherited, therefore, is the potential to rearrange these minigenes into antibody genes. Learning that genes may change locations and that minigene rearrangements may occur in three separate human chromosomes implies that the classical concept "gene are in fixed locations along the length of a chromosome" is not entirely true.

Chimeric Antibodies

Recent developments in the use of recombinant DNA techniques have shown the feasibility and potential of making recombinant antibody genes, cloning them, and reintroducing them into hybridoma cells that will then secrete recombinant antibodies possessing novel properties. Neuberger and colleagues have developed a cell line that secretes a chimeric human-mouse antibody molecule. It contains a human epsilon (ε) heavy-chain constant region attached to a mouse variable region.

Potential applications of such *chimeric antibodies* include cancer therapy and the treatment of autoimmune conditions, such as multiple sclerosis and systemic lupus erythematosus. It is thought that antibodies may be able to destroy specific immune cells that mount an attack on the body's tissues. In a related fashion, the antibodies may eventually be used for immune suppression in individuals who require organ transplants.

Table 9.1. Inherited immunological diseases

Disease	*Mode of inheritance*[1]
Ataxia-telangiectasia	AR
Cartilage-hair-hypoplasia	AR
Chronic granulomatus	X-LR
Complement-factor deficiencies	AR
Severe combined immune deficiency diseases (SCID)[2]	
Adenosine deaminase deficiency	AR
Purine nucleoside phosphorylase deficiency	AR
Other types of SCID demonstrated	X-LR, AR
DiGeorge's syndrome	F
Hypogammaglobulinemias	
Burton's type	X-LR
Agammoglobulinemia	X-LR
Immunologic amnesia	AR
Thymic dysplasia (Nezelof's syndrome)	AR
Wiskott-Aldrich syndrome	X-LR
Lymphoproliferative disease	X-LR

Note: As of 1985, there are at least twenty-four known primary immunodeficiency diseases. Some of the more common are listed in this table [1]AR–autosomal recessive, X-LR–sex linked recessive, F–familial, tendency inherited but mode of transmission undetermined. [2]Severe because the condition is rapidly fatal.

Natural Immunity and Acquired Immunity

Natural Immunity

'Natural immunity' of an animal is also known as innate immunity, native immunity or inherited immunity. This relates to a general or non-specific type of resistance which prevents infection by different kinds of pathogens. The extent of this natural immunity differs in different organisms. For example, man can easily get mumps while cats and dogs are immune to this disease. This level of natural immunity

may vary not only between species but also between races or strains, and sexes. This may also be controlled by nutrition, hormones and many other factors.

Acquired Immunity

'*Acquired immunity*', develops during the life time of an individual and refers to the immunity, which a specific individual displays against a specific pathogen. This is frequently related to the presence of antibodies or interferon in the blood. Acquired immunity based on antibodies is the most efficient type of acquired immunity and it may be either (i) *actively acquired* or (ii) *passively acquired.*

Actively acquired immunity

Actively acquired immunity may be either natural or artificial. Actively acquired natural immunity results from any infection from which a person recovers. During the infection, antibody production for that specific pathogen is stimulated so that when there is a subsequent infection by either the same or antigenically related pathogen, the antibodies assist in the body's defense.

Actively acquired artificial immunity is the most common method of immunization or vaccination. The immunogens are injected in the body in controlled quantity of stimulate the production of immunoglobulins. Killed and attenuated strains of bacteria and viruses are now used widely for immunization against many diseases (e.g., typhoid, small pox, poliomyelitis, yellow fever, measles, etc.) Attenuated organisms produce mild infection and induce natural immunity. Vaccines are also being developed now by a variety of other methods involving recombinant DNA technology.

Passively acquired immunity

'*Passively immunity*' may also be acquired either by natural or artificial means. While passive immunity acquired by natural means, involves the transfer of antibodies from mother to her unborn child, through the placenta during the later part of pregnancy, passive immunity of the artificial type refers to the original production of antibodies in some other individual (human or lower mammal) followed by injection of these antibodies with the help of a needle or syringe. Several drug manufacturing companies are involved in the large scale production of antibodies in horses and cows by active immunization.

Inherited Immune Deficiencies

Mutations in genes that encode cytokines or T cell receptors impair cellular immunity, which primarily targets viruses and cancer. However, because T cells activate the B cells that manufacture antibodies, abnormal cellular immunity (T cell function) causes some degree of abnormal humoral immunity (B cell function). Defects in the genes encoding antibody segments would impair immunity mostly against bacteria.

We are presently aware of more than twenty types of inherited immune deficiencies. The types classified as *severe combined immunity deficiency* (SCID) affect both branches of the immune system, humoral and cellular. About 50 percent of SCID cases are sex-linked, and therefore seen in boys only. In a less severe form, the boy lacks B cells but has T cells. Before the advent of antibiotic drugs, these boys died before the age of ten years of overwhelming bacterial infection. In a more severe form of sex-linked SCID, lack of T cells leads to death by eighteen months. Severe thrush (a fungal infection), chronic diarrhea, and recurrent lung infections usually kill these children.

A young man named David Vetter taught the world about the difficulty of life with SCID years before AIDS appeared. David had an autosomal form of the illness that caused him to be born without a thymus gland. His T cells could not mature and activate B cells, leaving him defenseless in a germ-filled world. Born in Texas in 1971, he spent his short life in a vinyl bubble, awaiting a treatment that never came. As David reached adolescence, he wanted to leave his bubble. An experimental bone marrow transplant was unsuccessful–soon afterwards David began vomiting and developed diarrhea, both signs of infection. David left the bubble but died within days of a massive infection.

For Laura Cay Boren, the first four years of life were an endless bout of severe infections. Her autosomal recessive form of SCID was caused by a lack of an enzyme called adenosine deaminase (ADA). Laura was luckier than David. Enzyme replacement and gene therapy have allowed her to live a near-normal life. About 15 percent of all SCID cases are due to ADA deficiency.

Another autosomal recessive form of SCID shows genetic drift. Affected children have a tiny thymus gland and very few lymphocytes. Among the general population, 2 in a million newborns have this SCID variant, but among the Navajo and Apache peoples in the southwestern United States, 1 in 3,340 newborns are affected. Table 9.2. summarizes some inherited immune deficiencies.

Table 9.2. Some inherited immune deficiencies.

Disorder	*Phenotype*	*Defect*
Sex-linked recessive		
X-linked immuno-deficiency	Severe viral, bacterial, and protozoan infections	Absent glycoprotein on T cell surfaces
Hyper IgM syndrome	Excess IgM, deficient IgA and IgG; disorganized lymphoid tissue; large tonsils; immune system destroys blood cells	Antibody protection can't switch from IgM to IgA and IgG
Duncan disease	Great risk of Epstein-Barr virus infection, causing lethal mononucleosis, death by age 20; B cell cancer; too few antibodies and red blood cells	Progressive loss of T cell function
SCID X2	No lymphoid tissue; deficient IgG; at high risk for viral infections	Lack of CD4 and CD8 helper T cells
Agammaglobulinemia, Bruton type (T^+ B^- SCID)	Bacterial infections	Lack of B cells
Agammaglobulinemia, Swiss type (T^- B^+ SCID)	Very small thymus; at high risk for viral, bacterial, fungal infections	Lack of T cells
Autosomal recessive		
ADA deficiency	Bacterial and viral infections	Lack of ADA poisons T cells, which cannot active B cells
Agammaglobulinemia	Thymocytes do not mature; no lymphocytes in bloodstream or lymph nodes; small thymus; no antibodies; diverse infections	Defect in protein kinase linked with T cell cytokine receptor
Immune defect due to absence of thymus	Viral infections	No thymus gland; no T cells; deficient antibodies
Autosomal dominant		
Immunodeficiency with defective leukocyte and lymphocyte function	Skin and eye infections	Defective antibody production

Table 9.3. Autoimmune disorders

Disorder	*Signs and Symptoms*	*Antibodies Against:*
Glomerulonephritis	Lower back pain	Kidney cell antigens that resemble strep bacteria antigens
Grave's disease	Restlessness, weight loss, irritability, increased heart rate and blood pressure	Thyroid gland antigens near thyroid-stimulating hormone receptor, causing overactivity
Juvenile diabetes	Thirst, hunger, weakness, emaciation	Pancreatic beta cells
Hemolytic anemia	Fatigue and weakness	Red blood cells
Myasthenia gravis	Muscle weakness	Receptors for nerve messages on skeletal muscle
Pernicious anemia	Fatigue and weakness	Binding site for vitamin B on cells lining stomach
Rheumatic fever	Weakness, shortness of breath	Heart cell antigens that resemble *Streptococcus* bacteria antigens
Rheumatoid arthritis	Joint pain and deformity	Cells lining joints
Scleroderma	Thick, hard, pigmented skin patches	Connective tissue cells
Systemic lupus erythematosus	Red facial rash, prolonged fever, weakness, kidney damage	DNA, neurons, blood cells
Ulcerative colitis	Lower abdominal pain	Colon cells

Adenosine deaminase deficiency

The severe combined immunodeficiency diseases, like *adenosine deaminase deficiency* (ADD), are characterized by the absence of all adaptive immune functions. ADD is a severe, but rare, recessive disorder of the human immune system. It occurs once in 200,000 liver births. It is also one of the few genetic diseases that can be "cured" ADD patients can be transfused with compatible bone marrow cells. The problem is finding his to compatible bone marrow for transplantation.

Adenosine deaminase (ADA) is an enzyme that converts adenosine to inosine in the purine salvage pathway. The ADA enzyme is essential to the production and maintenance of immunological T cells and B cells. The absence or continued loss of T and B cells results in immunodeficient children who die at very young ages. ADA deficiency, depending on the level within different cell types, may produce a clinical disease within ten days and up to five years after birth.

The ADA gene is located in the long arm of chromosome 20, region 1, band 3.1 (20q13.1). This gene is genetic interest for two reasons. First, it is a leading candidate for the first attempt at gene therapy (placing a normal ADA into the patient's own bone marrow cells). Second, for the study of its promoter gene sequence because this gene contains a promoter region that does not contain the CAAT or TATA DNA sequences used as signals in many genes for the beginning of mRNA synthesis. The promoter region for this gene is rich in G and C. The gene is 32,000 bases in length and contains twelve exons. The exon, or coding portion, is only 1,500 bases in length; 95 percent of these bases are committed to introns.

Ataxia-telangiectasis

Patients with recessively inherited ataxia-telangiectasis produce lower than normal levels of IgA. These people do not reject skin grafts in the normal time period; they suffer from progressive ataxia (lack of muscle coordination), brain abnormalities, pulmonary infections, and cancer.

X-Linked Immunological Diseases

There are at least four known X-linked inherited immunological diseases: *Wiskott-Aldrich syndrome, Burton's hypogammaglobulinemia, chronic granulomatus,* and the X-linked recessive type of the *combined immune deficiency agammaglobulinemias.* The Wiskott-Aldrich syndrome begins in infancy with recurring expression of otitis media (inflammation of the middle ear), thrombocytopenia (decrease in the number of blood platelets necessary for blood formation), eczema, and recurrent infections. In Burton's syndrome and combined immune deficiency disorders, patients lack sufficient IgG, IgA, and IgM. The lymph nodes are poorly defined or absent and plasma cells are absent. These persons cannot be immunized against bacterial infections of any type. These defects are congenital and the patients must receive repeated immunoglobulin injections to survive.

If depression of the immunoglobulins is very severe, a child may not live outside a sterile component, as in the case of David.

Autoimmune Diseases

To this point information has been presented to show that humans are absolutely dependent on the production of a wide range of antibodies for survival. However, it is known that on occasions the immune system can itself cause serious disease or undesirable consequences, like seasonal allergy attacks.

The autoimmune disease caused by autoantibodies (antibodies made against self) may be organ-specific, as in *Hashimoto's thyroiditis* antibodies attack the thyroid gland) and type I diabetes (antibodies attack beta-insulin producing cells of the pancreas), or systemic, as in *systemic lupus erythematosus* (SLE) where the autoantibody is carried in the patient's serum and reacts with many or most tissues of the body. Some of the more common target organs for autoantibody attack are the pancreas, stomach, kidney, and adrenal and thyroid glands. Nontarget-specific autoantibodies are associated with the rheumatoid disorders and involve the skin, muscles, and joints. Rare cases of male infertility involve the production of autoantibodies against the sperm, which causes the sperm to clump together. In cases of pernicious anemia, it is believed that autoantibodies interfere with the normal uptake of vitamin B^{12}. The high incidence of Hashimoto's thyroiditis and of circulating thyroid antibodies in Down's and Turner's syndrome patients suggests an association between autoimmunity and chromosome aneuploidy. Autoimmune diseases are more common in females than in males and incidence increases with age. For example, in rheumatoid arthritis, the most common autoimmune disease, 90 percent of patients are women.

Acquired Immune Deficiency Syndrome (AIDS)

Information gained from studies on patients with genetic abnormalities of the immune system proved essential in unraveling the immunologic abnormalities in AIDS patients. AIDs is caused by a retrovirus identified in the United States as HTLV-III (human T-cell lymphotrophic virus). (HTLV-I and II cause human leukemias and lymphomas). Because of an international dispute on who first discovered this virus, this single virus has three separate names in the literature–HTLV-III, LAV, and ARV. Recently, U.S. scientists adopted a fourth name, HIV, for *human immune deficiency virus*, to simplify reference to this virus. The AIDS virus attacks the body's immune system, leaving it susceptible to a spectrum of opportunistic infections (e.g., pneumonia caused by *Pneumocystis carinii*, a generally harmless protozoan) and cancers (e.g., Kaposi's sarcoma).

Human infection may be initiated by the free AIDS virus or by virus carried in infected cells. Once inside the body, the virus attacks a variety of immune-system cells that carry a cluster differentiation recoptor antigen type 4(CD4), a glycoprotein, molecule in the outer membrane. This molecule defines certain T lymphocytes (T4), monocytes, and macrophages. Macrophages normally interact with the T4 lymphocytes, stimulating the protective functions of the lymphocytes. However, the AIDS virus first attacks the CD-4 bearing monocytes and macrophages. Infected macrophages infect the T4 lymphocytes (T4L) and after a latency period, viral replication kills the T4L. This reduces the T4L population, thereby critically reducing the protection of the human immune system. Each T4L normally produces about one thousand progeny, but an infected T4L produces as few as ten. These ten, on reaching the bloodstream and after being antigen stimulated, begin producing virus and die.

Genetic structure of the AIDS virus

Perhaps the T4 molecule is the most significant molecule associated with HIV infection. By interacting with the outer envelope of the HIV, T4 may provide entrance to the cell. The membrane of the viral envelope contains many molecules of a glycoprotein (a protein attached to a sugar molecule). Each molecule of glycoprotein consists of two subunits–gp41 and gp120. The gp120 subunit interacts with T4 molecules of T4-carrying cells. After gp120-T4 contact, a cellular-membrane vehicle is formed around the virus and the virus is translocated from outside to inside the cell. Many retroviruses have the same three genes that code for the envelope protein (*env*), the core proteins (*gag* or group specific antigens), and for reverse transcriptase (*pol*). The three genes are flanked by long terminal redundancies, called LTRs. The function of the LTRs is to help control viral-gene expression. The difference between retroviruses in general and the HIV is that HIV contains at least four additional genes to help control gene expression. The four additional genes are called *tat*, which encodes a protein for regulation of mRNA transcription and translation of the gag and pol genes, *art/trs* that encodes a second regulatory protein for control of transcription and translation of the gag and pol genes, *sor* that encodes a protein that appears to be necessary for free HIV infection, and *orf* that encodes for a protein that appears to suppress HIV replication. The genome of the HIV is as follows:

Investigators, using information about the molecular structure of the HIV, are trying to produce a protective vaccine. However, a single vaccine to protect against HIV may not be possible because of two major problems: first, the HIV readily mutates, producing new infective strains of HIV and second, how can vaccine be produced that protects against free viruses in the body and viruses within infected cells? Conservative estimates are that an HIV vaccine will not be available through 1995.

AIDS transmission

AIDS is transmitted through sexual contact, transfusions of blood or blood products, by sharing contaminated hypodermic needles, and from infected mother to child at or before birth (infected mother means the carries the virus, regardless of whether she demonstrates AIDS symptoms). Ending the fist quarter of 1989, the National Center for Disease Control, Atlanta, Georgia, reported over 90,000 AIDS cases. Over half of these people have died. The U.S. Surgeon General stated that about 1.5 million people in the United States are infected but have not yet expressed the disease. The World Health Organization forecasts there will be from fifty million to one-hundred million HIV carries by 1993.

Evidence of AIDs-related deaths in the United States can be traced to the 1960s. However, there already existed within the U.S. population a genetic resistance as well as a susceptibility to the disease. For those that HIV infected, who contracted the disease, and who died, natural selection acted against their genotypes. For survivors of the infection or for those infected but who do not express the disease, natural selection favoured their genotypes. The process of natural selection is slow, but consistent. Many infected people will die prior to a natural stabilization of a highly resistant population.

Science cannot wait for natural selection because the cost in human life is too high. The search to control the HIV virus using a protective vaccine or effective drugs that are able to block the life cycle of the virus is underway. This search will be successful as it has been for many other epidemics.

Hypersensitivities: The Allergies

Perhaps the best known of the immunological diseases are *allergies*, or *hypersensitivities*. When an individual has been immunologically sensitized to an antigen, exposure at a later date can lead to a booster effect of the immune response that results in tissue damage. Some allergic reactions are immediate: sensitivity to insect stings or to such drugs as penicillin may bring on a severe reaction that, if not promptly treated, can lead to death. Less serious or delayed allergic reactions occur in persons sensitized to grass pollens, animal danders, house dust, and so on.

Immunoglobulin E (IgE) is most important with respect to allergic reactions to various summertime antigens, such as ragweed. If ragweed pollen is injected into the skin, the person becomes locally sensitized. The reaction of antibody to ragweed pollen results in a chain of acute physiological and pharmacological reactions. As the ragweed antigen combines with IgE, a special cell–a mast cell–release histamine. Histamine immediately dilates the blood capillaries and alters cell permeability in various tissues. This, in turn, causes a localized reddened skin reaction (wheal and flare). If the IgE is beneath the bronchial mucosa and adjacent to the smooth muscles in the chest area, an antigen-antibody reaction releasing histamine may cause an immediate bronchial constriction, stimulating an asthma attack.

Major Histocompatibility Complex: Transplant Rejection

A most important aspect of the human immune system is to recognize self from nonself. Each lymphocyte carries a unique tag of self-identity as a kind of reference point. The human immune system recognizes an enormous variety of antigens. The B cell is able to respond directly to a single antigen, but the T cell is activated only if that antigen is displayed on the surface of a lymphocyte. Because a self-identity marker is also present on the lymphocyte, the T cell must distinguish between self and nonself and proceed with its immune function. If the T cells cannot distinguish between an antigen and self, these cells may begin to attack the body tissue.

What are the molecules that serve as tags of individual identity? How many genes are involved in their production? To answer these questions, it is necessary to discuss yet a third class of proteins involved in immune response to an antigen. The first two classes of essential immune proteins are the immunoglobulins and T-cell receptor proteins. These are membrane-bound antigen-specific proteins analogous to the membrane-bound immunoglobulins of the B cells. The third class of proteins is coded for by a relatively large gene cluster called the *major histocompatibility complex* (MHC).

The MHC proteins were discovered in tissue-grafting experiments. In these experiments, unless the donor and the recipient of a graft were genetically identical, as in the case of identical twins or mice of an inbred strain, the graft was generally rejected. The graft was rejected because the recipient mounted an immune response to the donor's MHC proteins. Graft rejection implies that unrelated individuals almost always express different sets of MHC genes.

Three Gene Classes of the MHC

An MHC appears to be present in all vertebrates. It is composed of three families, or classes, of genes as defined by the proteins they produce in mice and humans. The three gene classes in mice and humans are referred to as class I, class II, and class III. Each of the three genes are multiallelic, having between fifty and one hundred different alleles in the general population. Proteins produced by the variant class-I and class-II alleles differ in their ability to present antigens and to initiate immune response. Class-III genes code for complement factors essential to the immune system.

Class-I and class-II alleles code for their respective class-I and class-II glycoproteins, which are anchored within the cell membrane and extend outside the membrane analogous to the immunoglobulins. The regions or domains of the class-I and class-II proteins, which extend beyond the cell membrane, permit the immune system to recognize self from nonself. One entry into the body, proteins, nucleic acids, etc. of any species or microbe are compared to self. When found to be foreign, the immune response is set in motion.

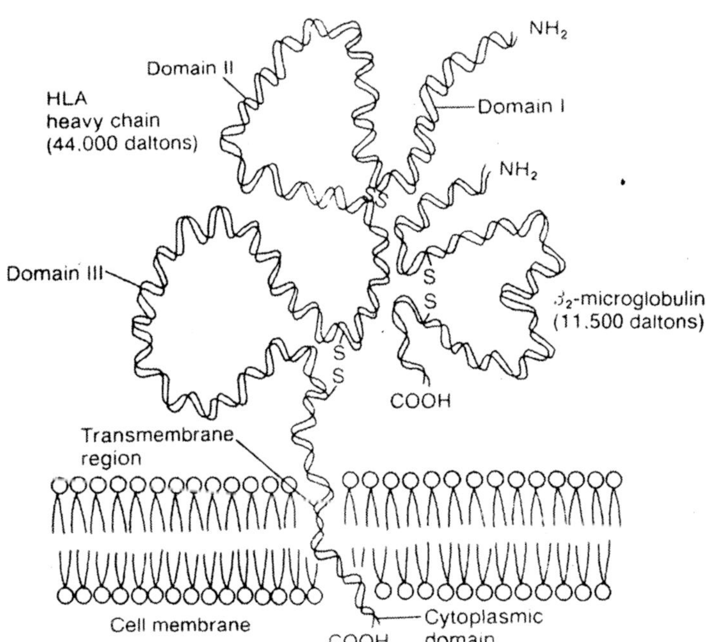

Fig. 9.8. Diagram of HLA class-I antigens.

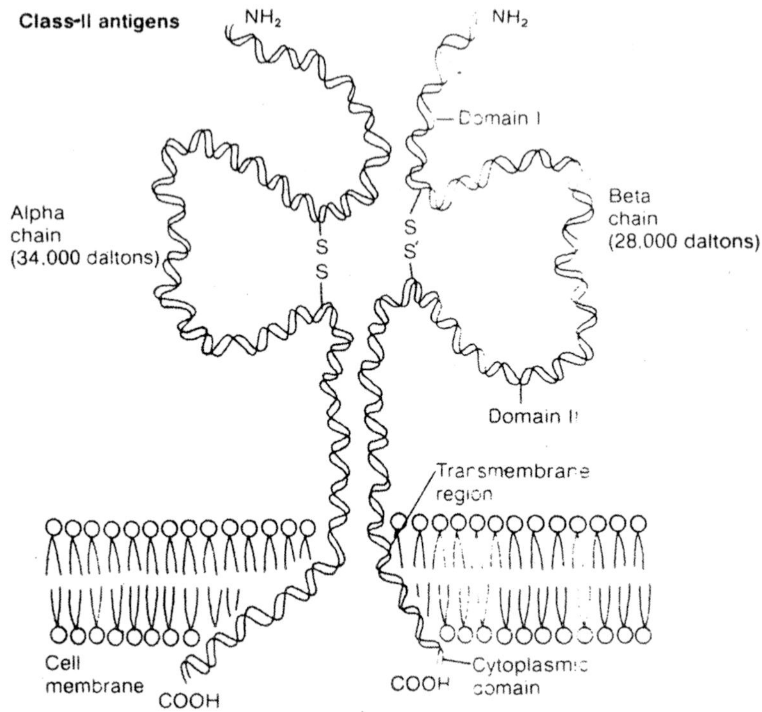

Fig. 9.9. Diagram of HLA class-II antigens.

In human organ transplantation, the recipient and donor, unless they are identical twins, will have distinct sets of class-I and class-II proteins on their cells. Both the donor organ and recipient individual will recognize each other's tissues as foreign because each person has a different combination of alleles in the MHC region that produce different class-I and class-II proteins. For this reason, *the proteins produced by the genes of the MHC region are referred to as MHC antigens.* Antisera has been produced against the MHC gene proteins and are used to determine an individual's genotype prior to organ transplant. One of the mysteries of the immune system yet to be unraveled is the mechanism by which the foetus, which is an *allograft* (has a different genetic constitution than its mother), is able to survive the immunologic defenses of the mother. Why does the mother's immune system accept the development of this foreign body within the uterus? Foetal development defies the basic tenets of transplant immunology.

Genetics of the MHC-HLA

In mice, the MHC is called the H2 complex and is located on chromosome 17. It consists of four million DNA base pairs. In humans,

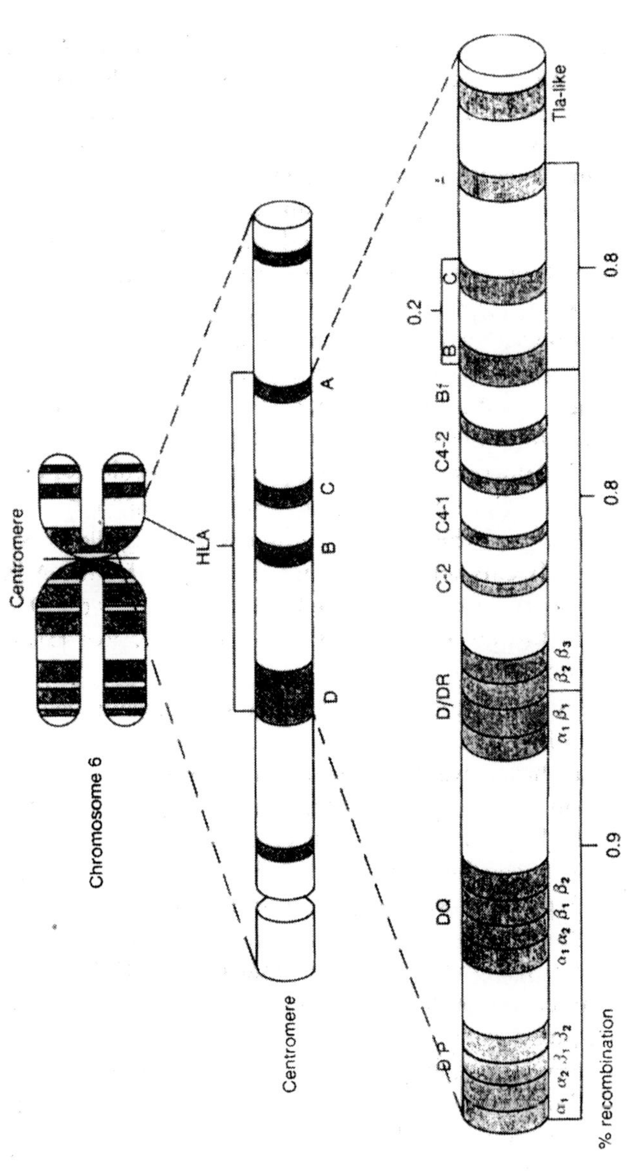

Fig. 9.10. HLA gene map complex.

the MHC is called the HLA system (human leukocyte antigen). The genes HLA-A, B, C, D/DR (D related) are located in close proximity to each other in the short arm of chromosome 6. The HLA complex consists of about six-million DNA base pairs. Recently, scientists at Harvard and Stanford Universities published a three-dimensional structure of class-I HLA protein. The investigators found that the string of amino acids that make up the HLA molecule is folded into a mass that is shaped like a "flat-bottomed ravine." Each wall of the ravine is made up of amino acid coiled into a helix. The area of the ravine is just wide enough to accommodate a foreign antigen. Evidence suggests that killer T cells, while straddling the ravine walls, interpret the foreign protein's sequence of amino acids, thus programming themselves to attack any similar foreign proteins. Once the programming process is understood, investigators should be able to manipulate the immune system with synthetic proteins, which may then act as a vaccine or as immune suppressor molecules for therapy in cases of autoimmune diseases and in transplant rejection.

The HLA system is highly polymorphic with many alleles existing for each locus. Twenty alleles have been identified for the A locus, forty-two for the B locus, eight for the C locus, twelve for the D locus, and twelve for the DR. The alleles for each of the five HLA loci present are codominant.

The group of HLA alleles on a single chromosome is called a *haplotype*. One HLA haplotype is inherited from each parent. Most individuals are heterozygous for some of the HLA alleles because there are so many alleles in the population for each HLA locus. Because the genes of the HLA system are so close together in the HLA region, crossing over between the A and B or B and D loci occur at a frequency of less than 1 percent. With such low frequency of chromosome exchange in the HLA region, parental haplotypes are almost always inherited intact. It is this very feature that makes HLA typing the method of choice for determining organ transplant compatibility and in cases of disputed paternity. Form a random sample of a population, only two persons among thirty thousand to fourty thousand will have identical HLA antigens. That is, the chance that any two people at random will have the same HLA antigens is .007–.005 percent.

HLV Antigenic Typing

Using a specific antisera (antibodies in serum formed against a given antigen) a cell preparation can be serologically typed according

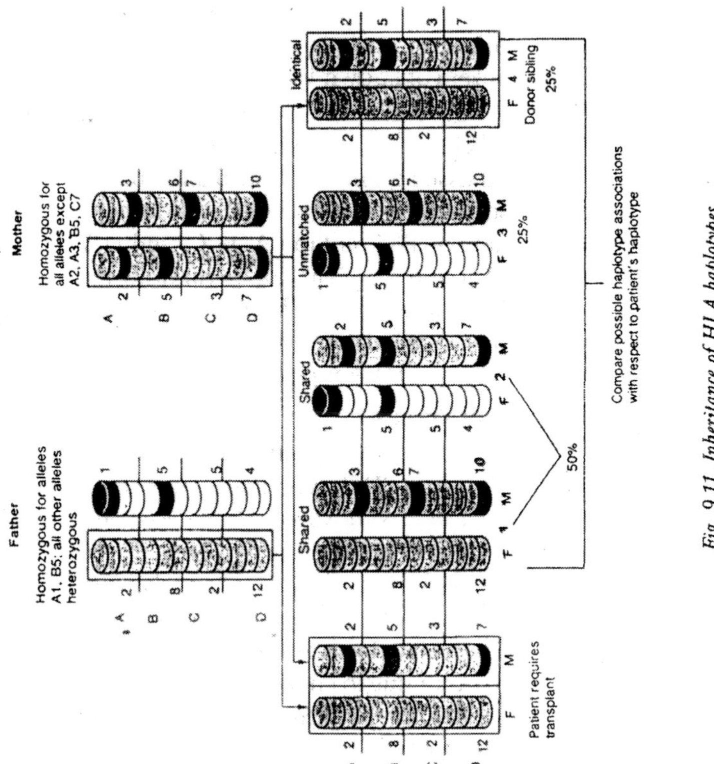

Fig. 9.11. Inheritance of HLA haplotypes.

to antigens it carries. This is called a *microcytotoxicity assay* because cells subjected to a series of known antisera in the presence of complement will be damaged or lysed if they are sensitive to a particular sera. The HLA-A, B, and C antigens are detected using the microcytotoxicity assay. Lymphocytes for HLA antigen typing are incubated in the presence of complement with a human antisera containing HLA antibody against either antigen A, antigen B, or antigen C. Lymphocytes bearing the specific antigen are damaged and will allow entry of an added dye. These stained cells are easy to locate. Undamaged cells remain unstained–they do not carry the specific antigen.

Closely related to the HLA-D region is a series of serologically defined antigens that are present on the surface of B lymphocytes and are formally designated as "D-related," or DR antigens. HLA-DR antigens are found not only on B cells, but also on monocytes, macrophages, and some bone marrow cells. DR types are also determined by the microcytotoxicity typing assay using a person's B lymphocytes and antisera of known DR specificity.

DNA probes are now available for six HLA class-II genes–HLA-D α and β, for HLA-DQ α and β, and DP α and β–and are being used in paternity cases and tissue typing for transplantation. By subjecting small aliquots of a suspension of lymphocytes to individual antisera made against the multiple alleles of the A, B, C, and D/DR loci, the phenotype of the donor and recipient can be carefully matched prior to organ transplant. In addition to the HLA typing for use in identifying transplant compatibility between donor and recipient, researchers have recognized an association of HLA types with various human diseases.

HLA Relationships of Disease

In 1971, McDevitt and Bodmer published the first report of an association between a given HLA type and a human disease. They reported that persons affected with systemic lupus erythematosus (SLE) demonstrated a specific HLA antigen with a greater frequency than nonaffected persons. Since then, an HLA antigen relationship has been established for over fifty human diseases. In each case, the incidence of a given HLA type associated with a specific disease is greater than would be expected to occur due to chance. However, because a given antigen is inherited, it does not mean the disease will be demonstrated. The association only means that there is an increased risk for demonstrating the disease compared to those who do not have the antigen.

Of all the HLA antigen-disease association, the most striking is that of the spondylarthropathies with HLA-B27. The spondylarthropathies are several disease states characterized by arthritis of the spine and sacroiliac and axial joints. *Ankylosing spondylitis* (AS) is the prototype; other related conditions include Reiter's disease, reactive arthritis, and a subgroup of juvenile rheumatoid arthritis.

Ninety percent of AS Caucasian patients have the B27 antigen. This also means that 10 percent of such patients do not have the B27 antigen. Therefore, it may be said that the presence of B27 antigen predisposes or increases the risk that a person with the B27 antigen will demonstrate AS and that the B27 antigen itself is not the sole cause of AS.

Table 9.4. HLA associated diseases

Disease	*HLA antigen(s)*	*Patients with antigen (%)*
Acute lymphatic leukemia	A2	60
Active chronic hepatitis	B8	36
	DR3	79
Addison's disease	D3	70
Ankylosing spondylitis	B27	90
Celiac disease	B8	71
	D3	96
Diabetes, type I	DR3, DR4	95+
Graves' disease	B8	18
	D3	53
Hashimoto's thyroiditis	D5	19
Hemochromatosis	A3	71
Hodgkin's disease	A1	40
Insulin-dependent diabetes	D4	75
Multiple sclerosis	A3	33
	B7	35
	D2	59
Myasthenia gravis	B8	39
	DR3	40
Pernicious anemia	D5	25
Pemphigus	D4	87
Psoriasis	A1	39
	C6	70
Reiter's disease	B27	80
Rheumatoid arthritis (adult)	DR4	47
Systemic lupus erythematosus	D3	70
Ulcerative colitis	B5	80

Note : Less than half of the known HLA antigen associated diseases are listed.

The prevalence of B27 correlates well with the occurrence of AS in certain race groups. For example, Japanese have virtually no B27, and As is very rare in that population. However, for the few Japanese who do carry B27, the relative risk for the disease is considered to be extremely nigh. From table 9.4., it can be seen that few disease associations have been made with the HLA-A locus. Most of the associations are with the B and D/DR loci. The reasons are not known.

Blood Typing and Transfusions

Early beliefs and attempts at blood transfusions were extremely dangerous for the patient. In 1874, Hamlin reported that "white corpuscles of cow's milk were converted into red blood corpuscles when milk was used in transfusing humans." In 1890, Jenkins reported that a human was transfused with lamb's blood as treatment for typhoid fever. Aside from these curious attempts at transfusions, many deaths occurred as a result of transfusions, many deaths occurred as a result of transfusing the blood of one human into another. As a result, many countries once banned the practice of human blood transfusion. It took some 275 years and countless tragedies before Karl Landsteiner, in 1900, discovered and classified the "agglutinating" and "agglutinable" blood factors that accounted for transfusion problems. For this achievement, Landsteiner received the 1930 Nobel Prize in medicine.

Landsteiner showed that red blood cells vary with respect to specific protein antigens found on the cell's surface. He classified human blood types based on the two different kinds of antigens: antigen A appeared on the red blood cells of type A blood and antigen B appeared on the red blood cells of type B blood. In addition, he found that certain people have both antigens A and B (type AB blood), and others have neither antigen A nor antigen B. Blood cells with no A or B antigens were designated type O. Persons with type A blood were shown to have a factor in the serum portion of their blood causing the agglutination, or clumping, of the B cells when the two types were mixed. Those with type B blood were shown to have a factor that caused the clumping of type A cells. The serum of type A blood contain antibody B, and the serum of type B blood carries antibody A. Type O blood does not contain either antigen A or antigen B, but does contain both A and B IgM antibodies.

Table 9.5. Blood types of generally safe transfusion possibilities

Blood group			*IgM antibodies against*	*Can give blood to*	*Can receive blood from*
Phenotype	*Genome*	*Antigens*			
A	I^AI^A, I^Ai	A	B	A, AB	A, O
B	I^BI^B, I^Bi	B	A	B, AB	B,O
AB	I^AI^B	A, B	None	AB	All
O	ii	None	A, B	All	O

More than three million Americans receive blood transfusions each year. For safe administration of blood from donor to patient, it is important to determine at least the ABO blood group of each individual prior to a transfusion. This is done by cross-matching, or testing for compatibility of recipient and donor blood. A blood antigen-antibody coagulation test is simple to run and is readily available to determine whether a given donor's blood will be compatible with the recipient's blood.

It is generally accepted that the effect of the donor's antibodies, especially those of a type O donor, on the cells of the recipient is not important because the antibodies of the donor are immediately diluted during limited transfusion. However, in some cases, type O blood may contain potent anti-A or anti-B hemolysins that destroy the red

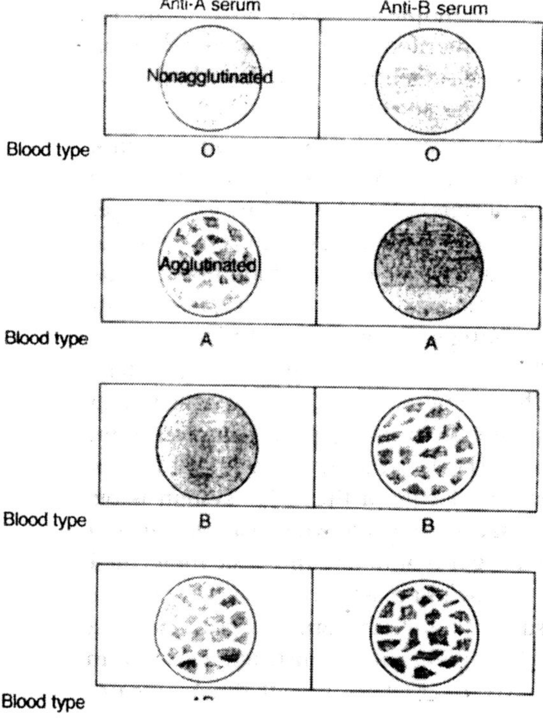

Fig. 9.12. ABO blood typing.

blood cells of the recipient. In theory, donors of group O can give blood to each of the three other types, as well as to their own type, because red blood cells of blood type O do not contain A or B antigens. Therefore, type O blood is referred to as the universal donor. In practice, it has been found best to use a donor of the same blood group as the recipient, and to use the universal donor only in an emergency.

Genetics of Common Blood Groups

ABO Blood Groups

For Landsteiner's human ABO blood groups, four separate phenotypes (A, B, AB, and O) and six genotypes (I^AI^A, etc.) can be postulated. The blood types are determined by the antibodies carried for use against foreign biological antigens. In 1910, von Dungren and Hirsfield worked out the genetics of the ABO system. Their work was elaborated on by Bernstein in 1925. The ABO blood groups' antigens are controlled by multiple alleles at a single locus. The A and B blood-group alleles are inherited without apparent dominance between the A or B genes (codominance); however, both A and B alleles are dominant to the gene for O.

It should be pointed out that on pure technical grounds, the common reference to ABO antigenic blood groups of humans is incorrect, but permissible. The expression 'antigen blood groups" implies that blood groups are recognized by the antigens a person possesses. For example, a person of blood type A has antigen A on his red blood cells. Persons of blood type O, by definition, have neither antigen A nor antigen B on their red blood cells. Thus, the system should be referred to as the AB-antigen blood system. There are other antigens on the red blood cells. One in particular, antigen H; should be added to the system making the common human blood system known as the ABH blood antigen system. It is now known that the basic antigenic molecule of the ABO system is built from a single precursor oligosaccharide onto which various sugars are enzymatically added. Which sugar is added to the basic molecule is determined by the H, A, and B alleles present. The H gene codes for an enzyme that attaches the sugar fucose to the terminal galactose of the oligosaccharide chain. The H antigen, once constructed, is common to all blood groups–A, B, AB, and O. Persons with the A gene produce an enzyme that attaches N-acetylgalactosamine to the terminal galactose of the oligosaccharide chain. Persons with the B gene produce an enzyme that attaches galactose to the terminal galactose of the oligosaccharide. Persons with both genes, A and B, are codominant and produce both antigens. Persons with type O blood have only the H antigen. Technically then, humans have antigenically an ABH blood system, but for historical reasons, the human blood groups are still referred to as the ABO blood antigen system.

A few people have been found with the homozygous hh genotype. Persons of hh genotype are not able to express their ABO blood groups. The A and B alleles of the ABO blood group gene may be present, but they are not expressed because galactose was not added to the precursor oligosaccharide. The homozygous hh genotype masks the real ABO phenotype; it is said to be epistatic. This rare condition is referred to as the *Bombay phenotype* because it was first recognized in that city. People having the genes for making the A or B antigens cannot by usual laboratory methods be distinguished from blood type O persons. The children of an hh person receiving the H gene from the other parent express the normal ABO blood type antigens.

Transformation of type B Blood into Type O

Investigators have recently transformed type B blood cells into type O by using the enzyme alpha galactosidase extracted from coffee bean. The enzyme cuts the galactose sugar molecule form the H antigen. In volunteers infused with the transformed blood, 95 percent of the converted RBCs were still in circulation after twenty-four hours and 50 percent were in circulation after one month. This compares well to RBC survival in regular transfusion. There were no recorded side effects to the volunteers. Investigators are now searching for an enzyme to cut off the N-acetylgalactosamine of antigen A, converting it to blood type O.

Lewis Blood Group

The *Lewis antigen gene* (Le) is responsible for the enzymatic addition of N-acetylgalactosamine to the H-determinant molecule. The Lewis antigens are not considered a permanent part of the RBC membranes as are the A, B, and H antigens. The Lewis antigens are soluble in the plasma; consequently, if the person is homozygous LeLe or heterozygous Lele they are found in saliva as well as the RBC. The homozygous lele makes no antigen.

Secretor Blood Group

During studies on the A, B and H antigens, some people were observed to have the A, B, and/or H antigens in the aqueous secretions of their eyes, nose, semen, and salivary glands. People able to secrete these antigens are referred to as *secretors*. The ability to secrete the

ABH antigens depends on the presence of a dominant *secretor gene* (Se). Secretors are of the genotype SeSe or Sese. Those of the genotype sese do not secrete the ABH antigens. The ability to secrete these antigens appears to have some relationship to whether a given blood type is associated with a given disease.

The MN Blood Groups

Blood groups other than ABO and Rh are rarely implicated in transfusion incompatibility or hemolytic disease of the newborn. A typical example is the MN blood group system, also determined by surface antigens. The basic features of the MN system are the following:

1. M individuals have red blood cells that react only with anti-M antibody.
2. MN individuals have red blood cells that react with both anti-M and anti-N antibody.
3. N individuals have red blood cells that react only with anti-N antibody.

Pedigree studies indicate that the M and N antigens are products of codominant alleles at a single locus. Individuals of genotype MM have blood type M, those of genotype MN have blood type MN, and those of genotype NN have blood type N.

The MN blood group system is of minor medical importance, because harmful anti-M and anti-N antibodies are rarely produced in humans, regardless of antigenic stimulation. Thus, donated blood is not routinely typed for MN, because neither the time nor the expense is warranted. The following section documents a few uses of blood groups beyond their role in transfusions.

Rhesus (Rh) Blood Group

The symbol Rh is taken from the word rhesus, a species of monkey (*Macaca rhesus*). The Rh antigenic factor is found on the blood cells of the rhesus monkey as well as in humans. Red blood cells carrying Rh antigen are Rh positive (Rh^+), and red blood cells not carrying Rh antigen are Rh negative (Rh^-). In the United States, 85 percent of the white and 92 percent of the black population are Rh^+. The Japanese, Chinese, and "pure" American Indians are 99 percent Rh^+. The Rh antigen of an Rh^+ person will elicit the production of antibodies in an Rh^- transfusion recipient. The first transfusion of Rh^+ blood into an Rh^- recipient usually sensitizes the individual, who produces low levels of antibody without a severe reaction. Once sensitized, the recipient can experience a severe reaction or even death on subsequent Rh^+ blood transfusions.

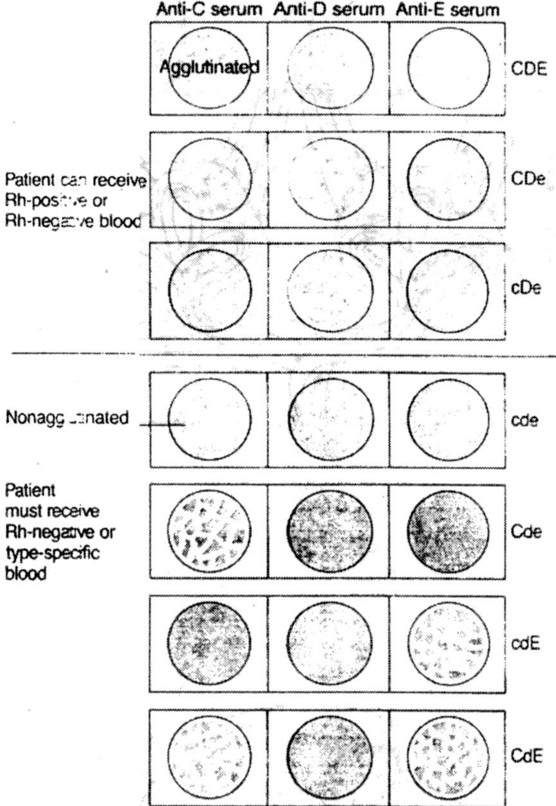

Fig. 9.13. Rh blood typing. Rh antigens using specific Rh and anti-C, anti-D, and anti-E serums.

Rh incompatibility

Mating between Rh^- women and Rh^+ men is often described as Rh incompatible. Rh incompatibility is estimated to occur in 13 percent of Caucasian marriages and in 5 percent of black marriages. In an Rh-incompatible mating, all offspring will be Rh^+ if the husband is homozygous Rh^+/Rh^+. If the husband is heterozygous Rh^+/Rh^-, there is a 50 percent chance the offspring will be Rh^+. Only on becoming pregnant with an Rh^+ foetus is an Rh^- female faced with the potential danger of the Rh sensitization (not all Rh^- women with Rh incompatible pregnancies become sensitized).

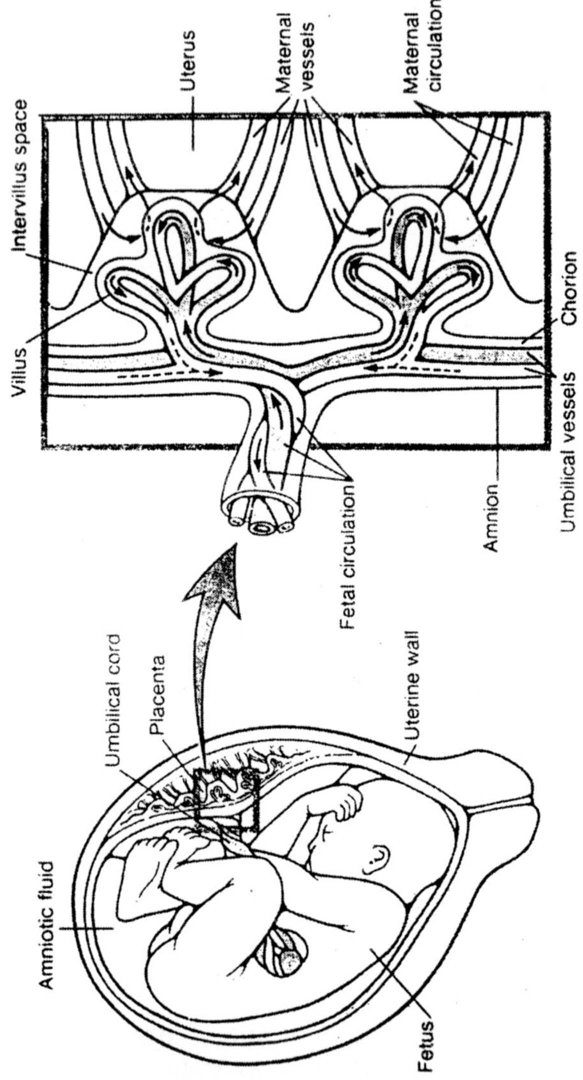

Fig. 9.14. The placenta and Rh incompatibility.

Rh sensitization during pregnancy

Each year an estimated two to three million pregnant women in the United States are at risk of becoming sensitized to the Rh blood factor. Those who are sensitized are exposed to the Rh antigen in the blood of an Rh positive foetus so that their immune system is able to rapidly produce the antibody against the Rh factor with the next exposure. Rh sensitization can be prevented by the injection of Rh immunoglobulin within seventy-two hours of the delivery of the baby.

In the Rh sensitization of the mother (often referred to as isoimmunization), the placenta plays an important role. The placenta forms where the fertilized ovum is implanted and becomes a highly vascularized network of blood vessels. An organ of both foetal and maternal origin, the placenta consists of blood vessels, vascular spaces, and small amounts of supportive tissue. The blood circulation system of mother and foetus are entirely separate, but there is an extensive contact surface since the foetal vessels extend into the chorionic villi, which in turn extend into the intervillus spaces that are filled with maternal blood. The possibility that red blood cells might cross this placental barrier was scarcely considered until the early 1940s.

In 1942, Levine suggested that foetal red blood cells make their way into the maternal circulation in sufficient amounts to stimulate isoimmunization of the mother. Later, Chown demonstrated the presence of foetal erythrocytes in the maternal circulation. It has now been demonstrated that 71 percent of all women with an Rh incompatibility have foetal red blood cells in their circulatory systems after delivery. However, it is also known that not all of these mothers are sensitized. Once a women becomes sensitized to the Rh factor,

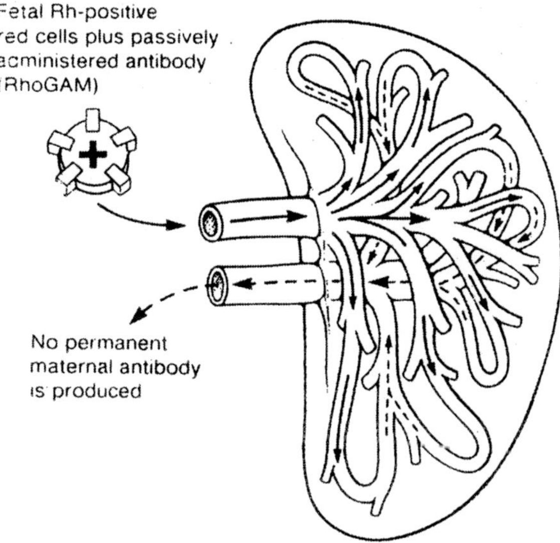

Fig. 9.15. Prevention of primary immune response to Rh(D) at delivery of an incompatible foetus.

future pregnancies will bring about more rapid antibody formation in response to the foetal Rh^+ cell. The antibodies cross the placenta into the foetal bloodstream and may result in infant hemolytic disease. Some of these infants are stillborn, some die shortly after birth with neurological defects, and others suffer permanent brain damage.

Erythroblastosis foetalis

Erythroblastosis foetalis (EF) is a hemolytic disease causing lysis of the red blood cells. If affects the erythropoietic system of the foetus and continues in the newborn. EF causes *hydrops foetalis.* The clinical symptoms of hydrops foetalis are jaundice, anemia, and enlargement of the liver and spleen due to the release of excessive amounts of bile pigment into the blood.

If the infant is severely jaundiced shortly after birth, the condition is referred to as icterus *gravis neonatorus.* Between 5 and 10 percent of the Rh-incompatible neonates develop *kernicterus,* a form of brain damage caused by an excessive amount of bilirubin in the blood resulting from lysed RBCs. During pregnancy, the excess bilirubin is handled by maternal circulation. At birth however, the newborn cannot

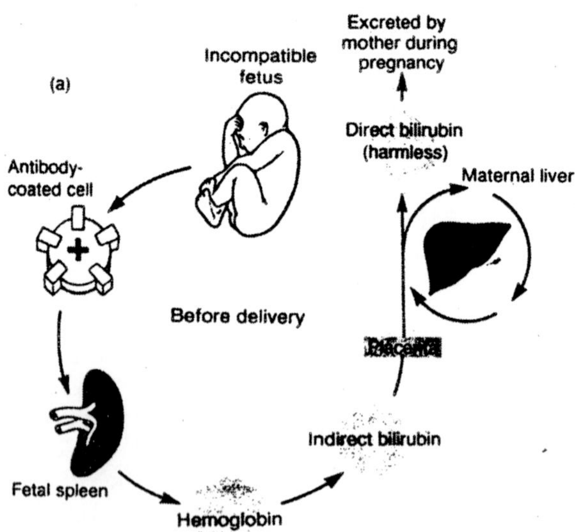

Fig. 9.16. The metabolism of Bilirubin. Before delivery, jaundice in the foetus does not occur because bilirubin produced by the breakdown of cells in the foetal spleen passes via the placenta to the maternal circulation.

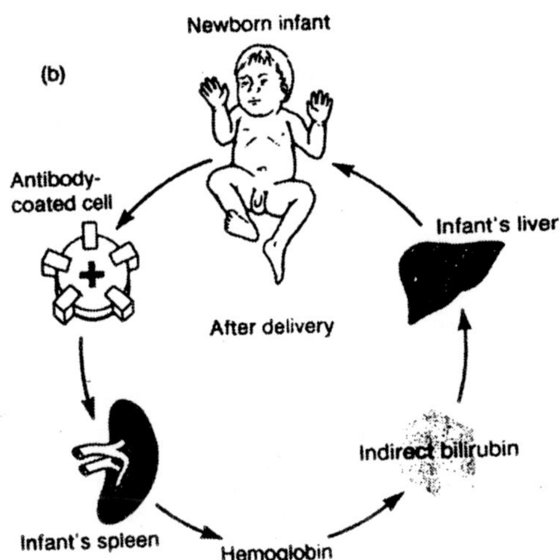

Fig. 9.17. The metabolism of Bilirubin. After delivery, the infant's liver cannot handle the amount of bilirubin and the child stays jaundiced with yellow-appearing skin and eye colour.

tolerate the volume of bilirubin due to continued RBC lysis. About 70 percent of the infants with kernicterus die within seven days of birth. Among those who survive, many have neurological handicaps. In fact, at one time kernicterus was responsible for 10 percent of all cases of cerebral palsy. In less severely affected infants, congenital anemia may dominate the clinical picture. Before treatment became available, neonatal mortality was 50 percent among infants born alive with hemolytic disease.

The Rh locus: multifactor or multiallelic

The knowledge of the Rh system has enabled researchers to solve the problem of erythroblastosis; however, the genetic and clinical terminology still remains confusing. Presently, three Rh factors are known, and they are produced from a very restricted region of the chromosome. Current Rh typing is based on the presence of three closely linked genes, each involved in the production of a separate antigen (a multiple-factor theory). The three Rh factors are C, D, and E. Factor D is responsible for the Rh^+ phenotype.

ABO protection in Rh incompatibility

It has been estimated that thirteen of every one hundred Caucasian marriages are Rh incompatible. Yet, of these thirteen, only one or two will be plagued with the Rh hemolytic disease. Why so few? The large difference between the number of children expected to have the disease and those actually found indicates that there is a protective mechanism operating in the Rh-negative mother that somehow inhibits the formation of Rh antibodies.

One form of protection is ABO incompatibility. In cases of ABO incompatibility, the mother and father are of different blood types. For example, if the mother is of blood type O and the father is of type B, a natural foetal antibody-antigen reaction is possible. If red blood cells carrying the antigenic B enter the maternal blood-stream, the anti-B antibodies she carries will complex with lyse and type B red blood cells. Should the father be blood type B. Rh^+, the lysis of the B cells as they enter the maternal bloodstream prevents the mother's immune system from making any anti-Rh antibodies. Children with Rh hemolytic disease usually have ABO compatible parents.

Utilization of Blood Group-Analysis Data

Through modern analysis of the ABO groups, a variety of legal and social problems can be dealt with more accurately. Some of the problems clarified are cases of disputed paternity, rape charges, or maternity mix-up. Blood analysis is also helpful for identification in kidnapping cases, in case of hit-and-run accidents, and in homicides. In these instances, geneticists and medical doctors are brought before the court to give testimony on the genetic basis of the findings in question. For example, the expert can, in most cases, present genetic blood group evidence for the exclusion of a father in a paternity suit. The expert reaches decision because of his knowledge of the Mendelian mode of inheritance of blood-group alleles. Blood-group testing can be used only in the exclusion of paternity. When the three well-known blood system (ABO, MNs, and Rh) are used together, there is about a 62 percent chance of excluding an accused man in a paternity suit. The use of all the fifteen known blood groups would eliminate a falsely accused man more than 90 percent of the time. With the use of histocompatibility leukocyte antigen (HLA) sera typing, the exclusion rate is brought to 98.6 percent, and with the use of HLA gene probes for the HLA-D and B regions and heteromorphic chromosome studies as presented later, the exclusion rate is over 99 percent. With new discoveries in the field of blood grouping, humankind has almost arrived at the time, predicted by Landsteiner, when blood grouping will be as specifically identifying as fingerprints.

ABO blood types and associated diseases

The word "association" as used with respect to the ABO blood groups and disease means that a given disease occurs in people of a given blood type more often than would be expected to occur by chance. The discovery of the association of blood groups and diseases dates to 1920. The more recent work stimulating research into blood group and disease association is that of Aird. He reported persons having gastric cancer were most often of blood type. Since Aird's initial article, numerous studies have been made on the association of the ABO blood groups and various diseases.

Table 9.6. Human blood types (ABO) and associated human conditions

	Blood type		
Condition	*A*	*B*	*C*
Cancers			
Stomach	X		
Cervix	X		
Pancreas	X		
Salivary glands	X		
Ovary	X		
Corpus of the uterus	X		X
Lungs (undifferentiated tumors)	X		X
Lungs (gland ulcer differentiation)	X		X
Lungs (squamous differentiation)	X		X
Skin (xeroderma pigmentosum)			X
Ulcers			
Duodenum (50 percent higher in O nonsecretors)			X
Gastric or peptic (highest in types A, B, and O nonsecretors and type O secretors)			X
Stomal or anastomotic			X
Diabetes mellitus			
Males only	X		
Females only			X
Manic depressive			X
Digitoxin-induced arrhythmias	X	X	
Erythroblastosis foetalis (mother, type O; child, type A or B)		(mother)	X
Ischemic heart disease	X	X	

Pernicious anemia	X		
Bronchopneumonia	X		
Rheumatic fever (carditis)	X	X	
Smallpox	X	X	
Bubonic plague	X		X
Adenovirus	X		
A_2 influenza virus			X
Myocardial infarction	X		
Cirrhosis of the liver	X		
Arteriosclerosis obliterans	X		
Gallstones	X		
Female venous thromboembolic disease (from oral contraceptives)	X	X	
Elevated serum cholesterol levels	X		
Electrophoretically slow-moving serum alkaline phosphatase		X	X
High carbohydrate diet		X	
High fat intake	X		

There are problems with attempts to associate a disease with a particular blood group. Although gastric cancer may be associated with blood type A, the same form of cancer can be found in individuals of blood types AB, B, and O. There are just relatively fever persons in these categories. Moreover, an association based on statistical evidence does not in itself represent conclusive evidence for a cause-and-effect relationship. There are a number of scientists who argue against a causal relationship for most, if not all, of the associations listed in table 7.5. The critics of an assumed causal relationship find fault with the methodology of the studies and therefore the conclusions of the studies. If there is a correlation (an association greater than due to chance), there may be a causal relationship. It is for this reason that blood types and associated conditions are reviewed in this chapter.

It is known that blood type antigens are inherited, but it is also known that the majority of the conditions listed in table 7.5 are not. What is suggested is that persons of specific blood types are more susceptible to certain conditions than persons of other blood types. Many associations have been reported for blood type A persons. Diseases associated most predominantly with blood type O are cancer of the skin, duodenal, gastric, and stomach ulcers, manic-depressive illness, erythroblastosis foetalis where the mother is of blood type O, and susceptibility to the A_2 influenza virus. Persons belonging to blood group B may find comfort in the reports to date that they are the least susceptible to blood group-associated disease. However, it may also be of interest to know only 10 percent of the white and 21 percent of the black population in the United States are blood type B.

Those reports indicating significant associations between a condition and a particular blood type also note that there was no association between a given blood group and certain other conditions. This means that certain diseases appear with equal frequency regardless of blood group, whereas certain other diseases appear to occur more frequently in persons of a particular blood group. At the moment, there is no explanation for why persons of one blood group should be more susceptible or resistant to given disease.

Application of Blood Groups

In addition to the importance of certain blood group system in transfusion and maternal-foetal incompatibility, blood groups have applications in studies of genetic linkage (the loci of several genetic disorders are linked with known blood-group loci), population studies, parentage exclusion, and criminology. Blood groups are useful in identification because, in many cases, the alternative blood groups are all reasonably common and any two unrelated individuals can be expected to differ in one or more blood groups. Indeed, the probability that any two randomly chosen Europeans will be identical for all known blood groups is less than 0.0003. Moreover, blood group frequencies may vary among populations. For example, among black Africans the chromosome with the highest frequency with respect to the Rh loci is *c D e*, whereas among Chinese it is *C D e*. Such observations are the basis of the use of blood groups in anthropology and population genetics to infer the genetic ancestry of populations.

Evidence based on blood groups is admissible in many courts in cases depending on identification of parentage. In this application the blood groups are interpreted according to two rules: (1) All allele present in a child must be present in one or both parents. (2) If a parent is homozygous for an allele, the allele must be present in the child. Suppose, for example, that the blood groups of mother, child, and two possible fathers are

Mother	O	Cc D–Ee	MN
Child	O	Cc dd Ee	M
Male 1	O	CC D–Ee	M
Male 2	A	cc dd ee	N

Because of the phenotype of the child, the sperm giving rise to the child must have carried I^O, either C or c (depending on whether the egg carried c or C, respectively), d, either E or e (depending on whether the egg carried e or E, respectively), and M. With respect to ABO and Rh, either male could have contributed such a sperm, as male 1 could be Dd and male 2 could be I^AI^O. However, male 2 cannot contribute a sperm carrying M, so he is excluded as the possible father. It is important to note that this analysis does not imply that male 1 is the father, but merely that he could be.

Application of blood typing to criminology is derived from the fact that the A and B antigens are extremely stable and can be correctly identified in dried blood or other substances after many years. A and B antigens can also be found in sweat, semen, and other body fluids. Secretion of these substances results from the presence of a dominant gene Se. Approximately 75 percent of Caucasians are secretors (that is, genotypically Se/Se or Se/se). Thus, in cases of rape, for example, the ABO blood type of the assailant can often be identified through examination of the semen with anti-A and anti-B antibodies.

10

Gene Therapy

Therapy, by definition, is the treatment of an individual. Therefore, the term *genetic therapy* (or *gene therapy*) is used within this chapter to mean any procedure that has as its purpose the prevention, reduction, or cure of a genetic disease. Although dramatic gains are being made in the areas of screening and counseling, lasting genetic therapy is still unavailable for most of the human hereditary diseases. However, in most cases a good diagnosis and discussion of available therapy will at least eliminate the agony that parents go through while looking for answers to What is it and can it be treated? Just knowing "what it is" can be of great relief to many parents.

There are no treatments or therapies currently available to *cure* any genetic disease, but in many cases, therapy will increase the affected person's life span and improve the quality of his or her life.

Genetic therapy for genetic-disease expression is called *euphenics*. Euphenics can be divided into three classes based on the type of therapy administered.

Class-I Genetic Therapy

Procedures and techniques are designed to offset the symptoms of the disease. The therapy involves organ transplantation, drugs, surgery, dietary manipulation, and the avoidance of substances that elicit symptoms. Brief summaries of the different uses of class-I genetic therapy follows:

Organ Transplantation

Organ and tissue transplants are possible ways of installing normal genes into genetically affected patients. The cells of the transplanted tissue contain the genes necessary to produce the required gene product for normal health. The human pancreas has been transplanted in cases of diabetes, and the kidney has been used in transplants for patients who demonstrated the inherited form of polycystic renal disease. Kidney transplants have been unsuccessful in the treatment of patients with Fabry's disease and cystinosis. Bone marrow transplants have also been performed on patients with thalassemia and various forms of immune deficiency diseases and in several lysosomal storage diseases, including Gaucher's and the Maroteaux-Lamy syndromes. However, bone marrow transplants are limited to diseases that do not affect the central nervous system because transplanted bone marrow cells do not cross the blood-brain barrier. Liver transplantation has been used in cases of severe familial hypercholesterolemia.

A major problem in transplantation therapy, even where it is successful, is obtaining scarce organs. In addition, there are problems with continued suppression of the immune system to hold the graft. "Successful" genetic transplants may lead to unforeseen complications later in life. In general, tissue transplantation has not made a major impact on patients demonstrating genetic diseases.

Drugs

Drugs are used in many cases where the genetic disease affects the central nervous system. In many cases, drugs have produced excellent results in normalizing mood and behavioural changes in inherited depressive disorders, in reducing excessive cholesterol in hypercholesterolemics, and in reducing pain and discomfort resulting from gout, migraine headache, and hereditary jaundice. Drugs may also be used to induce or increase the production of certain enzymes. It has been demonstrated, for example, that the drugs danazol and tamoxifen cause an increase in serum levels of alpha-antitrypsin. With respect to enzyme induction, in some types of hereditary jaundice, phenobarbital stimulates the production of a missing enzyme and relieves the symptoms of the disease. The drug 5-azacytidine, when administered to adult beta-thalassemia and sickle-cell anemia patients, turned on dormant genes to produce foetal hemoglobin, which boosted the patients' hemoglobin and red blood cell production. A search is underway to find other drugs that can also turn on normal, but silent, foetal genes to replace abnormal or missing adult gene products.

One inherited recessive disorder, *acrodermatitis enteropathica*, is effectively treated with the drugs diiodohydroxyquinoline and zinc sulfate. This disease is very striking in that the symptoms are scaling and blistering around the body orifices, the eyes, genitals, and anal areas. If left untreated, the affected person loses his or her hair, eyebrows, and eyelashes and passes bulky, putrid-smelling stools. Patients become withdrawn and die at an early age. The administration of diiodohydroxyquinoline brings about a dramatic improvement, hair grows, weight increases, and skin lesions heal. The action of the drug is not understood, but it has been suggested that there is an enzyme defect in the intestinal lining of such people. Because of the lack of enzyme activity in the intestinal lining, a "toxic" substance is not properly detoxified. The drug, binding to this substance, neutralizes the toxic effect, and normal body activities resume.

A wide variety of cytotoxic drugs are also used in cancer chemotherapy. Selected drugs preferentially kill tumor cell. Theoretically, with the use of these drugs a cancer should be disposed of before the host is damaged. However, cancer cells develop a drug resistance. Recent evidence suggests that this drug resistance is associated with a change in the chromosomes of tumor cells. It has been demonstrated that tumor-cell resistance to drug therapy is associated with irregular chromosome morphology and with the presence of extraneous small, paired chromosome fragments called *double minutes.* Some of the change in chromosome morphology is associated with the amplification of certain chromosome regions called *homogeneously staining regions* (HSRs). The presence of extra chromosomal material, double minutes, and HSRs, indicate that such tumor cells contain multiple copies of certain genes. Logically, protein products of these gene play a role in the cell's resistance.

A gene that codes for *multiple drug resistance* (mdr) has now been assigned to human chromosome 7. The gene codes for the production of P-glycoprotein, which is found in relatively large concentrations in human drug-resistant tumor cells. Because P-glycoprotein spans the cell membrane and was found to be structurally similar to some bacterial "pump" proteins (pump unwanted molecules outside the cell), it is speculated that drug-resistance genes–and mdr in particular–circulate the cytotoxic drugs out of the cancer cells before the drugs can kill. It is now believed that gene amplification produces an excess of the *glutathione peptide* that is present in drug-resistant colon-cancer cells. Glutathione neutralizes cell-damaging peroxides and hydroxyl radicals, which are agents that normally kill tumor cells. In cases where a drug or radiation is used to stop DNA replication, thereby stopping cell division of cancer cells, the cancer cell produces gaining an even better understanding of how cancer cells become resistant to different forms of therapy, more effective forms of therapy will be introduced.

Table 10.1. Inherited diseases and class-I genetic therapy

Disease	*Therapy*	*Results*
Acrodermatitis enteropathica	Diiodohydroxyquinoline plus zinc (e.g., zinc, sulfate)	Excellent
Adrenal congenital hyperplasia	Corticosteroids	Excellent
Adrenogenital syndrome	Cortisone induce salt loss	Good
Alpha$_1$-antitrypsin emphysema	No smoking (clean air)	Good
Apert's syndrome	Surgery	Poor to fair
Cystic fibrosis	Diet, antibiotics, mist	Good short-term life expectancy
Cystinosis	Diet (restricted methionine and cystine plus increase ascorbic acid)	Questionable
Cystinuria	High-fluid intake, alkaline, penicillamine	Fair to good for urolithiasis
Diabetes, type I	Insulin	Good
Diabetes, type II	Diet, insulin	Good
Diabetes insipidus	High-fluid intake, low salt	Fair
Familial goiter	Levothyroxine	Questionable
Familial hyperlipoproteinemia	Diet (restricted to short-chain fatty acid)	Fair
Familial xanthomatosis	Cholesterol diet plus drugs	Poor to fair
Favism (G-6-PD dificiency)	Avoidance of specific drugs and fava beans	Excellent
Fructosemia	Fructose-free diet	Good
Galactosemia	Galactose-free diet	Good
Glycinemia	Protein-restricted diet	Poor to fair
Gout (hyperuricemia)	Low-uric acid diet, colchicine plus allopunnol	Good for pain
Gynecomastia	Surgery	Cosmetically excellent
Hartnup disease	Nicotinamide	Fair to good
Hemochromatosis	Iron-restricted diet; venesections	Fair to poor

Hemophilia	Blood factor VIII injections	Good to excellent
Hereditary jaundice	Phenobarbital	Good
Hirschsprung's disease	Surgery	Good
Histidinemia	Histidine-restricted diet	Questionable
Huntington's disease	Drugs for behaviour control	Good
Hydrocephaly	Cranial shunt; head wrap	Good
Hypercholesterolemia	Low-cholesterol diet plus drugs	Fair to good
Hyperlipidemia	Low-fat diet	Fair to questionable
Hyperuricemia	Low-uric acid diet	Poor to fair
Hypospadias	Surgery	Cosmetically good
Ichthyosis vulgaris (alligator skin)	Retinoic acid	Fair to good
Intestinal polyposis	Surgery	Good if early
Lactose intolerance	Avoidance of milk and other dairy products	Excellent
Manic depression	Drugs	Fair to good
Maple syrup urine disease	Diet (restricted leucine, isoleucine, valine and methionine)	Poor to good
Marfan's syndrome	Surgery	Fair to excellent
Meningomyelocele	Surgery	Poor to fair
Methylmelonic acidemia	Vitamin B_{12}	Good
Migraine headache	Drugs	Fair to excellent
Mucopolysacchari-dosis VI	Bone marrow transplant	Under evaluation
Multiple carboxylase deficiency	Biotin	Good
Oroticaciduria	Uridine	Good
Osteogenesis imperfecta	Calcitonin	Fair
Osteoporosis	Calcitonin	Good
Pernicious anemia	Vitamin B_{12}	Excellent
Phenylketonuria	Phenylalanine-free diet	Good
Polycystic renal kidney disease (late)	Kidney dialysis; kidney transplant	Life saving
Prophyria, acute intermittent	IV 3 mg hermatin/kg body weight (experimental status)	Good (initial reports)
Pyridoxine depen-dency	High intake of pyridoxine	Fair to good

Ragweed hay fever	Desensitization	Fair to good
Renal tubular acidosis	Alkali	Good
Retinoblastoma	Surgery, radiation therapy plus cryotherapy	Life saving
Scurvy	Vitamin C (ascorbic acid)	Excellent
Short stature	Human growth hormone	Fair to good
Sickle-cell anemia	Cyanate injection; vitamin B_6 Nitrilosides	Fair to good in crisis Under study
Some solid tumors	Interferon	Poor to good
Spina bifida	Surgery	Poor
Tyrosinemia	Diet (restricted phenylalanine and tyrosine)	Questionable
Urea cycle disorders Argininosuccini-caciduria Ornithinuria Citrullinura	Moderate protein-intake restriction; diet supplemented by arginine	Under evaluation
Usher's syndrome (progressive deafness and blindness)	Vitamin A	Poor
Wilson's disease	Penicillamine restricted copper	Fair
Xeroderma pigmentosum	Avoidance of direct sunlight; use of medicated creams	Fair

Table 10.2. Inherited diseases and Class-II genetic therapy

Disease	*Therapy*	*Results*
Crigler-Naijar syndrome	Blood transfusions; phenobarbital	Fair; short-term life expectancy
Fabry's disease	Enzyme replacement	Questionable
Homocyastinuria	Methionine-restricted diet plus vitamin B_6[2]	Questionable
Rh incompatibilty	Gamma globulin injection	Good to excellent
Von Willebrand's disease	Blood factor VIII injections	Good to excellent
Wiskott-Aldrich syndrome	Blood platelet transfusion	Good

General Surgery

Surgery is most often used for the removal of an organ or an undesirable growth. In effect, surgery helps to relieve the patient's symptoms. For example, splenectomy (removal of the spleen) is quite effective in correcting the disorder of hereditary spherocytosis, a type of anemia. Removal of part or most of the large intestine also eliminates colon polyps that may become malignant. For achondroplasia, leg lengthening operations now lead to, on average, 30 cm of growth in the lower limbs.

Constructive cosmetic surgery has been very helpful in inherited disorders. It is frequently used in cases of cleft lip and palate, Apert's syndrome, gynecomastia (breast enlargement in males). Hypospadias (urethra opening on the underside of the penis), retinoblastoma (cancer of the eye), and some polygenic central nervous system disorders, such as spina bifida and various myeloceles.

Foetal Surgery

Foetal surgery has been developed for urinary-tract obstruction and hydrocephalus (water on the brain). The most successful surgical procedure has been the treatment of urinary-tract obstruction. The obstructed bladder or ureters are drained by temporary nephrostomy tubes implanted in the foetus. *In utero* drainage of foetal hydrocephaly to date has shown less long-term success. The decision to perform a surgical procedure on the foetus must be based on analysis of the total situation. A decision not to undertake foetal surgery may be made when the presence of other equally severe or noncorrectable anomalies are present.

Dietary Manipulation: Ingestion and Avoidance

The most successful treatment of inherited disorders to date appears to be diet regulation. This is not to say that diet therapy alone is adequate; rather, it reflects the present limitations in the treatment of inherited disease. Diet modifications may call for one of the following:

1. Low-level intake of dietary metals, such as low iron for patients with hemochromatosis (excess iron in the liver, heart, and pancreas) or low-copper intake for patients with Wilson's disease.
2. High-level intake of dietary metals, such as copper in patients with Menkes' disease (kinky-hair syndrome).
3. Low-level intake of various dietary amino acids. In phenylketonuria, a minimal level of phenylalanine has been established; in homocystinuria, cystine is kept low.
4. Elimination of dietary factors. In galactosemia and lactose sensitivity, milk sugars galactose and lactose, respectively, are eliminated from the diet; for glucose-6-phosphate. dehydrogenase

deficiency, fava beans are eliminated from the diet and various drugs, especially barbiturates, are eliminated from medications.

5. High-level intake of dietary vitamins. For the treatment of scurvy, vitamin C is used; for refractory rickets, vitamin D is used; and for pernicious anemia, vitamin B12 is recommended.

The success of dietary manipulation in treating a number of genetic disease suggests a close relationship between the gene and its environment and, more specifically, between the gene and nutrition. The administration of large doses of vitamins sometimes restores enzyme function, perhaps through establishing the lost affinity or association between an enzyme and its vitamin cofactor. For example, administration of vitamin B_6 offers some relief in homocystinuria and sickle-cell anemia. The most successful vitamin therapies to date are vitamin D for inherited refractory rickets and vitamin B_{12} for pernicious anemia and methylmelonic acidemia, although over twenty-five inherited diseases can now be treated with megadose vitamin therapy.

For many of the recessive metabolic inherited diseases, diet restriction is an effective treatment. For example, all six genetic disorders of the urea cycle respond to restricted protein ingestion and arginine supplementation. In phenylketonuria, the patient builds up phenylalanine, which causes brain damage. This patient is placed on a diet low in phenylalanine. In contrast to PKU is the disease oroticaciduria where, because of the lack of the enzyme orotic decarboxylase, orotic acid is not converted into uridine. High levels of orotic acid do not appear to be harmful, as is the case in high levels of phenylalanine, but the deficiency of uridine, which is a precursor to the synthesis of nucleic acids, is disastrous. In this case, adding uridine to the diet is an effective therapy.

For those defects (galactosemia, phenylketonuria, maple syrup urine disease, and tyrosinemia) that may be treated with some success, dietary management should begin as soon after birth as possible. Because early diet therapy in these disorders may be of significant help to the individual patient, it is important that the physician, who may only see one or two cases in his or her career, be able to obtain the information on both patient management and special diets. To this end, food and drug manufactures in the United States and Europe produce specially formulated diets for the treatment of particular inherited diseases, and a number of genetic centers provide assistance in diagnosis and management of patients, or in admission to a center if required. One specially formulated diet food is Lofenalac, a compound that is low in the amino acid phenylalanine, for use in treating phenylketonuria.

Class-II Genetic Therapy

Class-II genetic therapy involves the administration of human gene products–an enzyme or other proteins–to patients who demonstrate an enzyme or protein-factor deficiency.

Enzyme Therapy

Regardless of route of entry, enzymes, which are protein in nature, are quickly inactivated by the body's immune defense mechanisms. Because of this inactivation, enzyme therapy has not been successful and to date is not considered to be a practical means of genetic-disease therapy. Enzyme therapy has been tried with minimum success in the treatment of Fabry's disease, a glycolipid metabolic disease. With Hurler's and Hunter's syndromes (mucopolysaccharide diseases), unsuccessful attempts have been made to infuse patients with plasma from normal persons. It was hoped that in such treatment, the enzyme of the normal plasma would degrade the accumulated mucopolysaccharides in the patients. Hexosaminidase A enzyme therapy has also been tried in Tay-Sachs disease and Sandhoff's disease patients with similar lack of success. Arylsulfatase A therapy in metachromatic leukodystrophy patients is also unsuccessful. Children with argininemia (high levels of arginine) have been infected with Shope papilloma virus on the theory that a viral gene might allow for the production of the enzyme arginase, which would degrade the accumulated arginine. To date, these experiments have been unsuccessful. The administration of glucocerebrosidase to some Gaucher's disease patients (a recessive glycolipid metabolic disease) resulted in a temporary reduction in accumulated glycolipid levels. Recently, a capsule form of a fungus lactase enzyme has been marketed. The enzyme temporarily restores deficient human intestinal lactase enzyme activity in person who are lactase deficient. Ingestion of the enzyme allows these people to consume milk and milk-related products like ice cream, cheese, pizza, and many other popular foods.

There is at least one other bright spot in the investigations on enzyme therapy for genetic diseases. It is the use of alpha 1-antitrypsin.

Alpha 1-antitrypsin

Alpha 1-antitrypsin is an enzyme secreted by the liver and macrophages. It protects lung connective tissue from degradation by *proteases* (enzymes that break down proteins), particularly the enzyme elastase. An excess of elastase in the lungs leads to *emphysema* (severe breathlessness). Weekly administration of alpha 1-antitrypsin brings the level of this enzyme in the blood and lungs of emphysema patients to normal.

The alpha 1-antitrypsin gene is codominant, and at least twenty-five variants in the glycoprotein structure of the gene have been described. The more common variants of the gene are listed in table 8.3. The most common, or "normal," genetic variant is called Pi M (Pi stands for protease inhibitor). About 90 percent of the U.S. population is homozygous Pi MM and have the highest level of the serum enzyme. Heterozygous variants of this gene produce at least 30 percent of the normal level and the phenotype is basically unaffected. However, the homozygous Pi ZZ produces a very low level of the enzyme and the phenotype suffers from emphysema and other liver and pulmonary disorders. There are over 100,000 Pi ZZ patient in the United States who may benefit when broad-based enzyme therapy begins. Currently, the major problem is widespread administration is obtaining a sufficient supply of the enzyme. This problem should soon be overcome as researchers have sequenced the 324 amino acids making up the enzyme and have produced the enzyme via recombinant DNA techniques. An associated problem, however, is that the genetically engineered enzyme, for reasons not yet understood, is too quickly removed from the serum. Attempts are being made to deliver the enzyme into the lungs in an aerosol form. Until such time that the enzyme can be successfully maintained in the serum, the very best therapy is to stop smoking. The deleterious effects of cigarette smoking in alpha 1-antitrypsin-deficient individuals has been extensively documented. Unlike smokers with normal enzyme levels who run a 20 percent risk of developing lung disease, the risk of lung disease in smokers with the enzyme deficiency is over 70 percent and probably approaches 100 percent over time.

Table 10.3. Alpha 1-antitrypsin deficiency phenotypes

Genotype	*Affected U.S. population (%)*	*(%) Serum alpha 1-antitrypsin concentration (% relative to concentration in subjects with the MM phenotype)*
Pi MM	89.5	100
Pi MS	7.1	75
Pi MZ	3.0	57
Pi SZ	0.2	37
Pi ZZ	0.1	16

Protein Factors

Various factors required by the body, but missing because of a genetic error, can be successful replaced. Examples are blood factor VIII, the antihemophilic globulin required for blood clotting; calcitonin, a hormone necessary for the retention of bone calcium in such hereditary disorders as osteogenesis imperfectal the insulin hormone for diabetes; and anti-Rh globulin for prevention of Rh disease.

In both class-I and class-II genetic therapies, the procedures and treatment are compensatory and palliative rather than curative. That is, the use of class-I and class-II genetic therapies is aimed at modifying the consequences of a defective gene; such therapy does not cure the diseases. Class-I and class-II therapies do not alter the gene, only the environment. Class-III genetic therapy is gene therapy; it is a molecular attempt to cure genetic disease. Researchers are working to provide a means of curing genetic defects by manipulating the DNA of the gene.

Class-III Genetic Therapy

Class-III genetic therapy is the addition to a normal gene to cells carrying a defective gene. In theory, this third type of gene therapy is simple: insert a "good" gene into a somatic cell to compensate for a "bad" gene. In reality, somatic-cell gene therapy is biologically complex and raises many ethical questions. To begin with, the defective gene must be found from among the 100,000 or so other genes. The gene must then be cloned, placed into the cells of the proper tissue type, and finally induced to express normal function. The task is difficult, but not impossible. For example, in 1982 Wilson and colleagues established the amino acid sequence of normal *hypoxanthine-guanine phosphoribosyl transferase* (HPRT) enzyme from human RBCs. The enzyme consists of four protein subunits, each of which consists of 217 amino acids. Establishing the normal amino acid sequence of HPRT enzyme protein enabled Wilson and colleagues to identify the specific amino acid substitutions in mutant forms of the HPRT enzyme and to infer alterations that must have occurred in the DNA. They compared the amino acid data from three HPRT enzyme-deficient patients with missense mutations (one amino acid substituted for by another) who demonstrated gout with a missense mutation that did not affect the level of the HPRT enzyme in a Lesch-Nyhan patient. The substitution of asparagine for arginine at position 193 in the fourth individual resulted in a normal HPRT enzyme level, but the person developed the Lesch-Nyhan syndrome. While this method of mutant analysis has proved valuable in the description of pathogenic mutations, it is limited to those that result in the production of a protein product. In a study by Stout and Wilson, eleven of fifteen Lesch-Nyhan patients did not produce a protein.

Jolly and colleagues (1983) established the DNA sequence of the HPRT gene. The gene consists of nine exons and eight introns. Miller and colleagues have cloned the normal HPRT gene and inserted the DNA into murine retroviral vectors. The virus carrying the HPRT gene was used to infect HPRT-negative human cells in culture. The infected cells were cured of the genetic defect. The produced the HPRT enzyme at levels found in normal cells. Next, bone marrow cells of

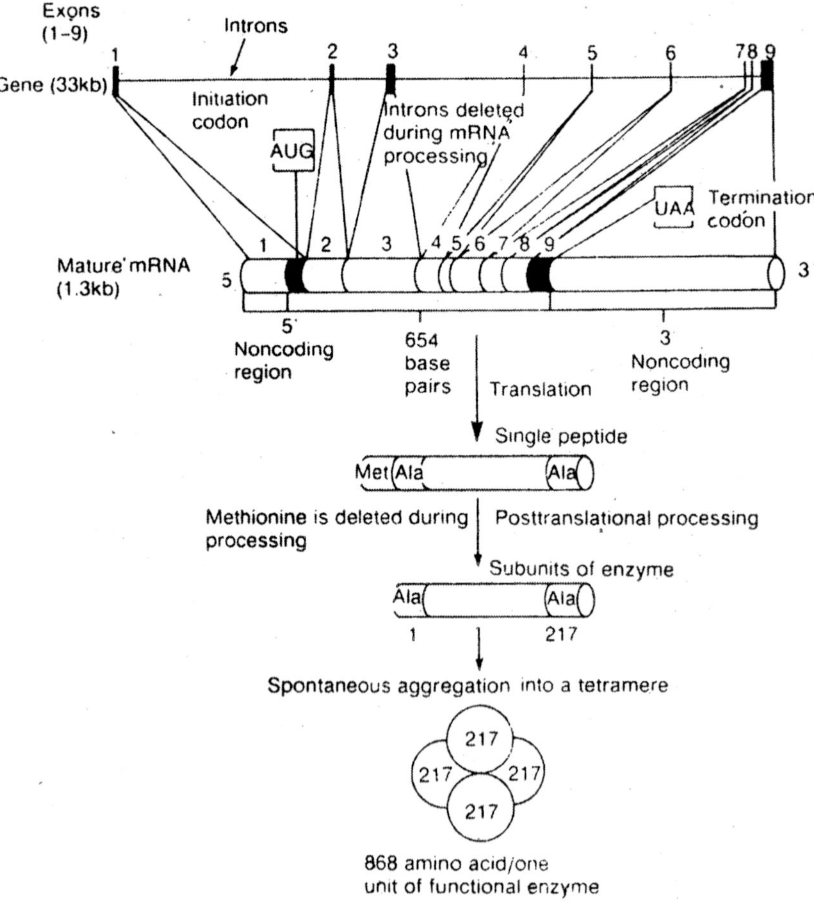

Fig. 10.1. Molecular reproduction of the HPRT gene.

mice were infected with the virus carrying the HPRT gene and replaced in the mice whose bone marrow cells were killed by radiation. The engineered bone marrow cells successfully populated the mouse bone marrow and produced the HPRT enzyme. The next step is an attempt to engineer human bone marrow cells. The plan is to remove bone marrow from the hip of a Lesch-Nyhan patient, infect it *in vitro* with the virus, and then transplant it back into the patient. There is some evidence from animal experiments that an injection of the engineered virus may be effective. However, at this time there is no way of controlling or knowing where in the human system the virus will or will not locate. The purpose of the initial experiment is to create a model by which gene transplantation or genetic engineering can be routinely accomplished. After it is accomplished, a gene transplant into a defective newborn may be an effective treatment.

In parallel experiments to those of Jolly and colleagues, the normal human HPRT gene was isolated, cloned, and made ready for the first human genetic engineering experiment. In the meantime, the DNA probes made to hybridize with the HPRT gene have made possible accurate prenatal diagnosis and carrier testing for the male foetus or adult suspected of having the defective HPRT gene.

TYPES OF GENE THERAPY

Studies in preparation for human gene therapy engage a variety of technologies, the latest being recombinant DNA technology. The different technologies to actually be used depend on how the physician wishes to correct the defective gene. Currently following forms of genetic therapy are envisioned.

Genes and Gene Regulation

Researchers developing gene therapy approaches face two main obstacles: identifying key genes and their mode of regulation and inserting genes into appropriate chromosomal sites. For each single-gene disorder, all of the following will be necessary:

1. Identification of the normal gene site and nucleotide sequence that has been altered and the protein it encodes.
2. Understanding of how the gene's expression is regulated.
3. Creation of vectors or physical methods that can be used to transport the normal gene into cells where the gene can be expressed.
4. Ensuring that the gene becomes stably integrated and expressed so that the protein product is present at levels sufficient to reduce or eliminate effects of the disorder.

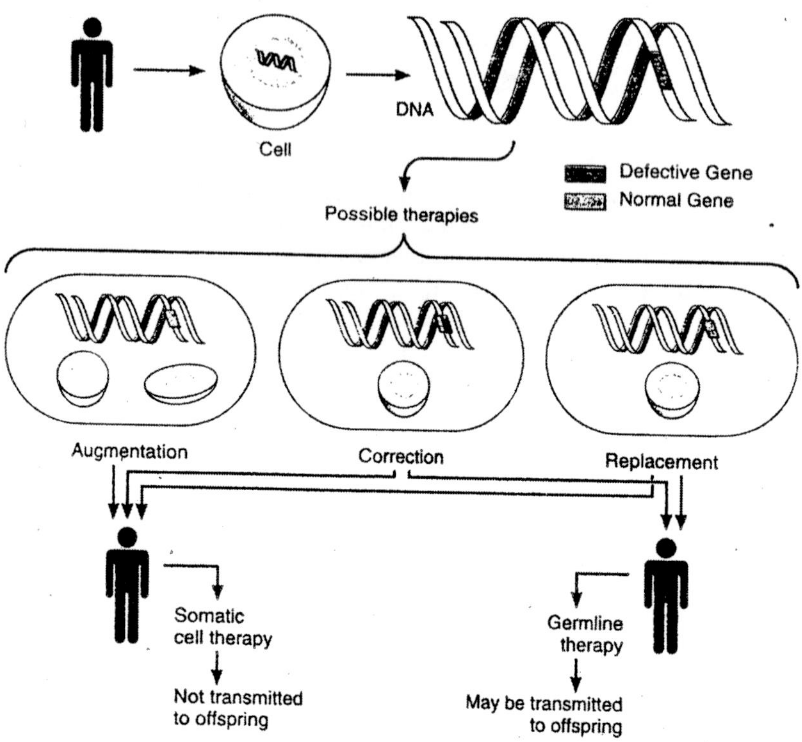

Fig. 10.2. Gene therapy consists of different strategies designed to modify the effects of a defective gene.

Gene Insertion

Since the early 1970s, the use of mouse and bird retroviruses as vectors for transferring normal human genes has been carefully investigated. *Retroviruses* are unusual viruses because their genetic material is RNA, not DNA. Once they infect a host cell, retroviruses are replicated through a mechanism that translates their RNA int. DNA. This, of course, is the opposite of the normal translation of genetic information, DNA into RNA, hence the prefix *retro-*. Human retroviruses cause several important diseases, including some cancers and AIDS.

In the laboratory, retroviruses have several advantages as vectors compared to others that have been studied. They are easy to manipulate and have a demonstrated ability to transfer foreign genes into cells and have the genes expressed. However, retroviruses are sometimes unstable, it is impossible to control where the genes they carry are inserted, and even if they do become integrated, the genes are not always expressed. Recipient human cells must be dividing before the gene carried by a retrovirus can be integrated in their chromosomes. For example, bone marrow cells divided rapidly and constantly to produce blood cells and may be good candidates for retroviral vectors, depending on the purpose of the gene insertion. Because of the cell division requirement, these vectors cannot be used for introducing genes into the mature cells–brain, nerve, muscle, or other cells–that make up most human tissues.

Despite these problems, it is evident that retrovirus vectors will play a prominent role in future developments. Other viral vectors are now receiving increased attention. For example, *herpesviruses*, which normally live inside nerve cells (and cause such problems as cold sores and genital herpes in humans) can perhaps be modified to serve as vectors for inserting genes into nervous tissue, physical methods, such as microinjection and high-velocity microprojectiles, are also being studied.

Through all of the difficulties have not yet been overcome, impressive progress has been made since the mid-1980s. Some forms of gene therapy now appear to offer great promise for ameliorating the effects of certain human genetic disorders.

Ethical Issues

The policy of the U.S. National Institutes of Health Recombinant DNA Advisory Committee (RAC) currently prohibits germline therapy experiments, but human somatic cell therapy research is taking place in laboratories throughout the country and the world. Most scientists and physicians view the projected uses of somatic cell therapies as extensions of accepted medical procedures, comparable to blood and marrow transfusions used for treating diseases. Also, few disagree with the justification that the use of somatic cell therapy is appropriate for treating a genetic disorders when no other treatment is available. Consequently, few specific objections have been raised based on ethical considerations.

In sharp contrast, many people, including ordinary citizens, scientists, politicians and others, have questioned the wisdom of creating genetic alterations that can be transmitted to future generations Is it acceptable or unacceptable to attempt to alter the germinal cells of an individual with a genetic disorder in such a way that the person can have children who will not have the disorder? Perhaps there would be general agreement that germline therapy might be allowed for single-gene disorders. However, most human traits are controlled by more than one gene, and little is known about the genetics of such traits.

With respect to humans, no clear consensus about germline therapies has yet emerged, in part because the research has not progressed to the point where serious concerns have arisen. However, because of the severe technical problems and a clear indication from most groups that they are not yet ready to accept experiments centered on human germline alterations, it seems unlikely that any research in this area will be approved in the near future.

Somatic Cell Gene Therapy

There are two general categories of somatic cell gene therapy, *gene substitution* and *gene augmentation.* As originally conceived, it was thought that *gene substitution* could be used to remove and replace mutant genes responsible for genetic disorders. However, there have been no successes in creating such substitution therapies. Alternatively, would it be possible to "correct" the abnormal DNA sequences of mutant genes? This might be accomplished by using established gene transfer methods to introduce a normal DNA sequence that would recombine, through complementary base pairing, with the mutant sequence. If the normal gene were then expressed, the needed protein would be available and the effects of the disorder would be reduced or eliminated. Some progress has been made in gene correction techniques and their application to research on animals such as mice.

Gene augmentation involves techniques that will equip cells with added genes that can produce the missing protein associated with the genetic disorder. Two basic methods are being developed; (1) a "standard" technique in which competent genes are inserted into cells of an afflicted individual, followed by introduction of the altered cells into the appropriate tissue where they are normally expressed, and (2) the insertion of the needed gene into "donor" cells that are then placed back into the body, where they produce the protein. This latter approach has been referred to as a "drug delivery system," and it offers exciting prospects. How does the donor cell system work? Vectors containing a normal gene encoding a missing protein–an enzyme produced by the liver, for example–are inserted into nonliver cells such as bone marrow cells, which produce white blood cells, or endothelial cells, which line the inside of blood vessels. These donor cells are then transferred into the body, where they become established, express the gene, and pump out useful quantities of the missing enzyme into the bloodstream. There have been notable accomplishments using both the standard and donor cell types of gene therapy.

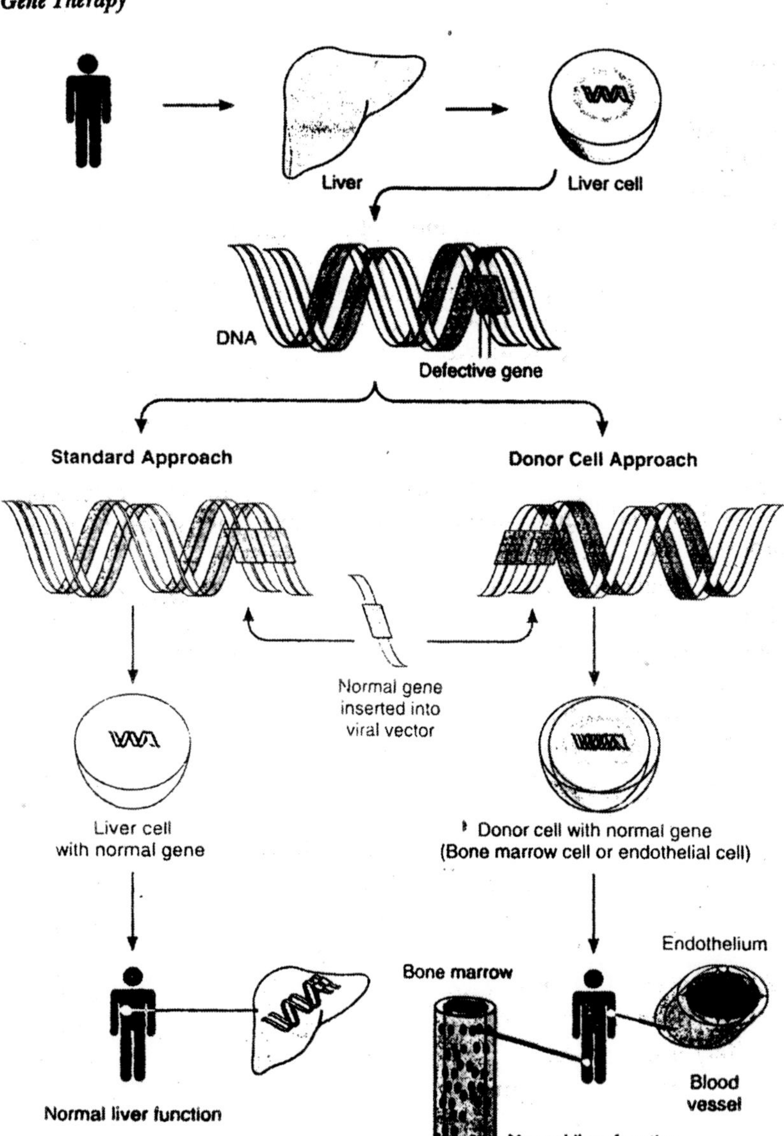

Fig. 10.3. Two gene augmentation approaches have been developed.

Animal Models and Case Study

For ethical and practical reasons, basic experimental research in gene therapy is not first conducted directly on humans. Given these constraints, how is it possible to make advances that may eventually lead to the treatment of humans? In general, new techniques and therapeutic approaches are developed and tested in the laboratory using cells obtained from various animals, including humans, which can be cultured (grown and amplified by cell division) in small containers. Subsequently, new methods are tested and analyzed using animals, primarily mice, rats, and rabbits. Once a technique that may have useful applications in gene therapy has been developed, it becomes necessary to conduct experiments on animals that have specific disorders that also occur in humans. Such a species is referred to as an *animal model*; that is, its disease serves as a model for learning about the nature of the condition in humans. The use of animal models for studying human diseases (for example, cancer) has a long and successful history. Only after exhaustive tests and demonstrations of safety will specific gene therapies ever be used on humans. Some very interesting research related to gene augmentation therapy demonstrates how this process works.

The familial hypercholesterolemia (FH) is a human genetic disorder characterized by high levels of cholesterol in the blood that often leads to premature death. The most common cause of FH is a deficiency in the number of protein receptors on liver cells that normally remove this substance from the blood. Is an animal model available for studying FH? What strategy might be developed for treating or curing FH? The answer to the first question is yes; the Watanabe rabbit has the same disorder. In the late 1980s, researchers provided information relevant to the second question by accomplishing the following:

1. Methods were devised for inserting foreign genes into the liver cells of various animals. This was notable because liver cell had previously been resistant to gene transfer.
2. Viral vectors were used to insert normal copies of the gene encoding the cholesterol receptor protein into cultured rabbit liver cells.
3. Once inside the rabbit cells, the cholesterol receptor gene produced functional receptor protein that removed cholesterol from the experimental system.

The next step will be to place the cells back into Watanabe rabbit in an attempt to cure the disease. For that to happen, the altered cells will have to become permanently established in the rabbit's body and continue to produce the receptor protein. Though these very exciting results have obvious implications for the treatment of FH, they represent only the first step in a long path leading to human clinical applications. What is required before similar types of gene therapy tests can be conducted on humans?

Authorization Procedures for Studies on Humans

Before conducting any experiments related to human gene therapy, scientists must obtain legal approval from several regulatory agencies. Research plans must first be reviewed by local institutional review boards and institutional biosafety committees. At the national level, approval is required from the RAC and the director of the NIH. In addition, the proposed experiment must be described in nonscientific language and published in the *Federal Register* so that members of the public can, if they wish, offer comments to the RAC and the NIH. The purposes of this review are to ensure that only qualified research teams conduct such tests, that all safety and ethical issues are fully considered and evaluated, and that the public has an opportunity to express its feelings.

The first experiment

How does the authorization process work in real life? Early in 1988, a request to conduct the first human gene therapy experiment in the United States was submitted by W. French Anderson of the National Heart, Blood, and Lung Institute and Steven Rosenberg and R. Michael Blaese of the National Cancer Institute. Their proposal stemmed from earlier investigations on a promising new approach for treating cancer. Specific immune cells–white blood cells called *lymphocytes*–that kill "foreign" cells, including cancer cells, normally exist in the body. In early studies, lymphocytes that had infiltrated the tumor (*tumor-infiltrating lymphocytes*, or TIL cells) of seriously ill patients were removed surgically and induced to grow and multiply in the laboratory. When sufficient numbers of cultured TIL cells had been generated, they were reinfused into the patient. TIL therapy caused significant tumor regression in about 50 percent of the 25 patients tested who had advanced cancers and had failed to respond to other treatments.

Why was TIL therapy effective only in some of he patients? To try to answer this question, Anderson, Rosenberg, and Blaese wanted to insert a single-gene "marker" (for resistance to the antibiotic neomycin) into TIL cells using a retrovirus vector and follow their progress once placed back into a patient. The marker would allow researchers to identify TIL cells with the gene by collecting blood samples periodically during treatment and then exposing all blood cells removed to neomycin; those with the marker would survive, while those without the resistance gene would perish. The marker would have no effect on either the TIL cells or their capacity for attacking cancer cells.

In accordance with the procedures described earlier, the proposal was approved by the various committees, but not without delays related to procedure (not all of the relevant preliminary data were made available to RAC members) and a large number of public hearings. A lawsuit alleging that the review process was not open to the public was eventually dismissed.

Finally, on May 22, 1989, a severely ill cancer patient received the first infusion of TIL cells that had been genetically altered to contain the foreign neomycin gene. Additional patients were treated during the following weeks. Results from the experiment were similar to those from earlier studies–some patients showed dramatic regression of their tumors, and others did not. The significance of the research lies in what was learned about the marker TIL cells introduced into the patient. They survived in the body, apparently attacked and killed tumor cells, and were unchanged by the inserted gene. The next step in this research was to seek approval to insert a gene for an active antitumor agent (a natural protein produced by the immune system) into TIL cells. Approval for such an experiment was granted in 1990, and results from new experiments may be available in 1991.

Though perhaps not technically a gene therapy experiment (the gene inserted into TIL was neutral, not therapeutic), this is considered the first such study in the United States. The TIL experiment opened the doors to further studies involving gene therapy. For example, the same group of researchers received approval to test the effects of a *tumor necrosis factor* (TNF) gene inserted into lymphocytes. TNF is a protein that shrinks tumors by cutting off their blood supply. The first patients were treated in 1990, and further tests are now under way. In 1990 and 1991, children with a rare, lethal immune deficiency known as ADA (because of a defective enzyme called *adenosine deaminase*) received transfusion of their own lymphocytes, which contained foreign ADA genes. One child, a 4-year-old girl, received cells on September 14, 1990. This is considered the first test of human gene therapy. Finally, in 1991, scientists succeeded in synthesizing normal dystrophin genes. Gene therapy trials involving this gene are expected to begin soon. Clearly, an increasing number of human gene therapy experiments can be expected in the next five years.

THE HUMAN GENOME PROJECT

What is known about the location of genes on human chromosomes? In a monumental achievement, the first partial map of the human genome was published in the journal *Cell* in 1987. This *genetic linkage map* is based on the location of gene loci on each chromosome relative to the positions of specific markers. Each marker provides a distinct reference point on the chromosome, and together they serve as a primitive map for identifying the approximate location

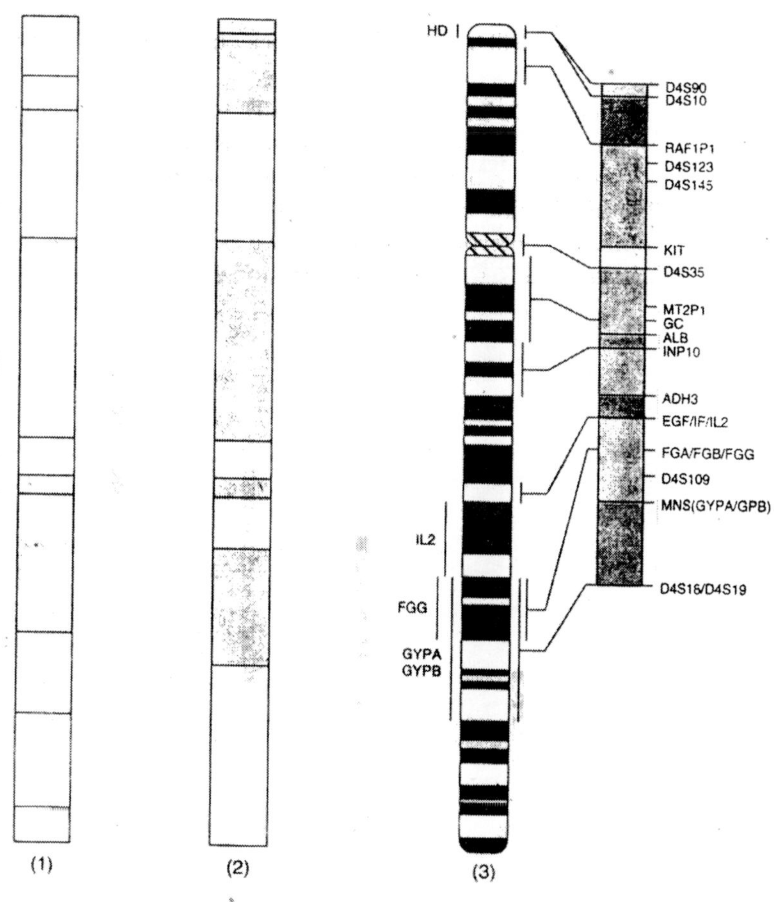

Fig. 10.4. Figure showing human genome map.

of various genes of interest. The blanks continued to be filled in and in October 1990, an undated human genome map was published in the journal *Science.*

A question on an even grander scale was raised during the mid-1980s: Could a high-resolution physical map consisting of the exact sequence of nucleotides in the entire human genome be drawn? A *physical map* shows the actual location of genes and also the distances between genes as determined by constant, identifiable landmarks. Figure shows the progression of human chromosome mapping. The crudest physical map uses chromosomal bands as landmarks. A map based on linkage markers represents a higher degree of resolution, and the ultimate physical map will be the complete nucleotide sequence of the entire genome.

In 1985, Charles De Lisi, director of the Department of Energy (DOE) Office of Health and Environmental Research, formally proposed a massive project designed to elucidate the complete sequence of the 3 billion nucleotide bases and 50,000 to 100,000 genes in the human genome. The enterprise became known as the *Human Genome Project,* and the original estimates for completion were \$3 billion and 15 years.

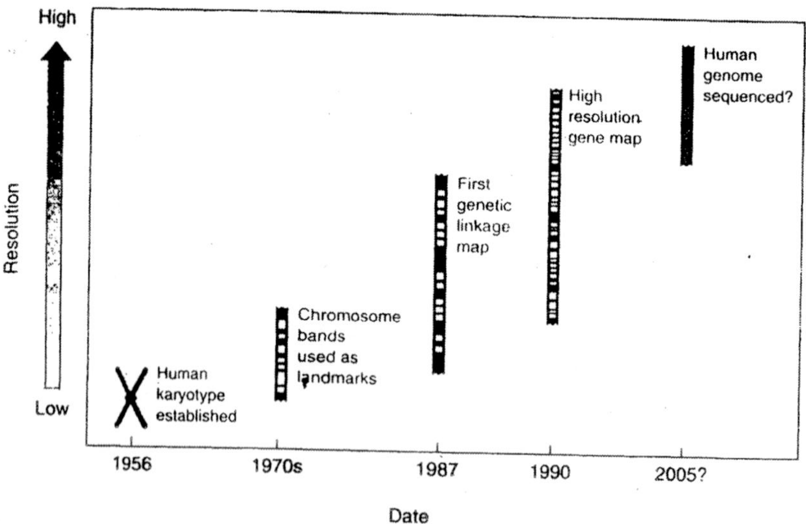

Fig. 10.5. Human genome map.

The initial reaction of the scientific community was mixed. On the one hand, there was great enthusiasm related to gaining access to the "Holy Grail of biology," as one molecular biologists put it. On the other hand, there was concern about the magnitude and cost of what promised to be biology's first "big science" project (that is, costing billions of dollars and involving large, organized, interdisciplinary teams of scientists) and the possible effects on traditional "small science." These concerns have not yet been completely dissipated, and funding of the project remains contentious. In 1989, the proposal was approved under NIH control (although the DOE will still play a research role). The official starting date was October 1, 1990, large-scale sequencing trials began in 1991, and the projected completion date is September 30, 2005, approximately 50 years after James Watson and Francis Crick published their seminal paper on the structure of DNA. The project is centered in the new Office of Human Genome Research at the NIH, and the same James Watson was named director in 1988. Thus, as he put it, he will now complete his journey "starting with the double helix and going up to the double helical structure of man."

Project Research

How will the Human Genome Project proceed? From 1990 to 1995, emphasis will be on developing new technologies required to complete a physical map. Model organisms such as *E.coli*, yeast, *Drosophila*, and mice will be "sequenced" (that is, the exact sequence of the nucleotide bases in their DNA will be determined). Interesting regions of the human genome–for example, those known to be near genes that cause important genetic disorders such as Huntington disease–will also be sequenced. During the next five years, technology for large-scale mapping will be developed, and some of the smaller human chromosomes will be sequenced. Finally, during the last five years of he project, sequencing of the entire genome will be completed.

Mapping the human genome will be the centerpiece of the project, but maps for other organisms, including economically important plants, will also be produced. The current human genome map contains the equivalent of one street marker for each 10 to 50 miles of DNA; a higher-resolution physical map, perhaps achievable as early as the mid-1990s, will reduce that distance between markers to 1 mile. The ultimate DNA sequence map will consist of the nucleotide sequences for each chromosome.

The precise timetable for the project will remain quite fluid since new technical advances can arise quickly and create opportunities for accelerating progress. One such development has already occurred. In December 1989, the prestigious journal *Science* named DNA polymerase the "molecule of the year" because of its role in a powerful new technique called the *polymerase chain reaction* (PCR). PCR came into widespread use in the late 1980s and early 1990s and was selected as the "major scientific development of 1989" in the same issue. Using PCR, short unique pieces of DNA can be amplified, their squences determined, and their precise chromosomal location established. In other words, single-copy DNA sequences–those that occur only once in the human genome–rather than linkage markers or other identifiers, can be used to define the basic landmark on the physical map. These sequence-tagged sites (STS) represent a common language and a new approach that may move the timetable for the physical map forward at least five years and perhaps more.

Importance of the Project

The knowledge gained from the Human Genome Project will be of immediate importance in diagnosing single-gene disorders that affect our species. Progress can also be expected in learning about the underlying nature of enormously complex polygenic (or suspected polygenic) disorders–such as alcoholism, mental illnesses, certain cancers, and heart disease–that affect millions of people but are poorly understood at present. When the project is completed, genes associated with polygenic traits may become apparent as well as the underlying mechanisms that regulat their activities. As with single-gene defects, a basic understanding of the geneitc defect should direct research toward methods for the treatment of various disordes. Certainly, many questions exist about the regulation of such genes, which seem to operate in some coordinated fashion to produce a given trait or genetic disorder. Farther down the road, the new information will also contribute to a greater understanding of gene control mechanisms in eukaryotes, human genome organization, normal and abnormal cellular growth and development, and evolutionary biology.

Besides being vital to the applied sciences, the project is of considerable importance to basic science. The genetic information in the genome repesents the fundamental instructions for the origin, development, maintenance, and eventual demise of an organism. All questions related to these topics can conceivably be addressed from the conceptual framework provided by knowledge of the genome.

However, deep knowledge of the genome does not guarantee that we will be able to understand complex processes such as the development of an organism from a fertilized egg. All the genes and their locations may be known, but key details for understanding how all of this information is processed may not be available. Nevertheless, the 1990s should be as fruitful for biological research as the previous decade, and almost certainly more spectacular.

The Future

With respect to humans, the future applications of some new genetic technologies offer exciting prospects for benefiting humankind. For other technologies the future remains uncertain. The molecular technologies we have described throughout this unit have revolutionized our understanding of many human diseases. This field of *DNA diagnostics*–the analysis of disease at the nucleic acid level–has progressed so rapidly that the time may be near when many genetic disorders, cancers, and viral diseases will be analyzed using rapid, automated, and inexpensive procedures. For some disorders, new technologies have already been used to determine the exact disease mechanism causing the phenotypic effect, and others will be identified shortly. Thus it may soon become possible to treat numerous disorders that diminish the quality of life or lead to premature death.

11

Human Genetic Disorders

Diagnosing human genetic disorders using new molecular techniques is not reaching an advanced state, Mutant alleles that cause a number of genetic disorders can now be identified in individuals before phenotypic effects become apparent. As a result, it has become possible to determine whether or not individuals are risk (that, it with a family history) for late-acting agenetic disorders, such as Huntington disease, will develop the disorder later in their lives. Also, these methods have been adopted for prenatal diagnosis for an increasing number of genetic disorders including Duchenne muscular dystrophy, cystic fibrosis, retinoblastoma (eye cancer), and Huntington disease.

Other complex disorders that appear to have at least a partial genetic cause, such as certain cancers diabetes, and cardiovascular disorders, may also become amenable to new DNA testing procedures. It is expected that such tests, when fully developed, will allow credible predictions of the risk for these and many other genetic disorders. This ability to predict the future accurately has severely challenged the ethics and values of people who may be directly or indirectly affected by the new genetic technologies. These individuals include scientists, physicians, politicians, parents and prospective parents, and all concerned citizens.

Common Baldness

This trait is the common form of hair loss in Caucasians, made famous because President John Adams and several of his illustrious descendants had the trait. Beginning at about age 20 in males and age 30 in females, hair begins to fall out at the sides near the front of the scalp, producing an M-shaped hairline. Recession of the hairline in front is usually followed by the development of a bald spot at the rear of the crown. Enlargement of the bald spot and continued recession of the front hairline lead to a final stage where only a peripheral fringe of scalp hair remains. The trait is a simple Mendelian trait due to an autosomal allele; however, the allele is *dominant* in males bur *recessive* in females. The trait is most common in Caucasians: Estimates are that at least 50 percent of Caucasian men and 25 percent of Caucasian women will be affected, although the extreme progression to a remaining fringe of hair with a bald pate occurs in only about one-third of affected men (affected women typically experience less hair loss than affected men). The trait is less common among Negroes than among Caucasian, and it is rare among Orientals.

The trait of common baldness illustrates several complications, First it is a *sex-influenced* trait, which refers to the fact that the trait is expressed differently in males and females (recall that the baldness allele is dominant in males but recessive in females). Extreme cases of sex-influenced traits, in which expression is limited to one sex, are known as *sex-limited* traits; example of sex-limited traits in humans are growth of facial hair, which ordinarily occurs only in males, and development of breasts, which ordinarily occurs only in females. Common baldness also illustrates *variable age of onset*; individuals who have the baldness genotypes from the time of birth, but the trait does not occur not occur until much later, and its time of occurrence is variable, beginning in the early 20s for some men but later for other men. The trait also exhibits *variable expressivity*, which means that it is expressed to different degrees in different people. The severity of expression of common baldness may vary from a slightly receding frontal hairline to an almost completely bald pate. The same genotype may be expressed differently in different individuals because of differences in *genetic background* (i.e. the genotypes at other loci that also influence the trait) or because of differences in nongenetic or environmental factors. The extreme end of variable expressively is known as *incomplete penetrance*, in which individuals who have a genotype associated with a particular trait do not express the trait at all. Finally, common baldness illustrates *variation among populations* in its incidences. As noted, the trait is common among Caucasians, less common among Negroes, and rare among orientals. Variation among populations is a characteristic feature of many inherited traits, and additional example will be found in later chapters of this book.

Chin Fissure

The most common type of chin fissure is a perpendicular furrow in the middle of the chin, and it is due to an autosomal dominant allele. The trait is variable in expression (ranging from a chin dimple to a Y-shaped furrow), but it has almost complete penetrance in males (i. e., all males who carry the allele will have some sort of chin fissure). Females are affected only about half as frequently as males, so the trait is sex influenced. Chin fissures have a variable age of onset; although sometimes present at birth, they more commonly appear in childhood or early adulthood. Among people of German extraction, approximately 20 percent of males and 10 percent of females have the trait.

Ear Pits

These harmless and usually shallow pits may occur in one ear or both. Ear pits are due to an autosomal dominant allele, which is variable in expression (one ear or both affected, shallow or deep pits), has incomplete penetrance (only about half of those individuals who are heterozygous or homozygous for the dominant allele will have ear pits), and is sex influenced (twice as prevalent in females as in males). The incidence of the trait varies among populations–1 percent in Negroes, Caucasians, 5 percent in American Negroes, and said to be "very common" in Chinese.

Darwin Tubercle

The Darwin tubercle is a variable-sized thickening of cartilage near the upper rim of the ear, and ut is probably to an autosomal dominant allele (i. e., homozygote and heterozygote are both affected). The allele has a high penetrance and the trait is not sex influenced because it has approximately equal frequency in males and females. Prevalence varies among populations, being about 20 percent in Germany and about 50 percent in England and Finland.

Ear Cerumen

In the Japanese language, the gray, dry, and brittle type of ear wax is called "rice-bran ear wax," whereas the brown, stick, and wet type is called "honey ear wax" or "cat ear wax." The dry type is due to an autosomal recessive allele. Type of ear cerumen has remarkable variation among populations; the dry type (which occurs in homozygous recessives) has the following frequencies in various population : Koreans (93 percent), American black (5 percent). The trait illustrates two important phenomena. Fist, most Koreans and Japanese who have wet cerumen will actually be heterozygous for the dominant " wet " allele; they are thus heterozygous for the recessive "dry" allele. Such individual who are heterozygous for a recessive allele are said to be *carriers* of the allele. Second, ear cerumen illustrates what is called *pleiotropy*, which refers to the fact that genes can have effects on several or many traits and not just one trait. In the case of ear cerumen, the alleles influence secretion of certain glands in the ear canal; these very same alleles also influence the chemical makeup of secretions from certain other glands, notably those involved in perspiration. Thus, the alleles that affect ear cerumen are *pleiotropic*. They affect several traits (type of ear wax and chemical makeup of perspirations.

Congenital Ptosis

The word *congenital* means "present at birth," and congenital ptosis refers to a drooping eyelid usually affecting only one eye that is in most cases due to weakness in upward movement of the eyelid. The condition is usually present at birth and may in some cases require surgical correction. The trait is caused by an autosomal dominant allele with incomplete penetrance (about 60 percent of homozygous the trait). Although the exact prevalence of congenital ptosis is unknown, it is thought to be quite common.

Epicanthus

Epicanthus refers to a condition in which the inner corner of the eye is bridged by an arching fold of skin extending from the bridge of the nose and giving the eye an almond-shaped or "Oriental" appearance. The trait is due to an autosomal dominant allele that is extremely variable in its expression . The trait often tends to disappear with age, being present in about 20 percent of 1-year-old Caucasians but in only about 3 percent of those 12 years and older. Approximately 70 percent of adult Orientals have epicanthus, but it is not clear whether genetic basis as in other racial groups. (As a point of interest, epicanthus is a normal characteristic of the *fetus* in all racial groups.)

Camptodactyly

Camptodactyly is a minor hand abnormality usually expressed as an inability to completely straighten the little finger. It is due to an autosomal dominant allele with incomplete penetrance and variable expressivity, but its prevalence is not known.

Min-Digital Hair

Presence of hair on the middle segment of the fingers is thought to be due to an autosomal dominant allele. Homozygotes for the recessive allele lack hair on these finger segments, even if they otherwise have abundant body hair. The prevalence of the react is unknown,

Phenylthiocarbamide Tasting

The ability to taste this chemical is due to an autosomal dominant allele usually symbolized as T; the recessive allele is symbolized as t. Thus, genotypes TT and Tt are tasters and report that the chemical is bitter, whereas tt genotypes are nontasters and cannot detect the chemical. Approximately 30 percent of Western Europeans are nontasters. Tasting ability is of some interest because it illustrates that genes can influence almost any trait, including, in this case, our sensory perceptions.

S-methyl Thioester Detection

Certain odiferous substances are excreted into the urine by at least some people after eating asparagus. At one time, excretion was thought to be due to a single dominate allele, nonexcretors being homozygous recessives, but now it appears that the primary genetic variation involves the ability to smell the substances. Approximately 10 percent of Caucasians can smell the diluted asparagus-related substances.

Familial Emphysema

Emphysema refers to an over inflation and distension of the air sacs in the lungs marked by continuous shortness of breath (even when at rest), a wheezy cough, and increased blood pressure in the arteries of the lungs leading to an enlargement of the right side of the heart. Without proper treatment, the disease progresses to obstructive lung disease and eventual death due to respiratory failure or congestive heart failure. Onset of the disease usually occurs in middle age, and the life span of affected individuals is shortened by 10 to 30 years depending on the success of treatment.

Emphysema can be caused by environmental factors, such as heavy smoking, but one form of the disease, called *familial emphysema*, in inherited as an autosomal recessive. In most populations, the incidence of familial emphysema is about 1 in 1700 individuals; the frequency of heterozygotes, however, is much greater–about 1 in 20 individuals! The high frequency of heterozygotes is particularly important in this case because there is evidence that heterozygotes may also have an increased risk of lung disease. In one study of 103 patients with obstructive lung disease, for example, there were 5 homozygotes and 25 heterozygotes.

Familial emphysema is also known as α_1-*antitrypsin deficiency*. α_1-antitrypsin is a protein found in blood serum. Its function is to inhibit an enzyme (*trypsin*) that breaks down other proteins. The locus for α_1-antitrypsin called the Pi locus, is apparently on chromosome 2, and at least 23 different Pi alleles have been identified. One of these alleles, designated Piz, leads t an inactive form of a1-antitrypsin. This allele is the one responsible for familial emphysema because PzPiz homozygotes have an extremely high risk of homozygotes have an extremely high risk of developing the disease. AS noted, however, Piz heterozygotes may also have an increase risk.

Cystic Fibrosis

Among Caucasians, one of the most frequent simple Mendelian recessive diseases of childhood is *cystic fibrosis*, which is due to a recessive allele on the long condition itself (homozygous recessives) is about 1 in 2500 individuals, but about 1 in 25 individual is heterozygous for the allele. Although relatively frequent in Caucasians, cystic fibrosis is extremely rare in other racial groups.

It is difficult to determine the mode of inheritance of cystic fibrosis and other conditions caused by rare, simple Mendelian recessives. The problem is that not all makings between heterozygous (Aa × Aa) actually produce affected children. To show that a disease like cystic fibrosis is due to a simple Mendelian recessive, one has to show that ¼ of the children from *Aa* × *Aa* matings are affected. The difficulty in showing this is that *Aa* × *Aa* matings can be detected only if they have affected children. To see what a problem this causes, focus on families who have just two children; the proportion of such families having zero, one, or two affected children will be 9/16 , 6/16, and 1/6, respectively (using the formula give earlier). Only the latter two families would be detected, however, so the proportion of detected families who have one or two affected children would be 6/7 and 1/7, respectively. Among the detected families, therefore, the overall proportion of affected children is (6/7) (1/2) + (1/7(1) = 8/14 = 57 percent, which is nowhere near the expected 25 percent! This problem arises because not all *Aa* × *Aa* matings are *ascertained* (discovered), and the observed proportion of affected children must be corrected for this *ascertainment bias*. The correction that should be applied is beyond the scope of this book. However, in one study of cystic fibrosis, the corrected turned out to be 24.9 percent, confirming the autosomal recessive mode of inheritance.

Cystic fibrosis is characterized by the malfunction of the pancreas and other glands, resulting in production of abnormal secretions. Its chief symptoms are recurrent respiratory infections, malnutrition resulting from incomplete digestion and absorption of fats and proteins, and cirrhosis of the liver. Patients with cystic fibrosis have an accumulation of thick, sticky, honeylike mucus in their respiratory tract, which often leads to respiratory complications. The victim becomes a prime target for secondary infections such as pneumonia or bronchitis. Cystic fibrosis is a disease of childhood, and the symptoms may appear early. The disease usually leads to death in childhood or adolescence, but the lives of affected children can be prolonged somewhat by intensive respiratory and dietary treatment. Left untreated, 95 percent of affected children will die before age 5. With treatment, the average life expectancy of affected girls is more than 16 years. If treatment is begun before any appreciable lung damage has occurred, the child's chance of living to age 21 or older is now greater than 50 percent. The treatment includes a special diet, daily administration of antibiotics, administration of extracts of the pancreas of animals, and special daily lung care (sometimes including the flushing out of the lungs with an aerosol mist). The treatment tis continuous and expensive, however.

Sickle Cell Anemia

Among blacks, the most common disorder inherited as a simple Mendelian recessive is *sickle cell anemia*, due to an allele on chromosome 11. The hereditary defect shows up in the red blood cells–the cells that carry oxygen from the lungs to the tissues. More specially, the defect is in the oxygen-binding protein, *hemoglobin*, a major component of the red blood cells. (Hemoglobin is the protein that physically binds with oxygen molecules and transports them.) People homozygous for the sickle cell allele have a form of hemoglobin that tends to crystallize or stack together when exposed to lower than normal levels of oxygen. The crystallization of hemoglobin causes the entire red cell to collapse from its normal ellipsoidal shape into the shape of a half-moon or 'sickle'. In this form the red cell cannot carry the normal amount of oxygen. More important the sickled cells tend to clog the tiny capillary vessels, interrupting the nutrient blood supply to vital tissues and organs. Children with the disease tire easily, may be retarded in their physical development, and tend to be susceptible to infections of all kinds–owing to their general weakened condition caused by the severe chronic anemia brought on by the reduced amount of normal hemoglobin. At intervals the affected people experience sickle cell crises marked by fever and severe, incapacitating pains in the joints, particularly in their extremities, and in their chest, back an abdomen caused by the clogging of the blood supply to these vita areas. These painful episodes may last from hours to days to weeks. They may be provoked by anything that lowers the oxygen supply; overexertion, high altitude, respiratory ailments, and so on. Pregnancy in a woman with sickle cell anemia is particularly dangerous, not only for the mother because of the increased stress but also for the baby. The sickling of cells in the placenta can decrease or block the oxygen supply tot the baby and cause a spontaneous abortion, miscarriage, or stillbirth. (Technically, the expulsion of a fetus from the time of fertilization to three months of pregnancy is an *abortion*; expulsion between three and seven months of pregnancy is a *miscarriage*; expulsion of a nonliving fetus thereafter is a *still birth*).

The lifespan of people affected with sickle cell anemia is distressingly short. They frequently die in their teens or twenties, almost always before age 45. At present there is not cure for the disease, although the symptoms can be treated, transfusions can be given, and the pain can be decreased or relieved.

A confession terminology about sickle cell anemia has come into widespread use. The carriers of the gene, the heterozygotes, are often said to have the sickle cell "trait." These people are actually quite healthy. Rare cases are known, however, of heterozygotes suffering symptoms of the disease brought on by extreme and prolonged reduction of oxygen in the blood. The affected people who are homozygous for the sickle cell gene are said to have the sickle cell "disease." Sickle cell anemia is the condition that afflicts *homozygous* people. These distinctions should be kept in mind to avoid confusion the condition of the heterozygotes, who are normally healthy, with that of the homozygotes, who are not.

Sickle cell anemia is extraordinarily common as compared with other serious conditions that have an equally simple inheritance. In most of West Africa, about 1 per 100 children has the disease, and about 1 person in 6 is a carrier. Among American blacks, the frequencies are lower because only part of their ancestry traces back to West Africa; about 1 per 400 children has the disease and approximately 1 per son in 10 is heterozygous. This single disease causes the young to die in enormous numbers; the yearly toll worldwide has been estimated at 100,000. The reason the disease is so common is found in an infectious disease, *malaria*, and in the law of segregation.

The geographic distribution of the high frequency of sickle cell anemia in Africa almost coincides with the geographic distribution of falciparum malaria, so named because for the protozoan parasite that causes it (*Plasmodium falciparum*). This overlap occurs because the carriers of the sickle cell allele have an enhanced resistance to the disease. The parasites causing malaria are mosquito-borne; they are transferred from bloodstream to bloodstream by mosquitos and get transported throughout the body by inserting themselves into red blood cells. The red blood cells with abnormal hemoglobin tend to sickle when infested with parasites; they clump together and are thereby effectively removed from circulation and destroyed. It may also be that the parasite cannot infect these red cells as readily as the normal ones. Still other factors may be at work, but, in any case, the heterozygous carriers of the sickle cell gene tend to be more resistant to malaria than normal homozygotes. In one study, the risk of severe malarial infections in children who were carriers was found to be only half as great as the risk to homozygous normal children. Partly because of this, carriers of the sickle cell allele have an enhanced ability to survive and reproduce in an environment in which malaria is widespread. Inevitably, since the carriers have an advantage over even the normal homozygotes, many matings occur between two carriers, and the law of segregation foreordains that 1/4 will be homozygous normal (and therefore more susceptible to malaria than their parents), but 1/2 of the children will themselves be carriers and will perpetuate the sickle cell allele because of their advantage.

Almost as if to emphasize the connection between malaria and abnormal hemoglobin, another inherited blood disease is found in other parts of the world where malaria is common, particularly in the Mediterranean basin. In this case the abnormality is in the amount of hemoglobin. The Mediterranean condition is also inherited as a simple recessive, and homozygous recessive individuals suffer a sever and debilitating anemia that is frequently fatal in early childhood. The disease itself is called *thalassemia major.* The coextensive distribution of malaria and thalassemia was first noticed in about 1950, and this immediately led to the suggestion that the carriers might have a decreased susceptibility to malarial infections. The geographic correlation was subsequently shown to be true of sickle cell anemia and malaria in Africa as well. The essential correctness of the interpretation of the geographic correlation is now established beyond reasonable doubt. Thalassemia major is very common in certain Italian populations, sometimes reaching an incidence of 1 per 100; up to 1 in 6 normal people are heterozygous carriers. The carriers of the gene have at worst a mild anemia known as thalassemia minor. Thalassemia is also found in appreciable frequencies among other peoples in Mediterranean region–for example, among the Greeks, Sardinians, Armenians, and Syrians.

Tay-Sachs Disease

Like cystic fibrosis among Caucasians and sickle cell anemia among Negroes, Tay-Sachs disease is a simple Mendelian recessive that has an increased incidence in a particular population–in this case Jews. (Incidentally, many diseases are named after the person or persons who first recognized the symptoms as recurring clinical entities. In this case, Tay was a British ophthalmologist and Sachs an American neurologist; both described the disease in the 1880s, although they were not aware of its genetic basis). *Tay-Sachs disease*, formerly called "familial infantile amaurotic idiocy," is due to a recessive allele on chromosome 15. The normal an enzyme, called *hexosaminidase A*, which functions in the breakdown of a sugary-fatty substance (*gangliside* GM_2) in the central nervous system. Homozygotes for the Tay-Sachs allele lack a functional form of hexosaminidase A. The absence of this enzyme causes an abnormal accumulation of ganglioside GM2 in the central nervous system, which leads to blindness, seizures, and a complete degeneration of mental and motor function. The disease symptoms are usually evident by six months of age; the disease becomes progressively more severe and is invariably fatal within the first five years of life. Among Jews from Centra Europe (Ashkenazi), the incidence of the condition is about 1 per 4000 births . This incidence is about 100 times higher than that among non-Jews or among Jews from the Mediterranean basin (Sephardic); the incidence of Tay-Sachs disease among non-Jews is about 1 per 400,000. Approximately 1 in 30 Ashkenazi Jews is a carrier of the recessive allele, which emphasizes the importance of the disease in this population.

The Complementation Test

One final example of variation among population in the incidence of inherited disorders is albinism among the Hope Indians of Arizona and the Jemes and Zuni Indians of New Mexico. *Albinism* results from an inherited defect in the ability of certain specialized cells to produce normal amounts of the brown-black pigment melanin. Albinos therefore lack pigmentation Their hair, skin, and the iris of the eyes are very light of white, although their eyes look oink due to reflected light passing through the blood vessels. The condition is inherited as a simple Mendelian recessive. It is not as serious a disorder as many others, but albinos tend to have vision problems, sometimes including blindness, and are extremely susceptible to sunburn and prone to skin cancer because they have no melanin to protect them from the sun's ultraviolet rays. The incidence of the condition among the Indians mentioned is about 1 in 200, and about 1 person in 8 is a carrier of the albino allele. Among people of European ancestory, by contrast, the incidence is about 1 in 40,000

Glucose 6-Phosphate Dehydrogenase (G6PD) Deficiency

One well-known gene on the X chromosome controls the production of an enzyme, *glucose 6-phosphate dehydrogenase* (G6PD), which is involved in carbohydrate metabolism and is important in maintaining the stability of red blood cells, Individuals who have abnormally low amounts of this enzymatic activity are prone to a severe anemia that occurs when many of their red blood cells cannot function normally and therefore break down and are destroyed. The anemia can be provoked by a number of environmental triggers such as inhaling pollen of the broad bean *Vicia faba* or eating the bean raw, in which case the illness is known as *favism.* Such individuals are also sensitive to certain drugs such as naphthalene (used in mothballs), certain sulfa antibiotics (such as sulfanilamide), or the absence of the offending substances these individuals are completely normal, and they recover from the anemia when the agents are eliminated. G6PD deficiency is found in high frequency in people of Mediterranean extraction (10 to 20 percent or more of males are affected) and among Asians (about 5 percent of Chinese males are affected), and it occur in about 10 percent of block American males. As noted, the locus of G6PD is X linked, and well over 50 alleles coding variant forms of the enzyme are known. However, only a few of these alleles lead to a sufficiently defective form of G6PD to cause the drug sensitivity and anemia associated with G6PD deficiency.

Hemophilia

Recall that *hemophilia* is a bleeding disorder due to an X-liked recessive that results from the excessively long time required for the blood to clot following an injury. Affecting about 1 in 7000 males, it is much rarer than red-green colour blindness, yet it is probably as well known. This is partly because it affects the blood; and even though we now know that blood has no magical or hereditary properties, we still carry a vestige of the old beliefs in such expressions as "pure blood" and "blood lines" and "blood relation." A second reason hemophilia is so well known is that it occurred in many members of the European royalty who descended from Queen Victoria of England (1819-1901). She was a carrier of the gene, and by the marriages of her carrier granddaughters, the gene was introduced into the royal houses of Russia and Spain. Ironically, the present royal family of Great Britain is free of the gene because it descends from King Edward VII, one of Victoria's four sons, who was not himself affected .

Actually, there are several forms of hemophilia, and incidence of 1 per 7000 in males applies only to the special type exemplified in some of Queen Victoria's descendents. Blood clotting is a complex process involving more than a dozen ingredients known as *clotting factors*, and different genes are responsible for producing these various clotting factors. An abnormal form of any one of these genes can lead to an inherited form of excessive bleeding. Most hereditary hemophilias are caused by genes on the autosomes, but two forms are inherited as X-linked recessives. One of these X-linked forms, called *hemophilia B* or *Christmas disease*, involves clotting factor IX, and it is extremely rare. The other X-linked form, called *hemophilia A* or *Royal hemophilia*, involves clotting factor VIII. Hemophilia A is the more common X-linked form, and it is the one percent in some of the European royal families.

Where Queen Victoria's hemophilia-A mutation came from is not known, She herself was a carrier, but her father was completely normal and nothing in her mother's family suggests that the gene was present. The bet guess is that either the egg or the sperm that gave rise t Queen Victoria carried a new mutation. Perhaps it was in the sperm of her father, Edward Duke of Kent, because it is thought that mutations are somewhat more likely to occur in the sperm of older men, and he was 52 when Victoria was born. However it happened, Victoria was a carrier. She had nine children five daughters (two were certainly carriers, two were almost certainly not, and the remaining daughter may or may not have been as she left no children) and four sons (one a hemophiliac–Leopold Duke of Albany–and three normal). In the five generations since Queen Victoria, 10 of her male descendants have had the disease. Virtually all died very young. Those who survived childhood often died in their 20s or early 30s, usually from excessive bleeding following injuries.

The most famous of Victoria's affected male descendants is undoubtedly her great-grandson Alexis, the son of her grand daughter Alix (Empress Alexandra of Russia) and Tsar Nicholas II. Alexis was Alexandra's firstborn son and heir to the Russian throne. Unbeknown to Alexandra, she was a carrier of the hemophilia-A allele, and her son had inherited the disease. Anna Vironbova, one of Alexandra's favourite ladies-in-waiting, recounts.

The parents became preoccupied with the boy's health. At one point the Tsar observed in his diary how difficult it was to live through the worry resulting from Alexis's excessive and life-threatening bleeding from minor injuries and bruises that all children unavoidably experience. The boy survived his childhood, but some historians have argued that the Tsar's preoccupation with Alexis's health contributed to the neglect of the empire that ultimately brought on the Bolshevik Revolution in 1917. During the revolution, the Tsar, Alexandra, Alexis, and his four sisters disappeared. The fate of the family is unknown, but they are thought to have been machine-gunned to death in Ekaterinburg fortress on July 17, 1918, just 13 days before what would have been Alexis's seventeenth birthday.

Lesch-Nyhan Syndrome (HGPRT Deficiency)

As a final example of X-linked recessive inheritance we consider the rare condition known as *Lesch-Nyhan syndrome*, which is due to a deficiency of enzyme HGPRT. Persons with this condition, virtually all male, have two groups of symptoms. One results from excessive accumulation of *uric acid* in the blood. One normal function of HGPRT is to route certain cellular metabolites into DNA via the minor pathway of DNA synthesis. (A *metabolite* is any compound produced during the course of chemical processes in the cell; the totality of all these processes is called *metabolism*.) In the absence of HGPRT, these metabolites accumulate and are eventually broken down by other enzymes into uric acid. Accumulation of excess uric acid leads to hyperuricemia (elevated levels of uric acid in the blood), bloody urine, crystals in the urine, urinary tracts stones, severe inflammation of the joints (*arthritis*), and the extremely painful swelling and inflammation of joints in the hands and feel (particularly the big toe) known as *gout*. (Although gout is always associated with elevated levels of uric acid, the inheritance of most forms of gout is multifactorial; only a minority of patients with excessive uric acid and gout have HGPRT deficiency).

Symptoms of excessive uric acid can be treated with appropriate drugs, but Lesch-Nyhan syndrome has another group of symptoms that are of unknown origin and are not alleviated by treatment for excessive uric acid. These symptoms involve the nervous system and are associated with jerky, unwilled, and uncoordinated muscular movement. The most bizarre symptom of Lesch-Nyhan syndrome is extreme aggressive behavior, most often expressed as self-mutation of the hands and arms , usually by biting. At present, such self-mutation can be prevented only by appropriate binding of the hands and arms. Lesch-Nyhan syndrome is an extremely serious condition. Most patients die before the age of five, and almost all die before adulthood. Though rare, the Lesch-Nyhan syndrome is important because it involves HGPRT, which allows us to understand the basis of some of its symptoms. The condition also illustrates the profound influence that gene can exert on behaviour, even though the biochemical basis of this influence is not yet understood.

Ambiguous Sex

Although most human beings are unambiguously male or female, about 1 percent have abnormalities in sexual development. These include several of the abnormal sex-chromosome constitutions. Such as XO and XXY, but they also include a wide variety of developmental abnormalities in children who have normal sex chromosomes. Some of these conditions are known to be caused by mutations in particular genes; other conditions have extremely complex or unknown causes. Among the latter group is a particularly common class of abnormalities in sexual development known as hermaphroditism. *Hermaphrodites* are neither entirely male nor entirely female but have a mixture of male and female sex organs; that is, their sex is ambiguous. (The name *hermaphrodite* is a compound of the names of the god Hermes and the goddess Aphrodite.) A distinction should be made between the true hermaphrodites and false or *pseudohermaphrodites.* True hermaphrodites are, by definition, capable of producing *both* functional sperm and functional eggs so that, theoretically, they could fertilize themselves. Some animals, such as earthworms, and most flowering plants are true hermaphrodites. The disorder in humans is invariably pseudo-hermaphroditism–that is, false hermaphroditism. Indeed, pseudo-hermaphrodites in humans produce *neither* functional sperms nor functional eggs.

The overall incidence of pseudohermaphroditism in humans is about 0.2 percent, or 1 per 500 newborns. This makes it relatively common among abnormalities of sexual developments. Pseudohermaphroditism is not a single condition, however, and its causes are diverse. Homozygosity for any its causes are diverse. Homozygosity for any one of several different autosomal recessive mutations can lead to pseudohermaphroditism, but beyond this rudimentary knowledge the genetics of the disorder are not well worked out. The trait is extremely variable from patient to patient. In some cases the internal reproductive organs are of one sex, the external organs of the other. In other cases the reproductive organs are female on one side of the body and male on the other. Often *ovotestes* are found; these are mixture of ovarian and testicular tissue that result from the failure of either the inner or outer part of the embryonic gonad to degenerate.

Although the underlying causes of pseudohermaphroditism are diverse and generally not understood in detail, there are exceptions. Among these are the several abnormalities of the sex chromosomes to be discussed in the next chapter. Two types of inherited pseudohermaphrodites not due to sex-chromosome abnormalities are also relatively well understood, and the nature of the disorder in these cases illustrates the importance of the male hormone testosterone in sexual development. One of these types is a male pseudohermaphroditism known as the testicular-ferminization syndrome; the other is a female condition known as adrenogenital syndrome.

Testicular-Ferminization Syndrome

The *testicular-Ferminization syndrome* is an inherited condition of chromosomally XY individuals. The mutation responsible for it is on the X chromosome, and its expression in limited to males. Affected individual are chromosomally XY, yet they have few traces of masculinity. The external genitalia are female, a vagina is present but incompletely developed, and rarely there is a rudimentary uterus and fallopian tubes. Tissue recognizable as testicular can be found, usually in an abnormal position in the body cavity, and the Wolffian ducts, associated with male sexuality, are somewhat developed. Affected individuals are therefore chromosomally male but for the most part phenotypically female, and they are, or course, unable to bear children. Nevertheless, many do marry as women and have normal sexual relations with their husbands.

The syndrome is evidently caused by the inabiilty of embryonic tissues to respond to testosterone. The hormone itself is produced in quantities comparable to those in normal men. Much of the detailed information on testicular feminization comes not from studies of humans but from studies of a virtually identical X-linked inherited condition in the mouse. The particular defect in affected mice is very likely a defective receptor protein whose function is to recognize testosterone at the surface of cell membrane and transport it into the cells. Although affected mice do produce testosterone the hormone is not incorporated into the appropriate cells. Testosterone cannot exert its triggering effect in these cells, so they do not differentiate in the normal male direction.

Adrenogenital Syndrome

The *adrenogenital syndrome* is a form of pseudohermaphroditism in humans caused in some instances by a hereditary defect in the adrenal glands. These glands, one perched atop each kidney, produce a variety of important hormones, among them *cortisol*. This hormone is produced by converting *cholesterol*, chemical step by chemical step, into the cortisol molecule. Along this kind of chemical assembly line, each step adds some atoms or takes some off or chemically rearranges the preceding molecule. Each step in this sort of assembly pathway requires the presence of a specific enzyme molecule to accomplish the chemical reaction, and the genetic information enabling the cell to manufacture each of the enzymes is present in a different gene. In the adrenogenital syndrome, one enzyme required in an intermediate step along the pathway is not functional. Thus, the chemical pathway is blocked and cortisol cannot be produced. The defective enzyme is supposed to convert one of the intermediate molecules (*progesterone*) in the pathway, but the enzyme is nonfunctional due to homozygosity for a recessive mutation and the conversion cannot be accomplished. As a consequence, the intermediate molecule accumulates in cells of adrenal gland, much as half-finished automobiles would accumulate on an assembly line if one of the workers quit abruptly.

Progesterone and related molecules can not accumulate forever, of course. They are gradually broken down step by step along another chemical pathway by another series of enzymes into a class of molecules that is chemically similar to testosterone; one representative molecule in this class is called *aldosterone*. These breakdown products have an action in the body similar to that of testosterone, but not so strong. Nevertheless, the testosterone-like effects of the breakdown products cause a degree of *virilization* (masculinization) in affected females.

Females afflicted with the adrenogenital syndrome have a normal uterus and fallopian tubes; the vagina is usually small, and the clitoris may be enlarged, giving it the appearance of a penis. Externally, these people are male like, although the reproductive organs are small and not as well formed as in normal males. The most serious effects of the syndrome are not the physical abnormalities, however; of greatest concern is the lack of cortisol, which can lead to death. Treatment of the condition involves continual administration of cortisol to the patient. The presence of cortisol prevents the adrenals from starting cholesterol molecules along the chemical pathway to cortisol. This in turn, prevents the accumulation of the intermediate molecules and their breakdown into testosterone-like forms. Finally, the absence of these testosterone-like molecule prevents further masculinization.

XYY and Criminality

Another sex-chromosomal abnormality associated with few physical manifestations is the XYY conditions–technically 47,XYY. Affected males have 47 chromosomes; they have the full chromosome complement of normal males plus a second Y chromosome. XYY males tend to be tall, averaging over 6 ft. Beyond this, distinctive physical symptoms are practically nonexistent, so it is questionable whether the term syndrome really applies to this case because there is nor group of characteristic symptoms. Among newborn males, the incidence of the XYY condition is though to be approximately 1 in 950.

The behaviour of XYY males has been a matter of concern and controversy since 1965, when a Scottish study uncovered in incredibly high frequency of the XYY condition among imprisoned men convicted of violet crimes. More than 50 similar studies have now been carried out in prisons in Europe and the United States, and it had held true that the frequency of XYY men confined for crimes of violence is about 10 times greater than it is among a sample of newborn males. Many of these studies have concentrated on tall men; because XYY males tend to be tall, this will automatically create a bias, and the findings must therefore be interpreted with caution. Nevertheless, the higher than expected frequency of XYY males among tall criminal s has suggested to some the XYY males may undergo impaired psychosocial development, perhaps caused by their height itself, which may produce social stresses tending to evoke violent, aggressive behaviour, or perhaps caused by some subtle and visible behavioural effect of the extra Y.

One Danish study of criminality among XYY males is particularly important tan revealing because it is free of many of the biases in other studies. The Danish research involved an examination of all available records on almost all men 28 to 32 years old whose height was 184 cm (72.4 in) or more and whose birthplace was Copenhagen. A total of 4139 sex-chromosomal determinations were made; among these there were 12 XYY (incidence 1 per 345), 16 XXY (incidence 1 per 259–the XXY sex-chromosomal constitution will be discussed in few pages), and 13 individuals with other types of chromosomal abnormality. The information examined included criminal records and types of crime, socioeconomic status of the men's parents, score on a Danish army intelligence test, and an index of educational level. Parental socioeconomic status was lower among men with criminal convictions than among those without convictions, but no differences were found among XY, XYY, and XXY males. Thus, lower socioeconomic status is associated with criminality but not with sex-chromosomal constitution.

Other Danish data pertaining to chromosome and criminality are summarized in Table 6.2. Among XYY males, the rate of criminal convictions was 41.7 percent (5 out of 12), which is significantly greater than the 9.3 percent among XY males. (The term *significant* as applied to statistical tests means that an observed difference is greater than could plausibly be attributed to chance). Among XXY males the rate of criminal convictions (3 to 16) was somewhat greater than among XY males, but the difference is sufficiently small that it could be due to change. The findings relative to height are in the third column of Table 6.2. On the average, XYY males are taller than tall XY males by about 3.7 cm (1.5 in), and XXY males are taller than tall XY males by about 2.7 cm (1.1 in). Tallness alone is not related to criminality, however, because men with criminal records are somewhat shorter than their noncriminal counterparts.

One of the most important findings of the Danish study is summarized in the last two columns of Table 6.2 XYY and XXY males have a significantly lower score on the Army intelligence test and a significantly lower index of educational level than do XY males. For each sex-chromosomal class, moreover, men with criminal records are substantially proper in both measures of intellectual function than men without criminal records. Of course, this does not necessarily imply that men with lower levels of intellectual function actually commit more crimes; they may merely be more likely to get crimes; they may merely or more likely to get caught. In many case, the Danish data seem to imply that the high rate of criminality among XYY males is due to their moderately impaired mental function.

Some comments are in order about the nature of the crimes committed by the XYY males Two of the XYY males were habitual criminals. One had more than 50 convictions for petty larceny, auto theft, burglary, embezzlement, and so on, but no record of violent or aggressive behavior. The other habitual criminal, who was also mentally retarded and had spent most of his life in institutions for the mentally retarded, had convictions for arson, theft, burglary, and embezzlement, but, once again, no convictions for aggressive crimes against people. The other three XYY males were petty criminals. Two had each been convicted twice for petty theft. The other had turned in a false report about a traffic accident and had started a small fire. Overall, *the crimes of the Danish XYY males were not crimes of violence and aggression; they were mainly crimes involving property.*

The Danish study illustrates the danger of preliminary scientific reports being sensationalized by the popular press. At about the time of the original Scottish discovery of XYY males among exceptionally violent criminals, a pathological killer sneaked into a dormitory in Chicago and brutally murdered several student nurses. A newspaper claimed that the man had been 'born to kill' because he was XYY. This turned out to be untrue–but the false report was widely disseminated and believed. Several proposals were made for the mass screening of newborns for the detection of XYY males to provide them with special education and intensive psychological guidance to counteract their supposed "killer instinct." It is good that the screening proposals were never carried out because, in the atmosphere of the time, they could have been harmful by creating a self-fulfilling prophecy. That is to say, if physicians and family are convinced that XYY males (or any other category of people) have severe, aggressive behavioural problems, then they may interact with the individuals in such a manner as to elicit the very behavior they expect, thereby fulfilling their own prophecy. In light of the Danish study, the rate of criminality is related more to level of intellectual function then to sex-chromosomal constitution.

XXY: Klinefelter Syndrome

Although the XXX and XYY conditions have few and relatively mild physical manifestations, some sex-chromosomal abnormalities do have distinctive physical effects, chiefly affecting sexual development. Most frequent among these is the XXY condition (technically 47 XXY), which occurs in about 1 per 1000 newborn males. Approximately two-thirds of these arise from primary nondisjunction in the mother, leading to an egg carrying two X chromosomes, and there is a slightly enhanced risk of the condition among sons of older mothers. Males with the XXY chromosomes constitution have a distinctive group of symptoms known as *Klinefelter syndrome.* Characteristic of the syndrome are very small testes but normal penis and scrotum. Although about half the patients have some degree of mental retardation, growth and physical development are, for the most part, quite normal, although affected males tent to be tall for their age. Puberty does not occur normally: The testes fail to enlarge, the voice remains rather high-pitched, and pubic and facial hair remains sparse. In about half the cases there is some enlargement of the breasts. Males with Klinefelter syndrome do not produce sperm and hence are sterile; up to 10 percent of men who seek aid at infertility clinics turn out to be XXY. XXY males in the Danish study seem to have moderately impaired intellectual function. However some men with Klinefelter syndrome, thought sterile, lead completely normal and even distinguished lives.

XO: Turner Syndrome

Another abnormality of the sex chromosomes, more than the XXY condition, is the XO condition (The "O" used in the shorthand designation XO refers to the absence of the homologue of the X.). Individuals with this sex-chromosome constitution have 45 chromosomes altogether : 22 pairs of autosomes and a monosomic X. Thus, the XO chromosome constitution is technically designated 45, X. The frequency of XO among live-born females is about 1 per 5000, but this far under-represents the occurrence of the condition. As will be discussed later in this chapter, about 15 to 20 percent of spontaneously aborted fetuses that have a detectable chromosome abnormality are XO, making this one of the most common chromosomal abnormalities in spontaneous abortions. More than 1 percent of all recognized pregnancies in humans involves and embryo that is chromosomally XO, and approximately three-fourths of these are caused by abnormal sperm that lack a sex chromosome.

The small fraction of XO fruits that survive develop into individuals who have *Turner syndrome.* Affected females are of short stature and often have a distinctive webbing of the skin between the neck and shoulders. There is no sexual maturation. The external sex organs remain childlike, the breasts fail to develop, the pubic hair does not grow. Internally, the ovaries are infantile or absent and menstruation dose not occur. Although the IQ of affected females is very nearly normal, they have specific defects in spatial abilities and arithmetical skills.

Autosomal Abnormalities

Most autosomal trisomies--and all autosomal monosomies–are so severe that they are incompatible with life. The result of the chromosomal abnormality may be failure of blastocyst implantation or spontaneous abortion of an often grossly malformed embryo or fruits. Nevertheless, three autosomal trisomies are sometimes compatible with embryonic development and birth, although the trisomic children may be severely abnormal. The incidence of the three conditions among live-born children is summarized in Table 11.1. The most common autosomal trisomy among live-born children is trisomy 21, also known as Down syndrome after the physician who first described the condition over a century ago.

Table 11.1. Most frequent autosomal trisomies in Live-born Humans

Trisomy	*Name of syndrome*	*Approximate frequency*
13 (47,+13)	Patau syndrome	1 per 5000 live births
18 (47,+18)	Edward syndrome	1 per 6500 live births
21 (47,+21)	Down syndrome	1 per 750 live births

Trisomy 21: Down Syndrome

The cause of Down syndrome was for many years a complete mystery, but in 1959 it was discovered that affected children have 47 chromosomes including three copies of chromosome 21, the smallest of the autosomes. (In the early years after the discovery that patients with Down syndrome were trisomics, it was mistakenly thought that the chromosome involved was chromosome 22; one may still find reference to "trisomy 22: in connection with Down syndrome in older books and journals. However, it is now agreed that the trisomic chromosome in all children with Down syndrome is indeed chromosome 12). The correct technical designation of the Down karyotype is 47, + 21–the"+" preceding the "21" denoting an entire extra chromosome 21.

Children with Down syndrome are invariably mentally retarded. They are of short stature due to delayed maturation of the skeletal system, and their muscle tone is poor, leading to a characteristic facial appearance in older children and adults . Approximately 40 percent of children with Down syndrome have major heart defects. Many children with the syndrome also have epicanthus–a small fold of skin across the inner part of the eye where the tear glands are; this gives the eyes a slightly Oriental look, and the common name for the disease–*mongolism*–derives from it. (Use of the term *mongolism* for Down syndrome is now discouraged because of the unnecessary and misleading racial connotations.) Altogether there are 10 or so major symptoms of the syndrome, but the condition is somewhat variable.

Down syndrome is very frequent in all racial groups (about 1 in 750 live births). The condition has probably been present in human populations since the ancient past, long before recorded history, perhaps even before our species, *Homo sapiens*, had evolved. There is, for example, a condition in chimpanzees that is very similar in symptoms to Down syndrome. This is also caused by trisomy, and the banding pattern of the chromosome responsible is so similar to that of human chromosome 21 that the conclusion is inescapable that both chromosome must have descended from some common ancestral chromosome in a primate population in the dim past. The incidence of Down syndrome in present-day human population is, as noted, about 1 in 750 live-born children. Down syndrome dose not usually run in families and therefore most families that have had a child with Down syndrome have only a small risk of having a subsequent child with the same condition. About 97 percent of families with a trisomic-21 child are in this category. (However, as will be noted later, the risk of having a child with Down syndrome increases with the age of the mother.) The remaining 3 percent of families with an affected child have a much higher recurrence risk. In these cases one of the parents carries a chromosomal rearrangement that leads to a high risk of trisomy 21 in the children even though the parent is phenotypically normal. For now, concentrate on the 97 percent; in these families the trisomic- 21 children are the result of primary nondisjunction leading to a gamete with two copies of chromosome 21.

Actually, the abnormal is almost always the egg. Moreover, the risk of a pregnancy leading to a child with Down syndrome increases dramatically with the age of the mother. Overall, the risk of having a child with Down syndrome is about 1 in 750, but averaging obscure the maternal-age dependence. The risk is only about 1 per 700 in mothers 15 to 19 years old. The risk increases with age, gradually at first, so that the risk of occurrence in mothers aged 30 to 35, however, it is 1 per 750. Then it shoots up. In mothers aged 45 and over, the risk is as high as 1 in 16–about 6 percent! As a result, women over 40 have approximately 40 percent of the children with Down syndrome, although they have only 4 percent of all children. The *recurrence risk* of Down syndrome (i.e., the risk of a subsequent child having Down syndrome when one child is already affect) also increases with the mother's age; generally speaking, risk of a woman having a second child with Down syndrome is about three times the risk of a woman of the same age who has not previously had an affected child.

The reason for the greatly increased risk of trisomy 21 in the children of older mothers is not known. Suspicion centers directly on the meiotic process in females. The cells undergoing meiosis in the ovaries of a female embryo are arrested in late prophase I, and they stay arrested until just before the egg is ready to be ovulated. Thus a cell undergoing meiosis in a woman 45 years old has actually resumed the process after a hiatus of 45 years. This by itself may greatly increase the chance of nondisjunction or other errors. Perhaps, too, hormonal changes in older women may increase the risk of meiotic accidents. In any case, the increased risk with age is confined to the mother. The risk does not increase with the age of the father. It may seem strange that this can be stated so definitely. Older women are almost always married to older men, and you might think that the age effects would be so intertwined that you would be unable to decide what to attribute to whom. However, reliable statistical procedures have been devised to separate the age effects of mothers and fathers, and the result of the elaborate analysis is that the increased risk in older couples is due almost exclusively to the age of the mother.

Children with Down syndrome, as you might guess, have a shorter life expectancy than normal. Many have heart defects among their other multiple physical abnormalities, and they are particularly susceptible to chest infections and bronchial pneumonia. The critical year in children with trisomy 21, as in the normal children, is the first year of life. The average lifespan of children born with Down syndrome in 1932 was nine years. But the life expectancy of affected individuals has been steadily increasing as antibiotics and mass immunizations have decreased the risk and severity of infections. By 1947 the life expectancy was 12 years; by 1963 it was 16. Today, affected children between ages 5 and 7 have a life expectancy of up to 27 years; some survive to age 40 or 50.

Because of the increased life expectancy of individuals with Down syndrome, many of them enter the reproductive period. Males with Down syndrome are sexually incompetent and do not reproduce. Affected females are capable for reproduction, however, and approximately 50 women with Down syndrome have give birth; usually the pregnancy has resulted from rape or incest. Because of secondary nondisjunction, one expects approximately half the children to have Down syndrome. In fact, only about 20 percent of the children are affected; the rest are completely normal. The frequency of affected children is lower than expected because fetuses with trisomy 21 frequently undergo spontaneous abortion.

The increased life expectancy of individuals with Down syndrome has had another consequence. Today more people with trisomy 21 are living than at any time in the past, and this creates a social cost of tremendous proportions because affected individuals are often utterly dependent on others for their welfare. If an affected child is institutionalized, the social cost may be paid through taxes. But virtually everyone acknowledges that many public institutions are understaffed by under-trained and underpaid personnel. Therefore, many physicians urge personnel. Therefore, many physicians urge the parents to raise an affected child at home and, indeed, there is a national organization of parents of Down syndrome children who do raise their children at home. The substantial medical costs must then be borne by the family. Unfortunately, there is also a psychological cost. Tragically, many people unwisely look upon the birth of an abnormal child as a stigma, a mark of Cain, and they inevitably feel remorse and guilt. The parents themselves must resist these emotions. We should long since have stopped believing with the Puritans that the birth of an abnormal child is retribution for past and hidden sins.

Concern for the welfare of society and the parents must not replace compassion for the child. An affected child can benefit from special and intensive care and training, but usually this is not delivered. Sometimes the child does not even receive the attention and stimulation required by every human being to remain alert and interested, regardless of ability. Public institutions are often too overburdened to provide the care, and the parents are not themselves trained to be able to provide it. Moreover, the retarded child is usually not capable of frolicking with other, normal children and learning form them. Indeed, small children can be especially cruel and heartless when they ridicule the physical abnormalities or mental deficiencies of others. All this spells out a very sad tale, not only for children with Down syndrome but also for countless others with various kinds of physical and mental abnormalities.

In grappling with the social issue raised by children with Down syndrome, it is all too easy to forget that the children are human begins as much in need of love, security, and affection as others. They have interests and personalities of their own, which have been poignantly described by Down-syndrome specialists Smith and Wilson:

> Children with down syndrome usually take great pleasure in their surroundings, their families, their toys, their playmates. Happiness comes easily, and
>
> From D.W. Smith and A.A. Wilson, 1973, The Child with Down's Syndrome (Mongolism), Saunders, Philadelphia.

throughout life they usually maintain a childlike good humor. They are not burdened with the grown-up cares that come to most people with adolescence and adulthood... Life is simpler and less complex. The emotions that others feel seem to be less intense for them. They are sometimes sad, happy, angry, or irritable, like everyone else, but their moods are generally not so profound and they blow away more quickly... A child with Down syndrome, though slow, is still very responsive to his environment, to those around him, and to the affection and encouragement he receives from others.

TRISOMY 18: EDWARDS SYNDROME

Down syndrome is not the only autosomal trisomy compatible with the live birth of affected children, but it is by far the most frequent. Two other autosomal trisomies also occur in live-born children: trisomy 18 (i.e., 47 +18) and trisomy 13 (i.e. 47 + 13). *Trisomy 18*, also known as *Edwards syndrome*, is about eight times less frequent than Down syndrome; it affects about 1 in 6500 live-born children. As with trisomy 21, the incidence of trisomy 18 increases with the mother's age, but the increase is relatively small, far less than in Down syndrome. Children with Edwards syndrome have multiple congenital abnormalities, including severe mental and physical retardation. They have an elongated skull with low-set, malformed, sometimes pointed ears; their jaw and oral cavity are small. They carry their fingers in an abnormal position, with the second finger overlapping the third. Virtually all these children are born with heart defects. Approximately 65 percent of affected newborns are female, presumably because males with the syndrome are more likely to undergo spontaneous abortion. The life expectancy of affected children also indicates a greater severity in males: The life expectancy of males is about three months; in affected females it is about nine months.

TRISOMY 13: PATAU SYNDROME

Trisomy 13, also known as Patau syndrome, is equally as severe as trisomy 18. It occurs in about it about 1 in 500 live births. As in other autosomal trisomies, affected children are severely retarded, both mentally and physically. Children with trisomy 13 have a small skull and eyes; the ears are often malformed and deafness is common. Most have harelip and cleft palate. Many have malformed thumbs and extra digits. Nearly 70 percent have heart defects. Patau syndrome also has a slightly increased incidence with maternal age, but again the increase is far less than the age effect in Down syndrome. The sex ratio of affected children is about 1 : 1. Rarely do affected children survive more than three or four months after birth.

Most cases of trisomy 18 and trisomy 13 are sporadic. They are unpredictable and do not run in families, but there are a few high-risk families. Thus the situation insofar as occurrence and recurrence are concerned appears to be much like that in Down syndrome. The abnormalities caused by the extra chromosome in all three trisomies are major and multiple, physical and mental. Each trisomy is unique, though somewhat variable. Because certain abnormalities occur regularly in trisomy 13 but only rarely in trisomy 18, for example, these criteria can be used to distinguish between the two syndromes, even though there is a considerable overlap of physical and mental defects. And unlike the abnormalities caused by single-gene mutations, the abnormalities in the trisomies cannot be traced to some single biochemical defect in the cells. Rather, the trisomy syndromes result from the cumulative action of abnormal gene dosage. Consequently there is at present no hope of effecting a "cure" for the trisomies.

The autosomal trisomies invariably involve major physical malformations and mental retardation and are often associated with heart defects. This is true of other major developmental disturbances as well, not only of those that are genetic in origin but also of those that are purely environmental, such as those caused by infection of the fetus with German measles virus (rubella) during the first months of pregnancy. Evidently the heart and nervous system are often affected because their embryonic development is delicate and incredibly intricate; major developmental aberrations are unlikely to leave these systems untouched. The heart is a fist-sized, four-chambered, 10-oz wonder. It bears between 60 and 200 times a minute, depending on circumstances, and its 30 million rhythmic contractions a year force nearly a million gallons of blood through almost 60,000 miles of arteries, veins, and capillaries. Little wonder that the ancients thought the center of life, personality, and emotion to be the heart! But the brain is even more impressive. In the cerebral cortex alone, the complex outer layer–the seat of the senses, voluntary muscle control, consciousness and rationality–there are nearly 10 billion nerve cells known as neurons, with each neuron supported and nourished by about 10 neuroglial cells. Each neuron has between 4,000 and 10,000 extensions connecting it to other neurons. Most of these cells and connections develop before birth, but the process actually continues until about age two. It is no surprise that major developmental abnormalities will almost always disrupt this prodigious network. To expect otherwise would be like expecting an earthquake not to interrupt telephone service! (Incidentally, brain cells in an adult cannot divide. They gradually die off and are not replaced–an inevitable part of aging. People beyond puberty lose some 10,000 neurons every day).